GEOMETRY AND TRIGONOMETRY FOR CALCULUS

PETER H. SELBY

Director, Educational Technology
MAN FACTORS, INC.
San Diego, California

JOHN WILEY & SONS

New York • Chichester • Brisbane • Toronto • Singapore

For David Kneeland Andrews,

with thanks for many things.

Library of Congress Cataloging in Publication Data

Selby, Peter H.
   Geometry and trigonometry for calculus

   (Wiley self-teaching guides)
   Includes bibliographical references.
     1. Geometry, Plane—Programmed instruction. 2. Geometry,
Analytic—Programmed instruction. 3. Trigonometry—Programmed
instruction. I. Title.
QA455.S38
ISBN 0-471-77558-4

Printed in the United States of America
30 29 28 27 26 25 24

Editors: Judy Wilson and Irene Brownstone
Production Manager: Ken Burke
Editorial Supervisor: Winn Kalmon
Artist: Carl Brown
Composition and Make-up: Meredythe Miller

To the Reader

As you already know, if you have read *Practical Algebra* (a companion volume in the Wiley Self-Teaching Guides series) or studied algebra elsewhere, the study of algebra provides us with many useful techniques in problem solving. Building on what we have learned in arithmetic, it goes far beyond by exploring the real number system, introducing the use of letters to represent numbers, and teaching us how to formulate and solve equations and inequalities. We also learn in algebra how to work with polynomials, how to handle algebraic fractions, and how to multiply and divide terms containing exponents and radicals. Finally, we find out how to combine many of these operations in the solution of quadratic equations, problems involving ratio, proportion, and variation, and word problems.

In your study of algebra you have had a brief introduction to graphic methods in the solution of linear and quadratic equations as well as inequalities. However, the major emphasis in algebra is upon *numerical* solutions of problems. The major emphasis in this book will be upon the *graphic* representation of problems and upon their solution by the combined analytic methods of geometry and algebra.

The first four chapters will cover plane geometry and prepare you for trigonometry. (If you have studied geometry before, these chapters will act as a review.) The two chapters on trigonometry will introduce you to numerical trigonometry and some of the methods of trigonometric analysis. The next two chapters, which deal with analytics, will help you learn some of the beautiful methods of solution that evolve through the combined techniques of geometry and algebra. Finally, the chapter on limits will lead you to the very front door of calculus, which marks the beginning of advanced mathematics. In addition to representing the start of advanced mathematics, calculus also represents the final goal for many students in their study of mathematics since it gives them the final problem-solving tool they will need for most aspects of science and engineering. For those who wish to study calculus by themselves, the Self-Teaching Guide *Quick Calculus*, by Daniel Kleppner and Norman Ramsey, provides an excellent introduction.

Obviously such subjects as plane geometry, trigonometry, and analytic geometry cannot be treated fully in a single book that seeks to cover the principal mathematical topics that lie between algebra and calculus.

Hopefully, however, this Guide will familiarize you with the approaches and procedures — mainly geometric — necessary for the study of calculus.

To gain maximum help from this book you should be aware that although the subjects covered here follow a coordinated plan, and a consistent effort has been made to show their interrelationship and mutual dependence on one another, each of the major topics — synthetic geometry, trigonometry, and analytic geometry — is essentially complete in itself and therefore can be studied independently of the others to meet your individual needs.

As always, your goal should be *learning*, not speed.

La Jolla, California   Peter H. Selby
January, 1975

SELECTED REFERENCES

Hemmerling, E. M., *Fundamentals of College Geometry*, 2nd ed. (New York: John Wiley & Sons, 1970).

Allendoerfer, C. B. and Oakley, C. O., *Fundamentals of Freshman Mathematics* (New York: McGraw-Hill Book Company, 1959).

Drooyan, I., Hadel, W., and Carico, C. C., *Trigonometry — An Analytic Approach* (New York: Macmillan Publishing Co., 1973).

Fisher, R. C. and Ziebur, A. D., *Integrated Algebra and Trigonometry* (Englewood Cliffs, N. J.: Prentice-Hall, Inc., 1972).

Cooke, N. M. and Adams, H.F.R., *Basic Mathematics For Electronics* (New York: McGraw-Hill Book Company, 1970).

Protter, M. H. and Morrey, C. B., *College Calculus With Analytic Geometry* (Reading, Mass: Addison-Wesley Publishing Co., 1970).

REFERENCE CHART FOR SELECTED TEXTBOOKS ON SYNTHETIC GEOMETRY, TRIGONOMETRY, AND ANALYTIC GEOMETRY

Chapter in This Book	Hemmerling	Allendoerfer and Oakley	Drooyan, Hadel, and Carico	Fisher and Ziebur	Cooke and Adams	Protter and Morrey
1. Plane Geometry: Definitions and Methods of Proof	1,2					
2. Plane Geometry: Congruency and Parallelism	3,4					
3. Plane Geometry: Circles and Similarity	5,7					
4. Plane Geometry: Areas, Polygons and Locus	8, 9, 10					
5. Numerical Trigonometry	11	12	5,6	5	23,24,25,26	9,12
6. Trigonometric Analysis		13	2,3,4	4	27,28,29	9
7. Analytic Geometry	13	14		11		3
8. Conic Sections		14		11		8,11
9. Limits				10		4

Contents

CHAPTER ONE

Plane Geometry: Definitions and Methods of Proof

Whether you are studying geometry afresh or are now simply reviewing the subject, it will be worth your while to consider for a moment where this branch of mathematics came from, what it is about, and what you may hope to gain from its study.

Geometry had its origin long ago in the measurements by the Babylonians and Egyptians of their lands, the design of irrigation systems, and the construction of buildings and national monuments. The word geometry is derived from the Greek words *geos*, meaning *earth*, and *metron*, meaning *measure*. As long ago as 2000 B.C. the land surveyors of these people used the principles of geometry to reestablish vanishing landmarks and boundaries. In fact the ancient Egyptians, Chinese, Babylonians, Romans, and Greeks all used geometry for surveying, navigation, astronomy, and other practical occupations. The Greeks undertook to systematize the known geometric facts by establishing logical reasons for them and relationships among them. The work of such men as Thales (600 B.C.), Pythagoras (540 B.C.), Plato (390 B.C.), and Aristotle (350 B.C.) in organizing geometric facts and principles culminated in the geometry text *Elements*, written about 325 B.C. by Euclid. This truly remarkable and seemingly timeless text has been in use for more than 2,000 years.

Geometry is a science that deals with forms made by lines. A study of geometry is an essential part of the training of engineers, scientists, architects, and draftsmen. The carpenter, machinist, tinsmith, stonecutter, artist, and designer also apply the facts of geometry in their trades. In this book you will learn a great many basic facts about such geometric figures as lines, angles, triangles, circles, and various other two-dimensional shapes.

You also will learn a good deal about critical thinking and logical reasoning. You will be led away from the practice of blind acceptance of statements and ideas and encouraged to think clearly and precisely before forming conclusions. In fact, many consider the development of this type of thinking the chief benefit to be derived from the study of geometry. The process of reasoning is used to prove geometric statements. You will learn to analyze a problem in terms of the data given and the laws and principles accepted as true, and, by logical thinking, to arrive at a solution to the problem.

But before a statement in geometry can be proved we need to agree on certain definitions and properties of geometric figures. It is essential that the terms we use in geometric proofs have exactly the same meaning to us all. So in this chapter we will consider first such elements as defined and undefined terms, basic assumptions, some familiar geometric figures, methods of proof, and the axioms, postulates, and theorems fundamental to our investigation of congruent triangles, parallel lines, distances, angle sums, parallelograms, trapezoids, medians, and midpoints.

When you have finished this chapter you should be able to:

- recognize and use correctly the basic terminology associated with such geometric concepts as point, line, surface, line segments, circles, arcs, angles, triangles, and pairs of angles;

- understand and use the fundamental methods of geometric proof based on deductive reasoning, axioms and postulates, basic angle theorems, and procedures for determining hypotheses and conclusions.

POINTS, LINES, AND SURFACES

1. Just as in the study of language we accept some words as undefined in order to use them to define other words, so in geometry we accept certain terms as undefined. With them we can then begin the process of defining all other geometric terms. And although we cannot define these basic terms in any precise way, we can give meanings to them by means of descriptions. These descriptions should not, however, be thought of as definitions—at least not formal definitions, although they are sometimes referred to as connotative definitions. So despite the fact that we cannot define certain basic terms we will be using in our

    study of geometry, we can give meaning to them by _____ them.

    — — — — — — — — — — — — — —

    describing

2. The first term we will discuss is the term *point*. No doubt you have your own concept of what this terms means from your own reading, from common usage, and from discussion with others. Now, however, we are interested—as we will be with all the terms we will discuss—in its meaning in the context of geometry.

    In geometry a point has position only. It has no length, width, or thickness. It is *represented* by a dot, but is not the dot itself, just as a flag may represent a nationality but is not the nation itself. A point is designated (named) by a capital letter placed next to the dot.

.C

Thus, a *point* has position only. (True / False)
(Underline the correct answer.)

$\cdot B$

‒ ‒ ‒ ‒ ‒ ‒ ‒ ‒ ‒ ‒ ‒ ‒ ‒ ‒

True

3. A *curve* has length but no width or thickness. It can be *represented* by the path of a pencil on paper, chalk on a blackboard, or by a stretched piece of string—or in many other ways. You will best understand that there are many different curves if you think of each curve as being generated by a *moving point*. Thus,

A *straight line* is a curve generated by a point moving in the same direction.

A *curved line* is a curve generated by a point moving in a continuously changing direction.

A *broken line* is a combination of straight lines.

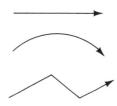

Our basic curve will be the straight line. It is designated by the capital letters of any two of its points, or by a small letter. Thus,

A straight line may be drawn between two points but is *unlimited in extent*; it extends in either direction indefinitely. Of course its representation (picture) cannot go on indefinitely. Another very important property of a straight line is that *it is the shortest distance between two points*. Also, when two straight lines intersect they intersect in (meet at) a point.

We think of a line as being generated by a _____ .

When two lines meet at a point they are said to _____ .

‒ ‒ ‒ ‒ ‒ ‒ ‒ ‒ ‒ ‒ ‒ ‒ ‒ ‒

moving point; intersect

4. A *surface* has length and width but no thickness; it is, therefore, two-dimensional. A surface may be *represented* by a table top, a blackboard, the side of a box, or the outside of a basketball. Again, these are *representations* of a surface but are not surfaces in geometric terms.

A *plane surface* or, simply, a *plane* is a surface such that a straight line connecting any two of its points lies entirely in it. A *plane* is a flat surface and might be represented by the top of a desk, a sidewalk, or a sheet of glass. *Plane Geometry* is the geometry that deals with plane figures, that is, figures that can be drawn on a flat or plane surface.

Hereafter in this book, unless otherwise indicated, a *figure* will mean a *plane figure*.

All surfaces are plane surfaces. (True / False)

————————————————

False. A sphere (ball) has a surface that is not plane (flat).

5. A *straight line segment* is the part of a straight line between two of its points, called the *endpoints* of the segment. It is named by using the capital letters of these endpoints or by a small letter.

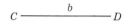

Thus *CD* or *b* may be used to name the straight line segment between *C* and *D*. We usually write the endpoints, like *CD*, to refer to the straight line segment itself and the small letter, like *b*, to refer to *how long* the segment is.

Here is another straight line segment: P———————a———————Q

To refer to the length of the segment we write _____ and to refer to the segment itself we write _____ .

————————————————

a; *PQ* (or *QP*)

6. The term *straight line segment* is often shortened to *line segment* or *segment*, or even *line*, if the meaning is clear. Thus, segment *AZ* (or simply *AZ*) means the straight line segment *AZ* unless otherwise indicated. Referring to a segment by its endpoints is quite useful.

Draw a line segment *XY* and label it two ways. *X* and *Y* will be its

_____ .

————————————————

X————r————Y ; endpoints. (You could use any small letter in place of *r* to refer to the length of *XY*.)

7. Now let's talk about dividing a segment into parts. If a segment is divided into parts, then:

    (1)   The whole segment equals the sum
          of its parts.

    (2)   The whole segment is greater than
          any part.

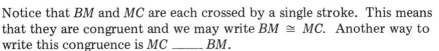

    Thus, in the illustration, if AB is divided into three parts, a, b, and c, then $AB = a + b + c$. Also, AB is greater than a or b or c (that is, $AB > a$, b, or c).
    If a segment is divided into two *congruent* parts:

    (1)   The two congruent parts have the
          same length. (Congruent means "same
          size and shape;" its symbol is $\cong$.)

    (2)   The point of division is the *midpoint
          of the segment.*

    (3)   A line that crosses a segment at its
          midpoint is said *to bisect the given
          segment.*

    Notice that BM and MC are each crossed by a single stroke. This means that they are congruent and we may write $BM \cong MC$. Another way to write this congruence is MC _____ BM.

    — — — — — — — — — — — — — —

    $\cong$

8. Since XD is shown as congruent to DY,
    we write $XD \cong DY$ and say that D is
    the *midpoint* of XY. D is also said to
    be the point of intersection of XY and
    PQ, that is, the point at which two
    lines cross each other or come together.
    If D is the midpoint of XY then PQ
    bisects XY.

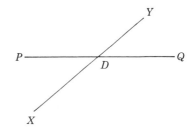

    A line that crosses a segment at its midpoint is said to _____
    the given segment.

    — — — — — — — — — — — — — —

    bisect

9. Now let's see how much you know about naming line segments and

points and finding lengths and points of line segments. In the figure below.

(a) Name each of the line segments shown.

_____

(b) Name the line segments that intersect at A. _____

(c) What other line segment can

     be drawn? _____

(d) Name the point of intersec-

     tion of CD and AD. _____

(e) Name the point of intersec-

     tion of BC, AC, and CD. _____

In the figure at the right,

(f) State the lengths of AB, AC, and AF.

_____

(g) Name two midpoints. _____

_____

(h) Name two line segments that are

     bisectors. _____

- - - - - - - - - - - - - - - -

(a) AB (or BA), AC (or CA), BC (or CB), CD (or DC), and AD (or DA)
(b) AB, AC, and AD
(c) BD
(d) D
(e) C
(f) $AB = 3 + 7 = 10$; $AC = 5 + 5 + 10 = 20$; $AF = 5 + 5 = 10$
(g) E is midpoint of AF, and F is midpoint of AC
(h) DE is bisector of AF, and BF is bisector of AC

10. It is time now to talk about circles. A
     circle is a closed curve all points of which
     are equidistant (the same distance) from
     a given point called the center. The sym-
     bol for a circle is ⊙, and for circles ⊚.
     Hence ⊙O stands for the circle whose
     center is O.
         The *circumference* of a circle is the
     distance *around* it. It contains $360°$ (360
     degrees).

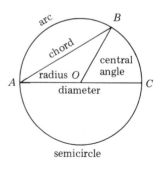

A *radius* is a line joining the center to a point on the circumference. It follows then, from the definition of a circle, that all radii (plural of radius) of a circle are congruent. (Remember, all points on a circle are the same distance from the center.) Thus in the preceding figure OA, OB, and OC are radii of $\odot O$ and $OA \cong OB \cong OC$.

A *chord* is a line joining any two points on the circumference. Thus AB and AC are chords of $\odot O$.

A *diameter* is a chord through the center of the circle. A diameter is twice the length of a radius and is longer than any chord not going through the center. Thus in the figure AC is a diameter of $\odot O$.

An *arc* is a part of the circumference of a circle. The symbol for arc is $\frown$. Thus $\overset{\frown}{AB}$ refers to arc AB. An arc of 1° is 1/360th of a circumference. (Note that arc $\overset{\frown}{AB}$ is different from chord AB.)

A *semicircle* is an arc equal to one-half of the circumference of a circle. A semicircle contains 180°. A diameter divides a circle into two semicircles. Thus, diameter AC cuts $\odot O$ into two semicircles.

A *central angle* is an angle formed by two radii. Hence the angle between radii OB and OC is a central angle. A central angle of one degree intercepts (cuts off) an arc of one degree. Therefore, in the figure at the right, if the central angle between OE and OF is 1°, then $\overset{\frown}{EF}$ is 1°.

Congruent circles are circles having congruent radii. Thus if $OE \cong O'G$, then circle $O \cong$ circle O'.

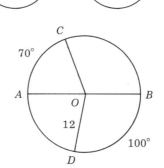

In circle O at the right find:

(a)   The lengths of OC and AB.

_____

(b)   The number of degrees in $\overset{\frown}{AD}$.

_____

(c)   The number of degrees in $\overset{\frown}{BC}$.

_____

(Note: Since AB is a diameter, $\overset{\frown}{ACB}$ and $\overset{\frown}{ADB}$ are semicircles.)

– – – – – – – – – – – –

(a)   Radius OC = radius OD = 12. Diameter AB = 24
(b)   Since semicircle ADB = 180°, $\overset{\frown}{AD}$ = 180° − 100° = 80°
(c)   Since semicircle ACB = 180°, $\overset{\frown}{BC}$ = 180° − 70° = 110°

11.   Now let's turn our attention to the subject of angles. An *angle* is the figure formed by two straight lines meeting at a point. The parts of the lines that are the sides of the angle are called *rays* and the point is its

vertex. The symbol for angle is ∠. The plural (angles) is ⩘. Thus rays *AB* and *AC* are the sides of the angle shown at the right. An angle may be named in any of the following ways:

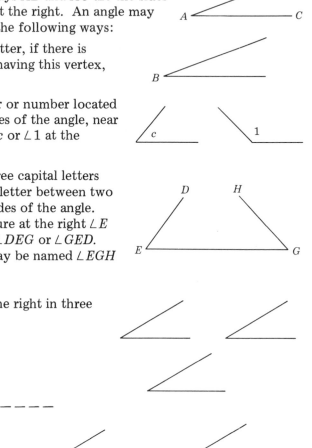

(1) By the vertex letter, if there is only one angle having this vertex, as ∠ *B.*

(2) By a small letter or number located between the sides of the angle, near the vertex, as ∠*c* or ∠1 at the right.

(3) By means of three capital letters with the vertex letter between two others on the sides of the angle. Thus, in the figure at the right ∠*E* may be named ∠*DEG* or ∠*GED.* Similarly ∠*G* may be named ∠*EGH* or ∠*HGE.*

Name the angles at the right in three different ways.

- - - - - - - - - - - - - - -

(You could, of course, use any other letters or numbers you wished as long as they correspond to the three methods.)

12. The size of an angle depends on the extent to which one side of the angle must be rotated or turned about the vertex until the turned side meets the other side. Thus the protractor at the right shows that ∠*A* is 60°. (A *protractor* is a simple device, similar to a ruler in concept but designed to measure, or help you lay out, angles ranging in size from 0° to 180°. You will find it handy to have one. They can be obtained from most bookstores or drafting supply stores.) If *AC* were rotated about

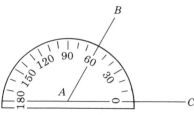

the vertex A until it met AB, the amount of turn would be 60°. In using a protractor it is easiest to have the vertex of the angle at the center and one side along the 0°–180° diameter.

The size of an angle does *not* depend on the pictured *lengths* of the sides of the angle. Thus, the size of ∠B at the right would not be changed if the pictured sides AB and BC were made longer or shorter.

There are various *kinds* of angles; that is, angles are given names according to certain characteristics of size. Thus:

An *acute angle* is one that is less than 90°.

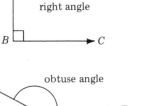

A *right angle* is an angle of 90°. (Note that we use the little square to indicate a right angle.)

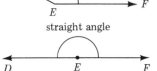

An *obtuse angle* is an angle that is greater than 90° and less than 180°.

A *straight angle* is an angle that equals 180°. (It is really just a straight line that we interpret as an angle.)

Name the angles below according to the classification we have established above.

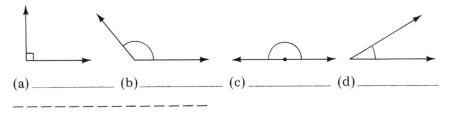

(a) _____ (b) _____ (c) _____ (d) _____

– – – – – – – – – – – – – – – –

(a) right angle; (b) obtuse angle; (c) straight angle; (d) acute angle

13. Below are a few more facts about angles with which you should be familiar.

(1) *Congruent angles* are angles that have the same number of degrees, that is, the same size. Thus, rt. ∠A ≅ rt. ∠B.

(2) A line that *bisects* an angle divides it
into two congruent parts. Thus, if AD
bisects $\angle A$, then $\angle 1 \cong \angle 2$. (Congruent
angles are shown by crossing their arcs
with the same number of strokes, hence
the arcs of $\angle 1$ and $\angle 2$ are crossed by a
single stroke.)

(3) *Perpendiculars* are lines that meet at
right angles. The symbol for perpendicular is $\perp$, and for perpendiculars $\perp$s.

(4) A *perpendicular bisector* of a given
segment is both perpendicular to the
segment and bisects it. Thus, if GH is
the $\perp$ bisector of EF, then $\angle 1$ and $\angle 2$
are right angles and M is the midpoint
of EF.

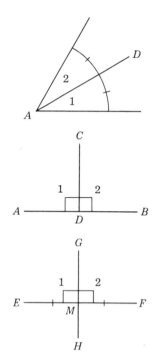

The following exercises will give you an opportunity to use some of the
things you have been learning about angles in the preceding frames.

(a) Name the obtuse angle in the
diagram. _____

(b) Name one acute angle in the
diagram. _____

(c) Find the value of (number of
degrees in) angle BOE.

_____

(d) Find the value of $\frac{3}{5}$ of a rt. $\angle$.

(e) In a half hour what angular
rotation is made by a minute
hand of a clock?

_____

(f) In the diagram shown find the
values of angles ADB and CDE.

_____

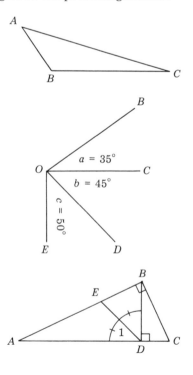

- - - - - - - - - - - - - -

(a) $\angle ABC$

(b) $\angle BAC$ or $\angle BCA$

(c) $a + b + c = 35° + 45° + 50° = 130°$

(d) $\frac{3}{5}(90°) = 54°$

(e) $\frac{1}{2}$ of $360°$ or $180°$

(f) $\angle ADB = 180° - \angle BDC = 180° - 90° = 90°$
$\angle CDE = 180° - \angle 1 = 180° - 45° = 135°$
(Note that $\angle 1 = 45°$ because angles $\angle ADE$ and $\angle BDE$ are marked congruent, hence each is one-half of $90°$.)

14. A *polygon* is a closed figure bounded by straight line segments as sides. The figure at the right is a polygon. Because it happens to have five sides it is also known as a *pentagon*, that is, a five-sided polygon.

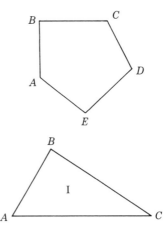

A *triangle* is a polygon having three sides. The symbol for a triangle is $\triangle$, and for triangles is $\triangle\hspace{-0.6em}\triangle$.

There are any number of different types of polygons, the most familiar of which probably are the four-sided figures (termed *quadrilaterals*) such as the square, the rectangle, the parallelogram, and so on. However, we will discuss these later. For the present we will concentrate our attention on the triangle.

A *vertex* of a triangle is a point at which two of the sides meet. (The plural of vertex is vertices.) A triangle may be named by naming its three vertices in any order or by using a Roman numeral placed inside it. Thus the triangle above is named $\triangle ABC$ or $\triangle I$. Its sides are AB, AC, and BC; its vertices are A, B, and C; and its angles are $\angle A$, $\angle B$, and $\angle C$.

Triangles are classified according to the congruence of their sides or according to the kinds of angles they have.

A *scalene triangle* is a triangle that has no congruent sides. Thus, in triangle ABC, $a \neq b \neq c$. (The small letter used for each side agrees with the capital letter of the angle *opposite* it.)

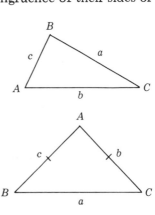

An *isosceles triangle* is one that has at least two congruent sides. Thus, in the triangle ABC, $AC \cong AB$ or $b = c$. The congruent sides are called the *legs* or *arms* of an isosceles triangle. The remaining side is the *base* (a). The angles on either

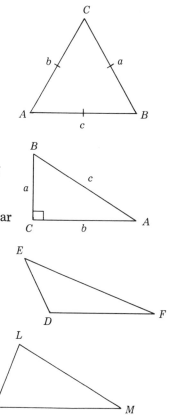

side of the base are the *base angles*, and the angle opposite the base ($\angle BAC$ here) is the *vertex angle*.

An *equilateral triangle* is one having three congruent sides. Thus, in the equilateral triangle ABC, $a = b = c$, that is, $BC \cong AC \cong AB$. An equilateral is also an isosceles triangle. (But notice that the isosceles triangle on the preceding page is *not* equilateral.)

A *right triangle* is a triangle containing a right angle. In triangle ABC, $\angle C$ is the right angle. Side c, opposite the right angle, is the *hypotenuse*. The perpendicular sides a and b are the *legs* or *arms* of the right triangle.

An *obtuse triangle* is one containing an obtuse angle. In triangle DEF, $\angle D$ is the obtuse angle.

An *acute triangle* is one having three acute angles. In triangle KLM, $\angle L$, and $\angle M$ are acute angles.

See if you can identify correctly the triangles shown below.

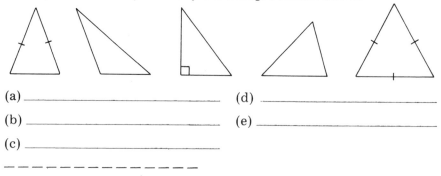

(a) _____   (d) _____

(b) _____   (e) _____

(c) _____

-- -- -- -- -- -- -- -- -- -- -- -- --

(a) isosceles triangle; (b) obtuse triangle; (c) right triangle; (d) scalene triangle (also an acute triangle); (e) equilateral triangle (also isosceles)

15. You should also be aware of some special lines in triangles that appear quite commonly in geometric constructions and problems.

An *angle bisector of a triangle* is a line (segment) that bisects an angle and extends to the opposite side. The segment BD, for example, is the angle bisector of $\angle B$, dividing $\angle B$ into the two congruent angles, $\angle 1$ and $\angle 2$.

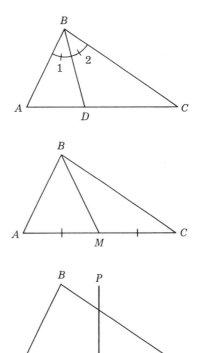

A *median of a triangle* is a segment from a vertex to the *midpoint* of the opposite side. BM, the median to AC, bisects AC, making $AM \cong MC$.

A *perpendicular bisector* of a side of a triangle is a line that bisects and is perpendicular to that side. PQ, the perpendicular bisector of AC, bisects AC and is perpendicular to it.

An *altitude* of a triangle is a segment from a vertex perpendicular to the opposite side. BD, the altitude to AC, is perpendicular to AC and forms the right angles 1 and 2. Each angle bisector, median, and altitude of a triangle extends from a vertex to the opposite side. (But notice that a perpendicular bisector does not necessarily pass through a vertex of the triangle.)

In an *obtuse triangle* the altitudes drawn to the sides of the obtuse angle fall outside the triangle. In obtuse triangle ABC (shaded), altitudes BD and CE fall outside the triangle. In each case a side of the obtuse angle must be extended.

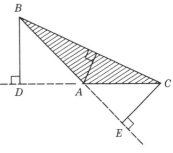

In the figure at the right, see if you can name the following:

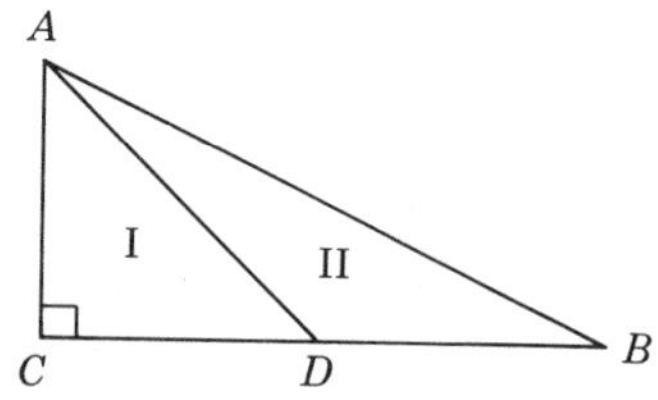

(a)   an obtuse triangle. _____

(b)   two right triangles. _____

and _____

In the figure at the right name:

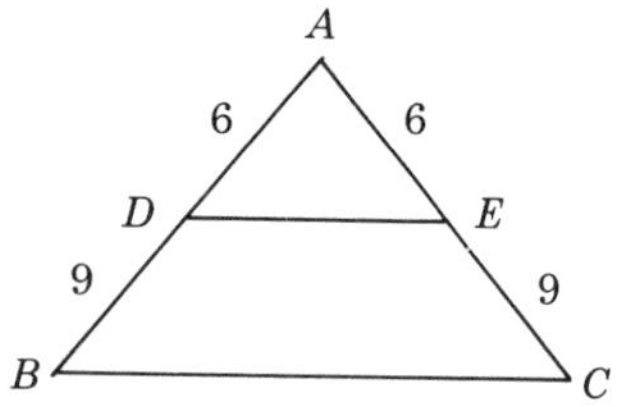

(c)   two isosceles triangles.

_____ and _____

(d)   the legs, base, and vertex angle of

each. _____

and _____

In the figure at the right name:

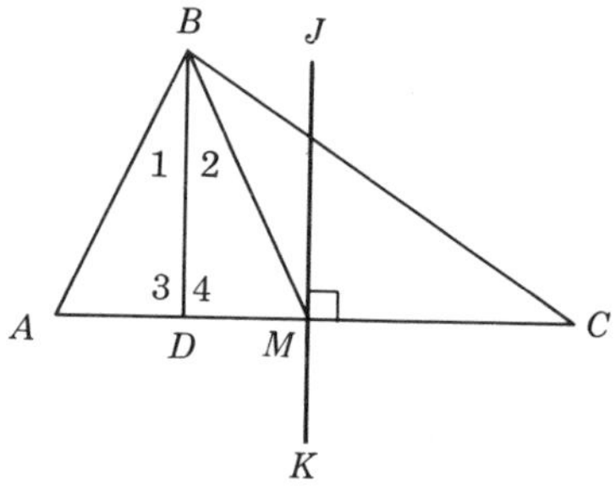

(e) BD if $\angle 3 \cong \angle 4$.

_____

(f) BM if $AM \cong MC$.

_____

(g) JK if $AM \cong MC$.

_____

(h) BD if $\angle 1 \cong \angle 2$.

_____

– – – – – – – – – – – – – – – –

(a)   Since $\angle ADB$ is an obtuse angle, $\triangle ADB$ (or $\triangle$II) is obtuse.

(b)   Since $\angle C$ is a right angle, $\triangle$I and $\triangle ABC$ are right triangles. In $\triangle$I, AD is the hypotenuse and AC and CD are the legs. In $\triangle ABC$, AB is the hypotenuse and AC and BC are the legs.

(c)   Since $AD \cong AE$, $\triangle ADE$ is an isosceles triangle. And since $AB \cong AC$, $\triangle ABC$ is an isosceles triangle also.

(d)   In $\triangle ADE$, AD and AE are the legs, DE is the base, and $\angle A$ is the vertex angle. In $\triangle ABC$, AB and AC are the legs, BC is the base, and $\angle A$ is the vertex angle.

(e)   Altitude

(f)   median

(g)   perpendicular bisector

(h)   angle bisector

16. In geometry pairs of angles of various kinds bear a useful relationship to one another. You will work with these relationships frequently, so it is important that you become aware of them.

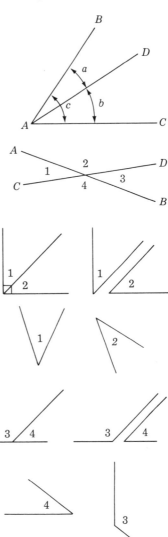

Adjacent angles are two angles that have the same vertex and a common side between them. Thus, as shown at the right, the entire angle c has been split into two adjacent angles, a and b. These adjacent angles have the common vertex A and a common side AD between them.

Vertical angles are two non-adjacent angles formed by two intersecting lines. Thus, $\angle 1$ and $\angle 3$ are vertical angles formed by the intersecting lines AB and CD. Similarly, $\angle 2$ and $\angle 4$ also are a pair of vertical angles formed by the same lines.

Complementary angles are two angles whose sum equals $90°$. They can be either adjacent or non-adjacent. In the first figure at the right angles 1 and 2 are adjacent complementary angles. However, in the other figures they are non-adjacent complementary angles. In all cases, $\angle 1 + \angle 2 = 90°$. Either angle is said to be the complement of the other.

Supplementary angles are two angles whose sum equals $180°$. Again, they may either be adjacent or non-adjacent. In the first figure at the right angles 3 and 4 are adjacent supplementary angles, hence their exterior sides lie in a straight line. However, in the other figures they are non-adjacent supplementary angles. In each case, $\angle 3 + \angle 4 = 180°$, and either angle is said to be the supplement of the other.

Name the pairs of angles shown below.

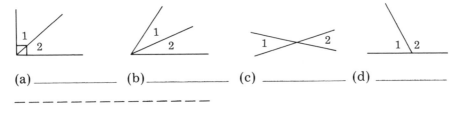

(a) _____ (b)_____ (c) _____ (d) _____

(a) complementary angles (also adjacent angles); (b) adjacent angles;
(c) vertical angles; (d) supplementary angles (also adjacent angles)

17. Now let's consider some *principles* that relate to pairs of angles.

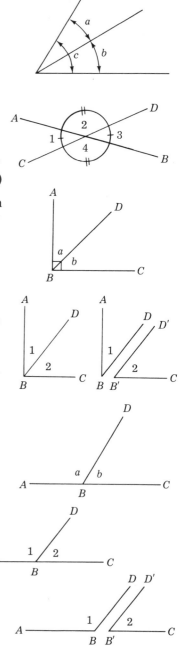

(1) If an angle of c degrees is cut into
     two adjacent angles of a degrees and
     b degrees, then $a + b = c$. Thus, if
     $a = 25°$ and $b = 35°$, then $c = 25° +$
     $35° = 60°$.

(2) Vertical angles are congruent. If
     AB and CD are straight lines, then
     $\angle 1 \cong \angle 3$, and $\angle 2 \cong \angle 4$. Thus if
     $\angle 1 = 40°, \angle 3 = 40°$; then $\angle 2 = \angle 4 =$
     $140°$. (Remember, the total number
     of degrees around any point is $360°$.)

(3) If two complementary angles contain
     a degrees and b degrees, then $a + b =$
     $90°$. Thus, if angles a and b are
     complementary and $a = 40°$, then
     $b = 50°$.

(4) Adjacent angles are complementary
     if their exterior sides are perpendicu-
     lar to each other. Thus in the figures
     at the right, angles 1 and 2 are com-
     plementary since their exterior sides
     AB and BC are perpendicular to each
     other. ($\angle ABD$ and $\angle D'B'C$ are not
     adjacent.)

(5) If two supplementary angles contain
     a degrees and b degrees, then
     $a + b = 180°$. Hence if angles a and
     b are supplementary and $a = 140°$,
     then $b = 40°$.

(6) Adjacent angles are supplementary
     if their exterior sides lie in the
     same straight line. Thus $\angle 1$ and
     $\angle 2$ are supplementary angles
     since their exterior sides AB and
     BC lie in the same straight line.
     ($\angle ABD$ and $\angle D'B'C$ are not
     adjacent.)

(7) If supplementary angles are congruent, each
of them is a right angle. Thus if $\angle 1$ and $\angle 2$
are both congruent and supplementary, then
each of them is a right angle.

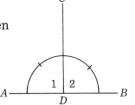

Apply the above principles relating to pairs of
angles to solve the following problems. (Use
your knowledge of algebra where needed.)

(a) If two angles are complementary and the larger is 30° more than

the smaller, what are the angles? _____

(b) Two angles are adjacent and form an angle of 135°. If the larger is
15° more than three times the smaller, what are the two angles?

_____

(c) If two angles are supplementary and the larger is three times the

smaller, what are the angles? _____

(d) What is the size of two angles if they are vertical and complemen-

tary? _____

— — — — — — — — — — — — — —

(a) Let x = smaller angle
$x + 30$ = larger angle
$x + (x + 30) = 90$, or
$x = 30°, x + 30 = 60°$.

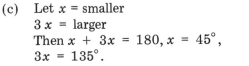

(b) Let x = smaller
$3x + 15$ = larger
Then $x + (3x + 15) = 135$, or
$x = 30°, 3x + 15 = 105°$.

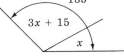

(c) Let x = smaller
$3x$ = larger
Then $x + 3x = 180, x = 45°$,
$3x = 135°$.

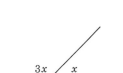

(d) Let x = each of the equal vertical angles
Then $x + x = 90$, or
$2x = 90, x = 45°$.

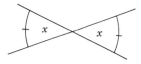

Before we proceed to the next section, on methods of proof, test yourself on your understanding of the material covered so far.

SELF-TEST

1. (a) Name the line segments that intersect at *E.* _____

    (b) Name the line segments that intersect at *D.* _____

    (c) What other line segments can be drawn? _____

    (d) Name the point of intersection of *AC* and *BD.* _____

    (frame 7)

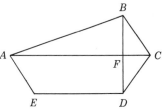

2. (a) Find the length of *AB* if *AD* is 8 and *D* is the midpoint of *AB*.

    _____

    (b) Find the length of *AE* if *AC* is 21 and *E* is the midpoint of *AC*.

    _____

    (c) Name two line segments that are bisectors if *F* and *G* are the trisection points of *BC* (that is, divide *BC* into three equal parts). _____

    _____   (frame 7)

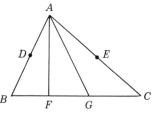

3. (a) Find *OB* if diameter *AD* = 36.

    _____

    (b) Find $\widehat{AE}$ if *E* is the midpoint of semi-circle $\widehat{AED}$. _____

    (c) Find the number of degrees in $\widehat{CD}$.

    _____

    (d) Find the number of degrees in $\widehat{AC}$.

    _____

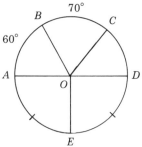

(e)   Find the number of degrees in

$\overset{\frown}{AEC}$. _____

(frame 10)

4. Name the following angles in the
diagram:

(a)   An acute angle at B. _____

(b)   An acute angle at E. _____

(c)   A right angle. _____

(d)   Three obtuse angles. _____

_____

(e)   A straight angle. _____

(frame 11)

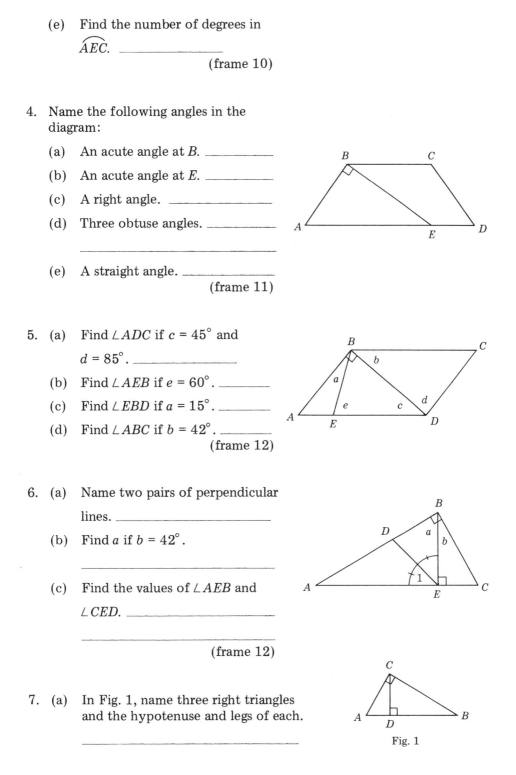

5. (a)   Find $\angle ADC$ if $c = 45°$ and

$d = 85°$. _____

(b)   Find $\angle AEB$ if $e = 60°$. _____

(c)   Find $\angle EBD$ if $a = 15°$. _____

(d)   Find $\angle ABC$ if $b = 42°$. _____

(frame 12)

6. (a)   Name two pairs of perpendicular

lines. _____

(b)   Find a if $b = 42°$.

_____

(c)   Find the values of $\angle AEB$ and

$\angle CED$. _____

_____

(frame 12)

7. (a)   In Fig. 1, name three right triangles
and the hypotenuse and legs of each.

_____

Fig. 1

(b)   In Fig. 2, name two obtuse triangles. _____

(c)   Name two isosceles triangles in the same figure. Also, name the legs, the base, and the vertex angle of each.

_____

_____

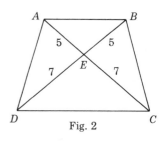

Fig. 2

(frame 14)

8.   State the relationship between each pair of angles:

(a) $\angle 1$ and $\angle 4$ _____

(b) $\angle 3$ and $\angle 4$ _____

(c) $\angle 1$ and $\angle 2$ _____

(d) $\angle 4$ and $\angle 5$ _____

(e) $\angle 1$ and $\angle 3$ _____

(f) $\angle AOD$ and $\angle 5$ _____

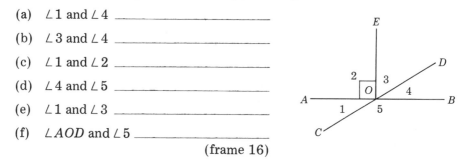

(frame 16)

Answers to Self-Test

1.   (a) AE, DE; (b) ED, CD, BD, FD; (c) AD, BE, CE, EF; (d) F
2.   (a) $AB = 16$; (b) $AE = 10\frac{1}{2}$; (c) AF bisects BG, AG bisects FC
3.   (a) 18; (b) $90°$; (c) $50°$; (d) $130°$; (e) $230°$
4.   (a) $\angle CBE$; (b) $\angle AEB$; (c) $\angle ABE$; (d) $\angle ABC$, $\angle BCD$, $\angle BED$; (e) $\angle AED$
5.   (a) $130°$; (b) $120°$; (c) $75°$; (d) $132°$
6.   (a) Since $\angle ABC$ is a right angle, $BC \perp AB$; since $\angle BEC$ is a right angle, $BE \perp AC$.
      (b) $a = 90° - b = 90° - 42° = 48°$.
      (c) $\angle AEB = 180° - \angle BEC = 180° - 90° = 90°$.
          $\angle CED = 180° - \angle 1 = 180° - 45° = 135°$.
7.   (a) $\triangle ABC$, hypotenuse AB, legs AC and BC.
      $\triangle ACD$, hypotenuse AC, legs AD and CD.
      $\triangle BCD$, hypotenuse BC, legs BD and CD.
      (b) $\triangle DAB$ and $\triangle ABC$
      (c) $\triangle AEB$, legs AE and BE, base AB, vertex angle $\angle AEB$.
      $\triangle CED$, legs DE and CE, base CD, vertex angle $\angle CED$.
8.   (a) congruent vertical angles
      (b) complementary adjacent angles

(c) adjacent angles

(d) supplementary adjacent angles

(e) complementary angles

(f) equal vertical angles

METHODS OF PROOF

Having learned something about such fundamental geometric elements as points, lines, and surfaces, we now are going to consider the method of logical reasoning by which we *prove* geometric facts. By logical reasoning we mean clear, orderly, rigorous thinking.

Basically there are two methods of reasoning: *inductive reasoning* and *deductive reasoning.* Inductive reasoning consists of observing a specific common property in a limited number of cases and then concluding that this property is general for all cases. Thus it proceeds from the *specific* to the *general.* Unfortunately, a theory based on inductive reasoning may hold for several thousand cases and then fail on the very next one. Having observed several thousand one-headed cows we might conclude that *all* cows were one-headed—until we visited the sideshow at the county fair and saw a two-headed calf on exhibit.

A more convincing and powerful method of drawing conclusions is called *deductive reasoning.* In reasoning deductively we proceed from the *general* to the *specific.* Starting with a limited number of generally accepted basic *assumptions* and following a series of logical steps we can prove other facts. Although the method of deductive logic pervades all fields of human knowledge, it probably is found in its sharpest and clearest form in mathematics. It is the principal method of geometry.

18. Deductive reasoning enables us to obtain true (or acceptably true) conclusions provided the statements from which they are deduced or derived are true (or accepted as true). It consists of the following three steps.

1. Making a *general statement* referring to a whole set or class of things, such as the class of dogs: All dogs have four feet.

2. Making a *particular statement* about one or some members of the set or class referred to in the general statement: All collies are dogs.

3. Making a *deduction* that follows logically when the general statement is applied to the particular statement: All collies are four-footed.

Deductive reasoning is known (in the field of logic) as *syllogistic reasoning* since the three types of statements above constitute a *syllogism.* In a syllogism the general statement is called the *major premise,* the particular statement is the *minor premise*, and the deduction is the *conclusion*. Thus, in the above syllogism:

1. The major premise is: All dogs have four feet.

2. The minor premise is: All collies are dogs.

3. The conclusion is: All collies are four-footed (quadrupeds).

Using a circle (as shown at the right) to represent each set or class helps illustrate the relationships involved in deductive or syllogistic reasoning.

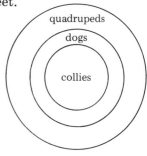

Using the above example as a guide, write the statement needed to complete each of the syllogisms below.

Major Premise (General Statement)	Minor Premise (Particular Statement)	Conclusion (Deduced Statement)
(a) All horses are animals.	This is a horse.	_____
(b) A king is a man.	_____	John is a man.
(c) _____	A square is a rectangle.	A square has congruent diagonals.
(d) Vertical angles are congruent.	$\angle a$ and $\angle b$ are vertical angles.	_____
(e) Complementary angles add up to 90°.	_____	$\angle c$ and $\angle d$ are complementary angles.

- - - - - - - - - - - - - - - - - - - -

(a)   This is an animal.
(b)   John is a king.
(c)   A rectangle has congruent diagonals.
(d) $\angle a$ and $\angle b$ are congruent.
(e) $\angle c$ and $\angle d$ add up to 90°.

19. Frequently the major premise appears as a conditional statement. Consider, for example, the statement,

"If I receive a passing grade on my exam, then I shall pass for the term."

This is a *conditional* statement because the word *if* implies a condition. The part of the statement following the word *if* is known as the *antecedent*, while the clause following the word *then* is called the *consequent*. If we assert (accept) the truth both of the conditional statement itself,

> "If I receive a passing grade on my exam, then I shall pass for the term,"

and the antecedent,

> "I receive a passing grade on my exam,"

then it will follow that the *consequent also will be true*,

> "I shall pass for the term."

Let's apply this to a geometric situation.

Example:
Accepting the conditional statement:

> If an angle is a right angle, then its measure is 90°.

And asserting the truth of the antecedent:

> ∠ABC is a right angle.

Affirms the truth of the consequent:

> The measure of ∠ABC is 90°.

In this example the conditional statement is another form of the definition of a right angle (see frame 12). This *if-then* relationship is the most common connective in logical reasoning. All mathematical proofs use conditional statements of this kind. The *if* clause, called the *hypothesis* or *premise* or *given*, is a set of one or more statements that will form the basis for a conclusion. The *then* clause which follows necessarily from the premise is called (as we learned in frame 18) the *conclusion*, or *consequent*.

Write the logical consequent of the two statements below.

1. If it is snowing, then it is cold outside.
2. It is snowing.

3. _____

‒ ‒ ‒ ‒ ‒ ‒ ‒ ‒ ‒ ‒ ‒ ‒ ‒ ‒ ‒

3. It is cold outside.

20. Odd as it may seem to you, it really doesn't matter what the contents of the first two statements (premises) are. So long as the first implies

the second, and the first statement is true, then the conclusion must be true. This is known as the *Fundamental Rule of Inference.*

A reasonable question to ask at this point is, "If we assert the truth of the *consequent* in a conditional statement, will this in turn affirm the truth of the *antecedent*?" Let's see.

Example: Consider this syllogism.

1. If it is raining, then it is cloudy.
2. It is raining.
3. Therefore it is cloudy.

We recognize this as a correct syllogism because the second statement asserts the truth of the antecedent in the first statement. However, suppose it appeared like this:

1. If it is raining, then it is cloudy.
2. It is cloudy.
3. Therefore it is raining.

Is this reasoning correct? (Yes / No)

————————————————

No, it is not. The second statement, instead of affirming the truth of the *antecedent* ("... it is raining"), asserts the truth of the *consequent* ("... it is cloudy"). The reasoning, therefore, is false, or incorrect.

21. Another rather common error in reasoning is that of *denying* the antecedent and assuming that this in turn has the effect of denying the consequent.

Example: Here is a correctly drawn syllogism.

1. If a person is a king, then that person is a man.
2. Joe Smith is a king.
3. Therefore, Joe Smith is a man.

Suppose, however, that instead of asserting the truth of the antecedent we *deny* the antecedent:

   Joe Smith is *not* a king.

This denial does *not* imply the truth of a denial of the consequent:

   Joe Smith is *not* a man.

As you can see from the diagram at the right, although Joe Smith *is not* an element in the set of kings, he *is* an element in the set of men. Hence reasoning following this pattern is *incorrect*.

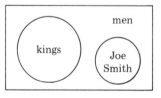

Indicate in each of the following problems whether the reasoning is correct or incorrect. (The symbol ∴ represents the word *therefore*.)

(a) If you are a good citizen, then you will vote.
     You are a good citizen.
     ∴ You will vote.   _____

(b) If you are a mother, you are a woman.
     You are a mother.
     ∴ You are a woman.   _____

(c) If you eat too much, you will get fat.
     You are fat.
     ∴ You eat too much.   _____

(d) If you marry, then your troubles will begin.
     You don't marry
     ∴ Your troubles don't begin.   _____

(e) If ∠a and ∠b add up to 90°, they are complementary.
     Angles *a* and *b* add up to 90°.
     ∴ They are complementary.   _____

– – – – – – – – – – – – – – – –

(a) correct; (b) correct; (c) incorrect (You may get fat *because* you eat too much, but not necessarily. There may be another reason. The error here is assertion of the consequent instead of the antecedent.); (d) incorrect (The error here is assuming that denial of the antecedent implies denial of the consequent.); (e) correct

22. Having considered some of the basic rules of logic and methods of valid reasoning—together with some of the common errors in reasoning—it is time we considered the building blocks of geometric reasoning known as *axioms* and *postulates*.

   The entire structure of proof in geometry must rest upon or begin with some unproved general statements, called *assumptions*. These are statements which we must *assume* or accept willingly as true in order to be able to deduce other statements. Assumptions are either axioms or postulates.

   An *axiom* is an assumption applicable to mathematics in general. Thus, the concept that "a quantity may be substituted for its equal in an expression or equation" applies to both algebra and geometry.

   A *postulate* is an assumption that applies to a particular branch of mathematics, such as geometry. Thus, the concept that "two straight lines can intersect in one and only one point" applies specifically to geometric figures.

It is essential that you learn the following axioms and postulates thoroughly! You will use them almost constantly when we get into proofs of theorems, so get to work on them *now*.

AXIOMS

23. *Axiom 1:*   Things equal (or congruent) to the same or equal (or congruent) things are equal (or congruent) to each other.

Thus the value of a dime is equal to the value of two nickels, since each value is 10¢. Or, given: $a = 5$, $b = 5$, $c = 5$, we can conclude that $a = b = c$.

Apply Axiom 1 to arrive at a conclusion with respect to the following data.

(a)   Given: $c = 15$, $c = d$ _____

(b)   Given: $f = k$, $g = k$ _____

(c)   Given: $\angle 1 = 20°$, $\angle 2 = 20°$ _____

(d)   Given: $\angle 1 \cong \angle 2$, $\angle 3 \cong \angle 1$ _____

- - - - - - - - - - - - - - - -

(a)   Since d and 15 each equal c, then $d = 15$.
(b)   Since f and g each equal k, then $f = g$.
(c)   Since $\angle 1$ and $\angle 2$ each equal $20°$, then $\angle 1 \cong \angle 2$.
(d)   Since $\angle 2$ and $\angle 3$ are each congruent to $\angle 1$, then $\angle 2 \cong \angle 3$.

24. *Axiom 2:*   A quantity may be substituted for its equal in any expression or equation. (Substitution axiom.)

Thus if $x = 7$ and $y = x + 2$, then by substituting 7 for x, $y = 7 + 2 = 9$. This amounts to evaluating an expression by substituting the value of one unknown to find the value of the other unknown, as you learned in your study of algebra.

What conclusion follows when Axiom 2 is applied below?

(a)   Evaluate $3a + 3b$ when $a = 2$ and $b = 4$.   _____

(b)   Find y if $2x + 3y = 60$ and $x = 15$.   _____

(c)   Given: $\angle 1 + \angle B + \angle 2 = 180°$
           $\angle 1 \cong \angle A$, $\angle 2 \cong \angle C$

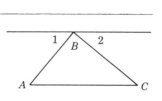

_____

- - - - - - - - - - - - - - - -

(a)   Substituting 2 for a and 4 for b we get $3(2) + 3(4) = 18$.
(b)   Substituting 15 for x we get $2(15) + 3y = 60$, $3y = 60 - 30$, $3y = 30$, $y = 10$.
(c)   Substituting $\angle A$ for $= 1$ and $\angle C$ for $\angle 2$ we get $\angle A + \angle B + \angle C = 180°$.

25. *Axiom 3:*   The whole equals the sum of its parts.

Thus the total value of a quarter, a dime, and a nickel is 40¢.

Apply this axiom to the following sets of data. Write your conclusions beside the figures.

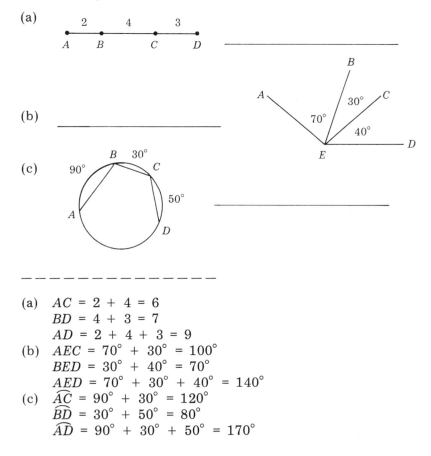

(a)

(b)

(c)

- - - - - - - - - - - - - - -

(a) $AC = 2 + 4 = 6$
      $BD = 4 + 3 = 7$
      $AD = 2 + 4 + 3 = 9$
(b) $AEC = 70° + 30° = 100°$
      $BED = 30° + 40° = 70°$
      $AED = 70° + 30° + 40° = 140°$
(c) $\overset{\frown}{AC} = 90° + 30° = 120°$
      $\overset{\frown}{BD} = 30° + 50° = 80°$
      $\overset{\frown}{AD} = 90° + 30° + 50° = 170°$

26. *Axiom 4:*   Any quantity equals (is equivalent to) itself. (Identity)

Thus $a = a$, $y = y$, $\angle C = \angle C$, $AB = AB$, and so on.

27. *Axiom 5:*   If equals are added to equals, the sums are equal. (Addition Axiom)

Thus we have the examples below.

$$7 \text{ nickels} = 35¢ \qquad\qquad a = a$$

$$\text{Add: } \underline{2 \text{ nickels} = 10¢} \qquad \text{Add: } \underline{ b = b }$$

$$9 \text{ nickels} = 45¢ \qquad\qquad a + b = a + b$$

28. *Axiom 6:*   If equals are subtracted from equals, the differences are equal. (Subtraction Axiom)

$$7 \text{ nickels} = 35¢ \qquad\qquad a = a$$

$$\text{Subtract: } \underline{2 \text{ nickels} = 10¢} \qquad \text{Subtract: } \underline{ b = b }$$

$$5 \text{ nickels} = 25¢ \qquad\qquad a - b = a - b$$

Now let's look at some examples showing the application of Axioms 4, 5, and 6.

Example 1: Given: $a = c$
    Find: Relationship of CE to DF.

1.	$a = c$	1.	Given
2.	$\underline{b = b}$	2.	Identity
3.	$a + b = c + b$	3.	Addition Axiom
4.	$CE \cong DF$	4.	Substitution

Example 2: Given: $\angle BAC \cong \angle DAE$
    Find: Relationship of b to c.

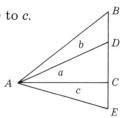

1.	$\angle BAC \cong \angle DAE$	1.	Given
2.	$a + b = a + c$	2.	The whole equals the sum of its parts.
3.	$\underline{a = a}$	3.	Identity
4.	$b = c$	4.	Subtraction Axiom

Apply Axioms 4, 5, and 6 to solve the following problems.

(a) Given: $d = a$
     Find: Relationship of ABC to BCD

(b) Given: $\angle AEC \cong \angle BED$
     Find: Relationship of a to c

- - - - - - - - - - - - - - - -

(a) 1. $d = a$ 1. Given
     2. $e = e$ 2. Identity
     3. $d + e = a + e$ 3. Addition Axiom
     4. $ABC = BCD$ 4. Substitution
     (*Note:* It may be well to mention here that the steps in a proof
     need not always appear in the same order. Often substitutions can
     be done early or later in a given problem. So don't be concerned
     if your answer doesn't *always* look like that given.)
(b) 1. $\angle AEC \cong \angle BED$ 1. Given
     2. $a + b = b + c$ 2. The whole equals the sum of its parts
     3. $b = b$ 3. Identity
     4. $a = c$ 4. Subtraction Axiom

29. *Axiom 7:*   If equals are multiplied by equals, the products are equal.
                   Also, doubles of equals are equal. (Multiplication Axiom)

     Thus if the price of a book is $5, the price of two books is $10.

30. *Axiom 8:*   If equals are divided by equals, the quotients are equal.
                   Also, halves of equals are equal. (Division Axiom)

     Thus if the price of 10 bricks is $4.00, then the price of one brick is:
     $\frac{10}{10} = \frac{\$4.00}{10}$, or 40¢.

     Below are examples of the application of Axioms 7 and 8.

     Example 1: Given: AB and AC are trisected, and $a = b$.
                   Find: Relationship between sides AB and AC.

     1. $a = b$ 1. Given
     2. $3a = 3b$ 2. Multiplication
                                Axiom
     3. $AB \cong AC$ 3. Substitution

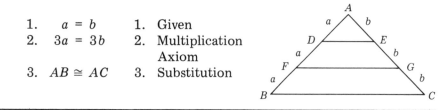

Example 2: Given: $\angle A \cong \angle C$, $\angle 1 = \frac{1}{2}\angle A$, $\angle 2 = \frac{1}{2}\angle C$.
Find: Relationship of $\angle 1$ to $\angle 2$.

1.	$\angle A = \angle C$	1.	Given
2.	$\frac{1}{2}\angle A = \frac{1}{2}\angle C$	2.	Halves of equals are equal
3.	$\angle 1 = \angle 2$	3.	Substitution

31. *Axiom 9:*   Like powers of equals are equal.

Thus, if $x = 6$, then $x^2 = 6^2$, or $x^2 = 36$.

Axiom 10:   Like roots of equals are equal.

Thus, if $y^3 = 8$, then $y = \sqrt[3]{8} = 2$.

POSTULATES

32. *Postulate 1:*   One and only one straight line (segment) can be drawn between any two points.

Thus, AB is the only straight line that can be drawn between A and B.

33. *Postulate 2:*   Two straight lines can intersect in one and only one point.

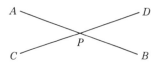

Only P is the point of intersection of AB and CD.

34. *Postulate 3:*   A straight line (segment) is the shortest distance between two points.

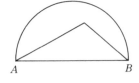

Straight line AB is shorter than either the curved or broken lines between A and B.

35. *Postulate 4:*   One and only one circle can be drawn with any given point as a center and a given line segment as a radius.

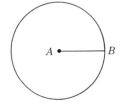

Thus, only circle A can be drawn with A as a center and AB as a radius.

36. *Postulate 5:*   Any geometric figure can be moved without change in size or shape.

Hence △ I can be moved to a new position without a change in its size or shape.

37. *Postulate 6:*   A straight line segment has one and only one midpoint.

Thus, only *M* is the midpoint of *AB*.

38. *Postulate 7:*   An angle has one and only one bisector.

Only *AD* is the bisector of ∠*A*.

39. *Postulate 8:*   Through any point on a line, One and only one perpendicular can be drawn to the line.

Therefore only *PC* ⊥ *AB* at point *P* on *AB*.

40. *Postulate 9:*   Through any point outside a line, one and only one perpendicular can be drawn to the given line.

Hence only *PC* can be drawn ⊥*AB* from point *P* outside *AB*.

Use the above postulates to help you decide whether each of the following statements is true or false. Write your answer and the postulate that supports it.

(a)   Two straight line segments can be drawn between points *A* and *B*.

_____

(b)   Both circles have *O* as a center and the same radius.

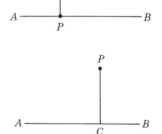

_____

(c) *M* and *M'* both are midpoints of the
    straight line segment *AB*.

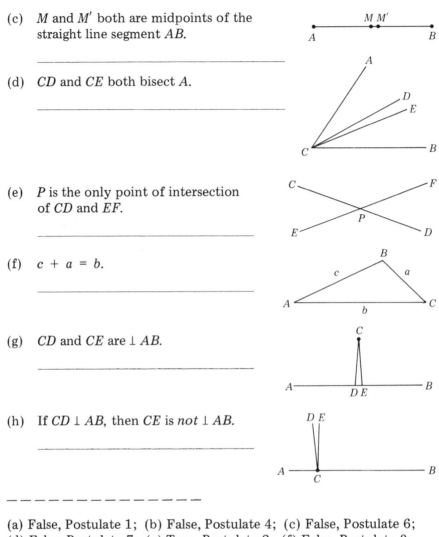

_____

(d) *CD* and *CE* both bisect *A*.

_____

(e) *P* is the only point of intersection
    of *CD* and *EF*.

_____

(f) $c + a = b$.

_____

(g) *CD* and *CE* are ⊥ *AB*.

_____

(h) If *CD* ⊥ *AB*, then *CE* is *not* ⊥ *AB*.

_____

– – – – – – – – – – – – – – –

(a) False, Postulate 1; (b) False, Postulate 4; (c) False, Postulate 6;
(d) False, Postulate 7; (e) True, Postulate 2; (f) False, Postulate 3;
(g) False, Postulate 9; (h) True, Postulate 8

BASIC ANGLE THEOREMS

41. A *theorem* is a statement to be proved. We are going to examine several
basic theorems, each of which requires the use of definitions, axioms, or
postulates for its proof. We will be using the term *principle* (sometimes
abbreviated as Pr.) to mean any of the important geometric statements,
such as theorems, axioms, postulates, and definitions. Later on we will
prove some of the principles but the main idea now is to become familiar
with their content.

Pr. 1: All right angles are congruent.

Thus, $\angle A \cong \angle B$. (See frame 47 for a proof.)

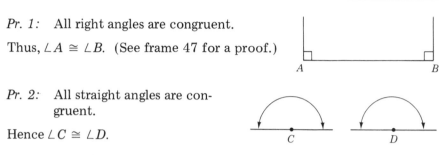

Pr. 2: All straight angles are congruent.

Hence $\angle C \cong \angle D$.

Pr. 3: Complements of the same or of congruent angles are congruent.

This is a combination of two principles:

(1) Complements of the same angle are congruent. Thus, $\angle a \cong \angle b$ since each is a complement of $\angle x$.

(2) Complements of congruent angles are congruent. Thus, $\angle c \cong \angle d$ since they are complements of the congruent angles $\angle x$ and $\angle y$.

Pr. 4: Supplements of the same or of congruent angles are congruent.

Again, this is a combination of two principles:

(1) Supplements of the same angle are congruent. Thus, $\angle a \cong \angle b$ since each is the supplement of $\angle x$.

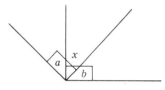

(2) Supplements of congruent angles are congruent. Therefore $\angle c \cong \angle d$ since they are supplements of the congruent angles $\angle x$ and $\angle y$.

Pr. 5: Vertical angles are congruent.

Hence $\angle a \cong \angle b$. This follows from Pr. 4 since $\angle a$ and $\angle b$ are supplements of the same angle, namely, $\angle c$.

Now let's see if you can apply the basic theorems contained in Principles 1 to 5 above. The following problems, like many of those we will use, combine the points we have just discussed and give you a chance to test your understanding of these points. Do your best to work them out without referring to the answer, but if you still need help don't hesitate to turn to the solution as a guide. We will start you out with an example.

State the basic angle theorem needed to prove $\angle 1 \cong \angle 2$ in each case.

Example: Given: *AB* and *AC* are straight lines.
      Prove: $\angle 1 \cong \angle 2$
Solution: Since *AB* and *AC* are st. (straight) lines, $\angle 1$ and $\angle 2$ are st. $\angle$s. Therefore, $\angle 1 \cong \angle 2$, since *all straight angles are congruent.*

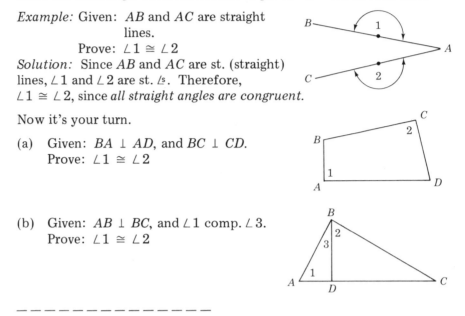

Now it's your turn.

(a)   Given: *BA* ⊥ *AD*, and *BC* ⊥ *CD*.
      Prove: $\angle 1 \cong \angle 2$

(b)   Given: *AB* ⊥ *BC*, and $\angle 1$ comp. $\angle 3$.
      Prove: $\angle 1 \cong \angle 2$

- - - - - - - - - - - - - - - - - -

(a)   Since *BA* ⊥ *AD* and *BC* ⊥ *CD*, $\angle 1$ and $\angle 2$ are rt. $\angle$s. Hence $\angle 1 \cong \angle 2$ because *all right angles are congruent.*
(b)   Since *AB* ⊥ *BC*, $\angle B$ is a rt. $\angle$, making $\angle 2$ complementary to $\angle 3$. And since $\angle 1$ is complementary to $\angle 3$, then $\angle 1 \cong \angle 2$. *Complements of the same angle are congruent.*

DETERMINING HYPOTHESIS AND CONCLUSION

42.   In frame 19, in connection with our study of the methods of proof, we considered conditional statements of the if-then form. For example, "If a horse is tired, he walks." Notice that we omitted the word *then*, which is perfectly proper to do since it is implied. The *if* clause we termed the *hypothesis*, and the *then* clause (the part following the comma) the *conclusion.*

    Another form of the same idea is the *subject-predicate* form. Thus, we could make the above conditional statement as follows: "A tired horse walks." The subject (A tired horse) is the hypothesis, and the predicate (walks) is the conclusion. The reason for going into all this is that the subject-predicate appears most commonly in geometric proofs. Let's look at another example of these two forms of conditional statement, one above the other for purposes of comparison.

Forms	Hypothesis (what is given)	Conclusion (What is to be proved)
Subject-Predicate Form A heated metal expands.	Subject A heated metal	Predicate expands
If-Then Form If a metal is heated, then it expands.	*If* Clause If a metal is heated	*Then* Clause then it expands

Identify the hypothesis and conclusion of each of the following statements.

(a)   If it's Tuesday, this is Belgium.

_____

(b)   If it's the American flag, its colors are red, white, and blue.

_____

(c)   Jet planes are the fastest.

_____

(d)   Stars twinkle.

_____

— — — — — — — — — — — — — —

	Hypothesis	*Conclusion*
(a)	If it's Tuesday	this is Belgium
(b)	If it's the American flag	its colors are red, white, and blue
(c)	Jet planes	are the fastest
(d)	Stars	twinkle

43. Now let's apply this general approach to some statements pertaining to geometric relationships so that you'll begin to associate this method of reasoning with the kinds of terms and situations we will be working with throughout the remainder of this chapter. First we will work with the subject-predicate form.

Determine the hypothesis and conclusion in each of the following statements.

		Hypothesis (Subject)	Conclusion (Predicate)
(a)	An equilateral triangle is equiangular.	_____	_____
(b)	A triangle is not a quadrilateral.	_____	_____
(c)	Perpendiculars form right angles.	_____	_____
(d)	Complements of the same angle are congruent.	_____	_____

- - - - - - - - - - - - - - -

(a)	An equilateral triangle	is equiangular
(b)	A triangle	is not a quadrilateral
(c)	Perpendiculars	form right angles
(d)	Complements of the same angle	are congruent

44. Let's turn our attention now to conditional statements of the if-then form.

Determine the hypothesis and conclusion of each of the following statements.

		Hypothesis (if-clause)	Conclusion (then-clause)
(a)	If a line bisects an angle, then it divides the angle into two congruent parts.	_____	_____
(b)	If two angles are right angles, they are congruent.	_____	_____
(c)	If a line divides an angle into two congruent parts, it is an angle bisector.	_____	_____
(d)	A triangle has an obtuse angle if it is an obtuse triangle.	_____	_____

- - - - - - - - - - - - - - -

(a)	If a line bisects an angle	then it divides the angle into two congruent parts
(b)	If two angles are right angles	(then) they are congruent

(c) If a line divides an angle   (then) it is an angle bisector
     into two congruent parts
(d) If it is an obtuse triangle   (then) a triangle has an obtuse
                                         angle

45. Another term you will need to be familiar with is *converse*. The *converse* of a statement is formed by interchanging the hypothesis and conclusion. To form the converse of an if-then statement, therefore, we simply interchange the if-clause and the then-clause. In the case of the subject-predicate form, interchange subject and predicate.

   Thus, the converse of "Rectangles are quadrilaterals (i.e. four-sided figures)" is "Quadrilaterals are rectangles." Similarly, the converse of "If a metal is heated, then it expands" is "If a metal expands, then it is heated." (Note that although the original statement is true in each of the foregoing examples, its converse is not necessarily true.) This leads us to the following two principles:

(1) The converse of a true statement is not necessarily true. Thus, the statement "Triangles are polygons" is true. Its converse, however, *need not* be true.

(2) The converse of a *definition* is always true. Thus, the converse of the definition "A triangle is a polygon of three sides" is "A polygon of three sides is a triangle." Both the definition and its converse are true.

In the following examples state whether the given statement is true, then form its converse and state whether this is necessarily so.

(a) A square is a rectangle.

(b) A right angle is smaller than an obtuse angle.

(c) An equilateral triangle is a triangle that has all equal sides.

– – – – – – – – – – – –

(a) Statement is true. Its converse, "A rectangle is a square," is not necessarily true since the sides of a rectangle do not all have to be of the same length.
(b) Statement is true. Its converse, "An obtuse angle is smaller than a right angle," is *not* true since by definition an obtuse angle is one that is greater than $90°$ (but less than $180°$).

(c)   Statement is true. Its converse, "A triangle that has all equal sides is an equilateral triangle" also is true since the original statement is a definition.

PROVING A THEOREM

46.   Let's summarize a few of the things we have learned about theorems. We know, for example, that a theorem is a statement to be proved. Obviously, therefore, it cannot be accepted as true until it *has* been proved. We also know that all theorems in geometry consist of two parts: a part that states what is given or known, called the *given* or *hypothesis,* and a part that is to be proved, termed the *conclusion* or *proof.* We have learned too that theorems can be written either as an if-then sentence, or as a simple declarative sentence, known also as subject-predicate form.

In frame 41 we worked with some elementary types of proof. Now we are going to consider a somewhat more formal (and more common) method of proof.

The formal proof of a theorem consists of five parts: (1) a statement of the theorem; (2) a general figure illustrating the theorem; (3) a statement of what is given; (4) a statement of what is to be proved; and (5) a logical series of statements supported by accepted definitions, axioms, postulates, and previously proved theorems. Although it is not considered part of the formal proof, it often is helpful to include a brief *analysis* or *plan* describing your approach to proving the theorem.

Write down, on a separate piece of paper, the five parts of a formal proof of a theorem and compare them with the answer shown below.

— — — — — — — — — — — — — —

(1)   A statement of the theorem.
(2)   A general figure illustrating the theorem.
(3)   A statement of what is given.
(4)   A statement of what is to be proved.
(5)   A logical series of statements supported by accepted definitions, axioms, postulates, and previously proved theorems.

47.   Actually there is no requirement that proofs be presented in formal form as we are going to do. They could be given just as conclusively in paragraph form. However, in paragraph form both you and others would have greater difficulty in following the line of reasoning. Long experience has shown that putting statements of proof in one column and the reasons justifying them in an adjacent column makes it easier

both for you and others to follow your line of reasoning. Below is an example of a formal proof.

THEOREM: *All right angles are*
                      congruent.

Given: $\angle A$ and $\angle B$ are rt. $\angle$s.
Prove: $\angle A \cong \angle B$
Plan: Since each angle equals 90°,
                 the angles are congruent,
                 using Ax. 1: Things congruent to the same thing are
                 congruent to each other.

PROOF: Statements   Reasons

Statements	Reasons
1. $\angle A$ and $\angle B$ are rt. $\angle$s.	1. Given
2. $\angle A$ and $\angle B$ each have 90°.	2. A rt. $\angle$ has 90°.
3. $\angle A \cong \angle B$.	3. Things congruent to the same thing are congruent to each other. (Axiom 1.)

Here are a few guidelines relating to formal proofs which you will find useful. Some of them are illustrated above.

(1)   Notice that the conclusion you are working toward is stated directly below the *Given* and is identified by *Prove*. Your objective is to try to reach this conclusion by making a series of *Statements*, each of which must be justified either by the fact that it is part of the *Given* or by the definitions or postulates that have been agreed upon.

(2)   Whenever the same reason appears more than once in a proof it is not necessary to restate it; just say "Same as 2" (or whatever statement it first appeared in).

(3)   Markings on the diagram should include helpful symbols such as square corners for right angles, cross marks for congruence, question marks for parts to be proved equal or congruent, and so forth (see frame 13).

(4)   The plan is advisable but is not an *essential* part of the proof. If included it should state the major methods of proof to be used.

(5)   The *Given* and *Prove* must refer to the figures and letters of the diagram.

(6)   The last statement is the one to be proved. Statements must refer to the figures and letters of the diagram.

(7)   A reason must be given for each statement. Acceptable reasons are: given facts, definitions, axioms, postulates, assumed theorems, and theorems previously proved.

When searching for reasons to confirm your statements remember to check the axioms and postulates we discussed in frames 23 through 40. These plus the definitions (frames 2 through 16) and the principles (frame 17) we have covered will be your main sources of support for your statements, at this point.

Incidentally, in case you still are in doubt as to why we need formal proofs to support what appear to be obvious conclusions, keep in mind what we said earlier, that one of the chief contributions of geometry was its development of deductive reasoning. And deductive reasoning involves a chain of reasoning from certain general, accepted definitions and assumptions to specific conclusions. It is the basic method of mathematics and one of the main things geometry teaches. Being "formal" in our proofs simply amounts to being clear and consistent. Also, what appears *obvious* from an intuitive viewpoint in a simple case may be quite difficult to prove with any exactitude in a less obvious case.

Now it is time for you to try your hand at a proof. Below is a partially completed theorem for you to work with. A plan is included to assist you in following the approach taken. Your job is to fill in the missing reasons.

THEOREM: *If two angles are complements of the same angle, they are congruent.*

A — a — B — b

Given: $\angle A$ and $\angle B$ are complementary to $\angle X$.

Prove: $\angle A \cong \angle B$

Plan: Using the subtraction axiom (Axiom 6) the same angle may be subtracted from the angles complementary to it. The remainders are congruent angles.

X — x

PROOF: Statements

Statements	Reasons
1. $\angle A$ and $\angle B$ are complementary to $\angle X$.	1.
2. $a + x = 90°$ $b + x = 90°$	2.
3. Hence $a + x = b + x$	3.
4. $x = x$	4.
5. $a = b$	5.
6. $\therefore \angle A = \angle B$	6.

Reasons: 1. Given
          2. Complementary angles are angles whose sum is a right angle. (By definition.)
          3. Things equal to the same or equal things are equal to each other. (Axiom 1.)
          4. Any number is equal to itself. (Axiom 4.)
          5. If equals are subtracted from equals, the differences are equal. (Axiom 6.)
          6. Definition of congruence.

48. A *corollary* is a theorem that is closely related to another theorem, an assumption, or a definition. Thus, we can state as a corollary to the above theorem: If two angles are complements of congruent angles, they are congruent. We could prove the corollary in the same way.

    Now go back and study the theorem in frame 47 carefully. Using it as a guide, write the *complete* proof of the following theorem.

THEOREM: *If two angles are supplements of the same angle, they are congruent.*

Given:

Prove: (diagram)

Plan:

PROOF: Statements   Reasons

Statements	Reasons
1.	1.
2.	2.
3.	3.
4.	4.
5.	5.
6.	6.

– – – – – – – – – – – – – –

Given: $\angle A$ and $\angle B$ are supplementary to $\angle X$.
Prove: $\angle A \cong \angle B$

Plan: Using the subtraction axiom, the same angle may be subtracted from each of the pair of supplementary angles. The
remainders are the required angles.

PROOF: Statements   Reasons

1. $\angle A$ and $\angle B$ are supplementary to $\angle X$.	1. Given.
2. $a + x = 180°$ and $b + x = 180°$	2. Supplementary angles are angles whose sum equals $180°$. (By definition.)
3. Hence $a + x = b + x$	3. Things equal to the same or equal things are equal to each other. (Axiom 1.)
4. $x = x$	4. Any number is equal to itself. (Axiom 4.)
5. $a = b$	5. If equals are subtracted from equals, the differences are equal. (Axiom 6.)
6. $\therefore \angle A \cong \angle B$	6. Definition of congruence.

49.   A corollary to the above theorem would be: If two angles are supplements of congruent angles, they are congruent.

Here is another theorem for you to prove, just for practice. On a sheet
of paper, write your list of statements and reasons.

THEOREM: *If two straight lines intersect, the vertical angles are
congruent.*

Given: The straight lines a and b
intersecting in the point P
and forming pairs of vertical
angles 1 and 2, 3 and 4.
Prove: $\angle 1 \cong \angle 2$ and $\angle 3 \cong \angle 4$

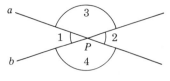

Hint: If $\angle 1$ and $\angle 2$ are supplements of the same angle then, by the
theorem you just proved above, they are congruent.

— — — — — — — — — — — — — — —

The above hint is also your plan, or analysis.

PROOF: Statements	Reasons
1. *a* and *b* are straight lines intersecting at *P* and forming vertical ∠s 1 and 2, 3 and 4.	1. Given
2. ∠1 and ∠3 are supplementary.	2. If the exterior sides of two adjacent angles lie in a straight line, then the angles are supplementary (frame 16).
3. ∠2 and ∠3 are supplementary.	3. Same as 2.
4. ∠1 ≅ ∠2	4. If two angles are supplements of the same angle, they are congruent. (Pr. 4)

In the next chapter we're going to go on to the subject of congruent triangles. But before we do let's review briefly what we have covered so far so you won't get lost.

What We Have Covered So Far

(1)   *Definitions of Basic Geometric Terms*—point, line, surface, line segments, bisectors; properties of circles, angles, polygons, triangles; pairs of angles and certain principles relating to angle pairs.

(2)   *Methods of Proof*—logical reasoning, inductive reasoning, deductive reasoning, assumptions, general statements, particular statements; the syllogism, major and minor premises, conclusion; conditional statements; the Fundamental Rule of Inference.

(3)   *Geometric Reasoning*—10 major axioms; 9 postulates; assumptions.

(4)   *Basic Angle Theorems*—explanation of a theorem; 5 basic principles or theorems applied.

(5)   *Determining Hypothesis and Conclusion*—subject-predicate and if-then forms of conditional statement; applying logical reasoning to geometric relationships; forming the converse of a statement.

(6)   *Proving a Theorem*—two-column form of proof; practice in proving some basic theorems.

Having agreed upon some basic definitions, established a method of proof, stated a number of necessary axioms, postulates, and principles, and had a little practice proving theorems, we are now ready for a test of your understanding of the material covered in this chapter. Take the Self-Test which follows.

SELF-TEST

1. Write the statement needed to complete each of the syllogisms below.

Major Premise (General Statement)	Minor Premise (Particular Statement)	Conclusion (Deduced Statement)
(a) Straight angles are congruent	$\angle A$ and $\angle B$ are straight angles.	_____
(b) Even numbers are divisible by 2.	_____	Numbers ending in an even number are divisible by 2.
(c) _____	Triangles are polygons.	Triangles have as many angles as sides.

(frame 18)

2. Write the logical consequent of the two statements below.

  1. If the sun is shining, then we will go on a picnic.
  2. The sun is shining.

  3. _____

(frame 19)

3. Indicate whether or not you feel the following reasoning is correct.

  1. If it is night, then it is dark outside.
  2. It is dark outside.
  3. Therefore it is night.   (Correct, Incorrect)

(frame 20)

4. Indicate if the following reasonsing is correct or incorrect.

  1. If a person is a child, he likes candy.
  2. Suzie Smith is not a child.
  3. Suzie Smith does not like candy.   (Correct, Incorrect)

(frame 21)

5. In each of the following state the conclusion that follows when Axiom 1 is applied to the given data.

(a) $a = 7, c = 7, f = 7$

_____

(b) $f = h, h = a$

_____

(c) $b = d, d = g, g = e$

_____

(d) $\angle 1 = 50°, \angle 3 = 50°, \angle 4 = 50°$.

_____

(e) $\angle 1 \cong \angle 2, \angle 2 \cong \angle 3, \angle 3 \cong \angle 4$.

_____

(f) $\angle 1 \cong \angle 4, \angle 2 \cong \angle 4, \angle 3 \cong \angle 4$.

_____

(frame 23)

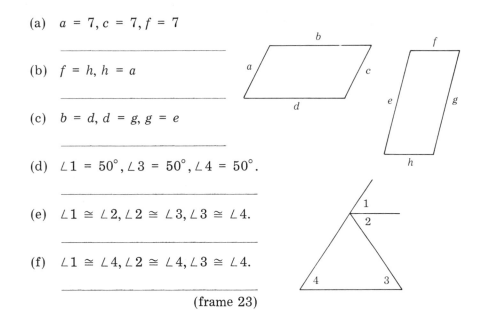

6. What conclusion follows when Axiom 2 is applied below?

(a) Evaluate $a^2 + 3a$ when $a = 10$. _____

(b) Does $b^2 - 8 = 17$ when $b = 5$? _____

(c) Find y if $x + y = 20$ and $y = 3x$. _____

(d) Find x if $x^2 + 3y = 45$ and $y = 3$. _____

(e) Find y if $x + y + z = 180°$, $x = y$, and $z = 80°$. _____

(frame 24)

7. State the conclusion that follows when
Axiom 3 is applied to the given data.

_____

(frame 25)

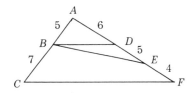

8. Apply Axioms 4, 5, and 6 below.

    (a) Given: $b = e$; find relationship of
         BA to DF.

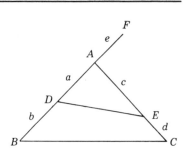

    (b) Given: $b = c$, $a = d$; find relation-
         ship of AB to AC.

(frame 28)

9. In the figure at the right, AD and BC are trisected. Answer the questions
    below by referring to the appropriate axiom.

    (a) If $AD \cong BC$, why is $AE \cong BF$?

    _____

    (b) If $EG \cong FH$, why is $AG \cong BH$?

    _____

    (c) If $GD \cong HC$, why is $AD \cong BC$?

    _____

    (d) If $ED \cong FC$, why is $EG \cong FH$?

    _____

(frame 30)

10. State the basic angle theorem needed to provide an answer to each of
     the following.

    (a) Why is $\angle 1 \cong \angle 2$?

    _____

    (b) Why is $\angle DBC \cong \angle ECB$?

    _____

    (c) If $\angle 3 \cong \angle 4$, why is $\angle 5 \cong \angle 6$?

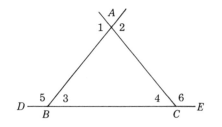

    _____

(frame 41)

11. Identify the hypothesis and conclusion of each of the following statements.

	Hypothesis	*Conclusion*
(a) If you like hot sauce, you will like Mexican food.	_____	_____
(b) If the figure is a pentagon, it has five sides.	_____	_____
(c) Angles equal to the same angle are equal to each other.	_____	_____
(d) Fat people are happy.	_____	_____
(e) A three-sided figure is a triangle.	_____	_____

(frames 42 and 43)

12. Indicate whether each of the following statements is true, then form its converse and state whether this is necessarily true.

(a) California is a state of the United States.

(b) A quadrilateral is a polygon.

(c) A house is a home.

(d) A man is a two-legged creature.

(frame 45)

13. Give the formal proof of the following theorem: *Straight angles are congruent.* (Hint: Note that each straight angle is equal to 180°, hence Axiom 1 applies.)   (frame 47)

Answers to Self-Test

1. (a) $\angle A$ and $\angle B$ are congruent.
   (b) Numbers ending in an even number are even numbers.
   (c) Polygons have as many angles as sides.

2. We will go on a picnic.

3. Incorrect. Step 2 asserts the truth of the consequent instead of affirming the truth of the antecedent, hence step 3 represents a conclusion that is not necessarily true. Therefore, the reasoning is incorrect.

4. Incorrect. Denial of the antecedent does not imply the truth of a denial of the consequent. Although Suzie Smith is not a child she can still like candy.

5. (a) $a = c = f$
   (b) $f = a$
   (c) $b = e$
   (d) $\angle 1 \cong \angle 3 \cong \angle 4$
   (e) $\angle 1 \cong \angle 4$
   (f) $\angle 1 \cong \angle 2 \cong \angle 3$

6. (a) 130
   (b) Yes
   (c) $y = 15$
   (d) $x = \pm 6$
   (e) $y = 50°$

7. $AC = 12$, $AE = 11$, $AF = 15$, $DF = 9$.

8. (a) $BA \cong DF$
   (b) $AB \cong AC$

9. (a) If equals are divided by equals, the quotients are equal. (Axiom 8.)
   (b) Doubles of equals are equal. (Axiom 7.)
   (c) If equals are multiplied by equals, the products are equal. (Axiom 7.)
   (d) Halves of equals are equal. (Axiom 8.)

10. (a) Vertical angles are congruent.
    (b) All straight angles are congruent.
    (c) Supplements of contruent angles are congruent.

11.

	Hypothesis	Conclusion
(a)	If you like hot sauce	you will like Mexican food.
(b)	If the figure is a pentagon	it has five sides.
(c)	Angles congruent to the same angle	are congruent to each other.
(d)	Fat people	are happy.
(e)	A three-sided figure	is a triangle.

12. (a) True. Its converse, "A state of the United States," is not necessarily true; it might be any of the 50 states.

(b) True. Its converse, "A polygon is a quadrilateral," is not necessarily true; a polygon can have any number of sides.

(c) True. Its converse, "A home is a house," is not necessarily true since a home to some people might be a boat or a cave.

(d) True. Its converse, "A two-legged creature is a man," is not necessarily true; it might be an ape.

13. Given: $\angle A$ is a straight angle.
    $\angle B$ is a straight angle.
    Prove: $\angle A \cong \angle B$

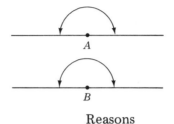

PROOF: Statements

1. $\angle A$ is a straight angle.
2. $\angle A = 180°$.
3. $\angle B$ is a straight angle.
4. $\angle B = 180°$.
5. $\angle A \cong \angle B$

Reasons

1. Given
2. Definition of a straight angle.
3. Given
4. Same as 2
5. Things equal (or congruent) to the same thing are equal (or congruent) to each other. (Axiom 1.)

CHAPTER TWO

Plane Geometry:
Congruency and Parallelism

In this chapter we will consider the geometric methods used to prove congruency in triangles. We shall also investigate the unique properties of parallel lines and of the geometric figures that contain parallel lines. When you have finished this chapter you will be able to:

- prove the congruency of triangles by five basic methods, and use the properties of isosceles and equilateral triangles to assist your proof;

- apply the basic properties of parallel lines and the relationship between the angles formed by a transversal in solving geometric problems and proving theorems;

- apply the relationships and special properties of the angles of a triangle and of a polygon to solve for unknown parts;

- recognize and use the properties of parallelograms, trapezoids, medians, and midpoints to solve geometric problems.

CONGRUENT TRIANGLES

1. *Congruent figures* are figures that have the same size and shape. In other words, they are exact duplicates of each other. Such figures can be made to coincide so that their corresponding parts will fit together. Thus, two circles having the same length radius are congruent circles.

   Congruent triangles are triangles that have the same size and shape. Thus, if two triangles are congruent, their corresponding sides and angles are congruent. And, once again, the symbol for congruency is ≅, which means "is congruent to." The congruent triangles ABC and $A'B'C'$ at the right and on the following page have congruent corresponding sides ($AB \cong A'B'$, $BC \cong B'C'$, and $AC \cong A'C'$) and congruent corresponding angles ($\angle A \cong \angle A'$, $\angle B \cong \angle B'$, and $\angle C \cong \angle C'$).

We designate this congruency as
$\triangle ABC \cong \triangle A'B'C'$ and read this as
"Triangle *ABC* is congruent to triangle
A-prime, *B*-prime, *C*-prime."
Note in the congruent triangles how
corresponding congruent parts can be
located: corresponding sides lie opposite congruent angles and corresponding angles lie opposite congruent sides.

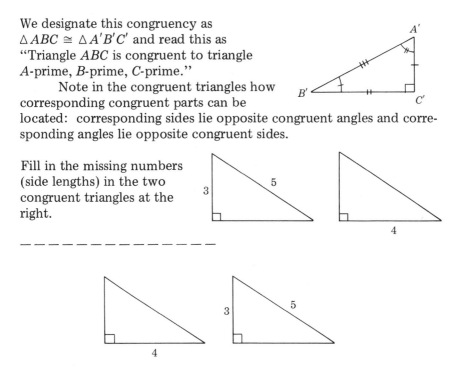

Fill in the missing numbers
(side lengths) in the two
congruent triangles at the
right.

2. Having decided what congruent triangles *are*, we're now going to consider four basic principles relating to congruent triangles, the last three
of which actually are methods of *proving* that triangles are congruent.

Pr. 1: If two triangles are congruent,
their corresponding parts are
congruent. (Or, corresponding
parts of congruent triangles are
congruent.)

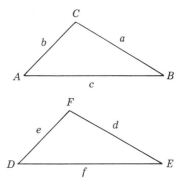

Thus, if $\triangle ABC \cong \triangle DEF$, then $\angle A \cong \angle D$,
$\angle B \cong \angle E$, $\angle C \cong \angle F$, $a = d$, $b = e$, and
$c = f$.

Pr. 2: Two triangles are congruent if
two sides and the included angle
of one triangle are congruent to
the corresponding parts of the
other triangle. (the side-angle-side
or SAS postulate)

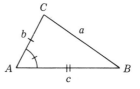

Thus, if $b = e$, $\angle A \cong \angle D$, and $c = f$, then $\triangle ABC \cong \angle DEF$.

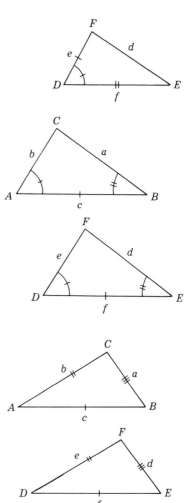

Pr. 3:   Two triangles are congruent if two angles and the included side of one triangle are congruent to the corresponding parts of the other triangle. (the angle-side-angle or ASA postulate)

Thus, if $\angle A \cong \angle D$, $c = f$, and $\angle B \cong \angle E$, then $\triangle ABC \cong \angle DEF$.

Pr. 4:   Two triangles are congruent if three sides of one triangle are congruent to three sides of another. (the side-side-side or SSS postulate)

Thus, if $a = d$, $b = e$, and $c = f$, then $\triangle ABC \cong \triangle DEF$.

From the following groups of three triangles select the two that are congruent in each case and state the congruency principle involved.

(a)

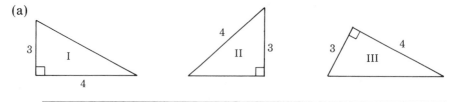

(b)

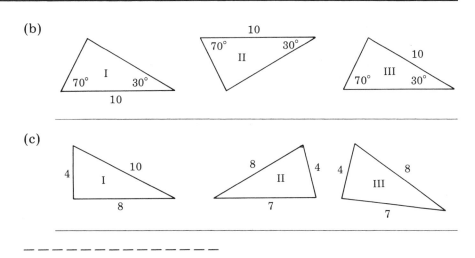

(c)

(a)   Triangles I and III, SAS.
(b)   Triangles I and II, ASA.
(c)   Triangles II and III, SSS.

If your answers are incorrect be sure you understand why before you go
on. These congruency principles are basic to the rest of the chapter.

3.   In the problems above you merely had to select the congruent triangles
     and then state the congruency principle (that is, ASA, SAS, or SSS).
     The problems below require you to *prove*, as simply as possible,
     that $\triangle I \cong \triangle II$ in each case.

Example: Given: $\angle 1 \cong \angle 4$
            $\angle 2 \cong \angle 3$
     Prove: $\triangle I \cong \triangle II$

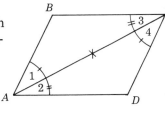

Solution: AC is a common side to both
triangles, and since the two sets of adjacent angles are congruent, then
$\triangle I \cong \triangle II$, ASA.

(a)   Given: $BE \cong EC$
               $AE \cong ED$
      Prove: $\triangle I \cong \triangle II$

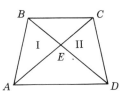

(b) Given: Isosceles $\triangle ABC$
          Isosceles $\triangle ADC$
          (see Chapter 1, frame 14 for
          definition of isosceles)
     Prove: $\triangle$ I $\cong$ $\triangle$ II

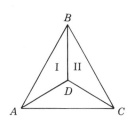

— — — — — — — — — — — — — — —

(a) $\angle 1$ and $\angle 2$ are vertical angles, hence con-
     gruent. And since the two pairs of
     adjacent $\angle$s are congruent, then $\triangle$ I $\cong$ $\triangle$ II,
     SAS.

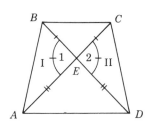

(b) BD is a common side of both triangles I
     and II. And since $\triangle$ ABC and ADC both
     are isosceles, then $AB \cong BC$ and $AD \cong$
     DC, hence $\triangle$ I $\cong$ $\triangle$ II, SSS.

4. Now let's exercise the idea of congruency of triangles in a slightly differ-
    ent way, but still using the congruency principles from frame 2.
        For each of the diagrams below state the *additional congruencies*
    needed to prove $\triangle$ I $\cong$ $\triangle$ II by the congruency principle indicated.

    (a) By SSS. _____

    (b) By SAS. _____

    (c) By ASA. _____

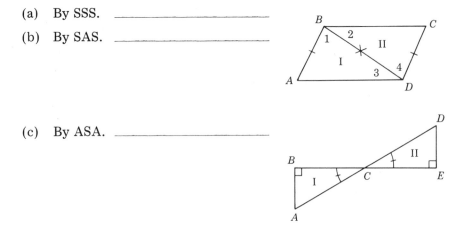

(d) By ASA. _____

(e) By SAS. _____

- - - - - - - - - - - - - - - -

(a) $AD \cong BC$. Since side BD is a common side and $AB \cong CD$, if
     $AD \cong BC$ then $\triangle$ I $\cong \triangle$ II by SSS.

(b) $\angle 1 \cong \angle 4$. Since side BD is common and $AB \cong CD$, if $\angle 1 \cong \angle 4$
     then $\triangle$ I $\cong \triangle$ II by SAS.

(c) $BC \cong CE$. Since $\angle$s B and E are right $\angle$s and the vertical angles are
     congruent, if $BC \cong CE$ then $\triangle$ I $\cong \triangle$ II by ASA.

(d) $\angle 2 \cong \angle 3$. Since B and D are right $\angle$s and $BC \cong CD$, if $\angle 2 \cong \angle 3$
     then $\triangle$ I $\cong \triangle$ II by ASA.

(e) $AB \cong DE$. Again, since $BC \cong CD$ and B and D are right $\angle$s, if
     $AB \cong DE$ then $\triangle$ I $\cong \triangle$ II by SAS.

5. Every valid geometric proof should be independent of the figure used
    to illustrate the problem. Figures are used merely as a matter of con-
    venience. Strictly speaking, before a congruency such as that shown in
    problem (c), frame 4, could be proved, it should be stated that: (1) A,
    B, C, D, and E are five points lying in the same plane; (2) C is between
    B and E; and (3) C is between A and D. However, to include informa-
    tion such as this which can be inferred from the figure would make the
    proof tedious and repetitious. In this book, therefore, it will be permis-
    sible to use the figure to imply such things as betweenness, collinearity
    of points (i.e., points lying in the same line), the location of a point in
    the interior or exterior of an angle or in a certain half-plane (i.e., points
    lying on the same side of a line), and the general relative position of
    points, lines, and planes.

    However, the student must be careful *not* to infer congruence of
    segments and angles, perpendicular and parallel lines, and bisectors of
    segments and angles just because "they appear that way" in the figure.
    Such things must be included in the hypotheses or in the developed
    proofs. With this caution in mind, let's proceed to an example of a
    proof in which we *use* the concept of congruency to prove congruence
    between two sides of two different triangles.

Example: Given: $BF \perp DE$
                    $BF \perp AC$
                    $\angle 3 \cong \angle 4$
            Prove: $AF \cong FC$
            Plan: Prove $\triangle$ I $\cong \triangle$ II

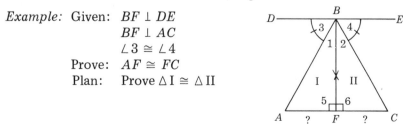

PROOF: Statements	Reasons
1. $BF \perp AC$	1. Given
2. $\angle 5 \cong \angle 6$	2. ⊥s form rt. ∠s. Rt. ∠s are ≅.
3. $BF \cong BF$	3. Identity
4. $BF \perp DE$	4. Given
5. $\angle 1$ is the complement of $\angle 3$. $\angle 2$ is the complement of $\angle 4$.	5. Adjacent ∠s are complementary if exterior sides are ⊥ to each other.
6. $\angle 3 \cong \angle 4$	6. Given
7. $\angle 1 \cong \angle 2$	7. Complements of ≅ angles are ≅.
8. $\triangle \text{I} \cong \triangle \text{II}$	8. ASA
9. $AF \cong FC$	9. Corresponding parts of △ are ≅.

Use this same general approach (that is, establishing congruency) to prove the following.

Given: AD bisects $\angle BAC$; $AD \perp BC$
Prove: D is the midpoint of BC.
Plan: Prove D is the midpoint by showing that $BD \cong DC$. Since these line segments will be congruent if $\triangle ABD \cong \triangle ACD$, the problem really is one of showing that these triangles are congruent.

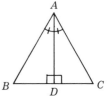

PROOF: Statements	Reasons
1. AD bisects $\angle BAC$.	1. Given
2. $\angle BAD \cong \angle CAD$	2. Def. of bisector of an angle
3. $AD \perp BC$	3. Given
4. $\angle ADB$ and $\angle ADC$ are rt. ∠s.	4. Def. of perpendicular lines
5. $\angle ADB \cong \angle ADC$	5. Right angles are congruent
6. $AD \cong AD$	6. Identity
7. $\triangle ABD \cong \triangle ACD$	7. ASA
8. $BD \cong DC$	8. Corresp. parts of ≅ triangles
9. D is the midpoint of BC.	9. Def. of midpoint of a line segment

6. Now let's try proving a congruency problem stated in words.

Prove: If the opposite sides of a quadrilateral (four-sided figure) are congruent and a diagonal is drawn, congruent angles are formed between the diagonal and the congruent sides.

The *hypothesis* is the portion of the above statement *before* the comma; the *conclusion* is the portion *after* the comma. From the statement we can derive the following information.

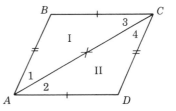

Given: Quadrilateral $ABCD$
        $AB \cong CD$, $BC \cong AD$
        AC is a diagonal
Prove: $\angle 1 \cong \angle 4$, $\angle 2 \cong \angle 3$
Plan: Prove $\triangle I \cong \triangle II$

PROOF: Statements

Statements	Reasons
1. $AB \cong CD$, $BC \cong AD$	1. Given
2. $AC \cong AC$	2. Identity
3. $\triangle I \cong \triangle II$	3. SSS
4. $\angle 1 \cong \angle 4$, $\angle 2 \cong \angle 3$	4. Corresp. parts of congruent triangles are congruent

Prove the following statement: *If the diagonals of a quadrilateral bisect each other, then its opposite sides are congruent.*

Given: Quadrilateral $ABCD$
      AC bisects BD
      BD bisects AC
Prove: $AB \cong CD$
      $BC \cong AD$
Plan: Show that $\triangle AEB \cong \triangle DEC$ and $\triangle BEC \cong \triangle AED$, hence the corresponding sides are congruent.

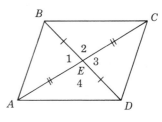

PROOF: Statements

Statements	Reasons
1. $AE \cong EC$ and $BE \cong ED$	1. Def. of a bisector
2. $\angle 1 \cong \angle 3$ and $\angle 2 \cong \angle 4$	2. Vertical angles are congruent
3. $\triangle AEB \cong \triangle DEC$ and $\triangle BEC \cong \triangle AED$	3. SAS
4. $AB \cong CD$ and $BC \cong AD$	4. Corresp. parts of $\cong \triangle$

ISOSCELES AND EQUILATERAL TRIANGLES

7. In frame 14 of Chapter 1 we defined isosceles and equilateral triangles and we have used some of their properties in several of the examples

and problems discussed thus far. However, there are a number of basic principles relating to such triangles which we will state now.

Pr. 1:   If two sides of a triangle are congruent, the angles opposite these sides are congruent. (Base angles of an isosceles triangle are congruent.)

Thus, in △ *ABC*, if *AB* ≅ *BC*, then ∠ *A* ≅ ∠ *C*.

Pr. 2:   If two angles of a triangle are congruent, the sides opposite them are congruent.

Thus, in △ *ABC*, if ∠ *A* ≅ ∠ *C*, then *AB* ≅ *BC*.

Pr. 3:   An equilateral triangle is equiangular.

Thus, in △ *ABC*, if *AB* ≅ *BC* ≅ *CA*, then ∠ *A* ≅ ∠ *B* ≅ ∠ *C*. (Pr. 3 is a corollary of Pr. 1.)

Pr. 4:   An equiangular triangle is equilateral.

Thus, in △ *ABC*, if ∠ *A* ≅ ∠ *B* ≅ ∠ *C*, then *AB* ≅ *BC* ≅ *CA*. (This is the converse of Pr. 3 and a corollary of Pr. 2.)

Now, how do we apply these principles? Let's start with Principles 1 and 3: In a triangle, congruent angles are opposite congruent sides.

Example: In the triangle at the right, state which congruent angles are opposite the congruent sides.
Solution: Since *AC* ≅ *BC* (they each have a length of 12), then ∠ *A* ≅ ∠ *B*.

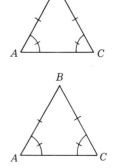

In the following three problems state which congruent angles are opposite the congruent sides.

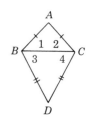

(a) ⎯⎯⎯⎯⎯⎯⎯⎯⎯⎯⎯⎯⎯⎯⎯⎯⎯⎯⎯⎯⎯⎯

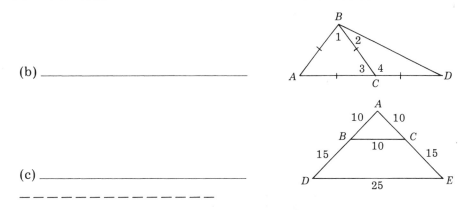

(b) _____

(c) _____

- - - - - - - - - - - - - - - -

(a) Since $AB \cong AC$, then $\angle 1 \cong \angle 2$; and since $BD \cong CD$, then
$\angle 3 \cong \angle 4$.

(b) Since $AB \cong AC \cong BC$, then $\angle A \cong \angle 1 \cong \angle 3$; and since $BC \cong CD$,
then $\angle 2 \cong \angle D$.

(c) Since $AB \cong BC \cong AC$, then $\angle A \cong \angle ACB \cong \angle ABC$; and since
$AE \cong AD \cong DE$, then $\angle A \cong \angle D \cong \angle E$.

8. Principles 2 and 4 we can apply as shown in the example below. We
can summarize these two Principles as follows: *In a triangle, congruent
sides are opposite congruent angles.*

Example: In the triangle at the right, state which
congruent sides are opposite congruent angles.
Solution: Since $\angle a = 55°$, and $\angle a = \angle D$, then
$BC \cong CD$.

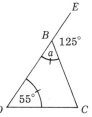

State which congruent sides are opposite congruent angles in each of
the problems below.

(a) _____

(b) _____

(c) _____

- - - - - - - - - - - - - - -

(a)   Since $\angle A \cong \angle 1$, $AD \cong BD$; and since $\angle 2 \cong \angle C$, $BD \cong CD$.
(b)   Since $\angle 1 \cong \angle 3$, $AB \cong BC$; and since $\angle 2 \cong \angle 4 \cong \angle D$,
      $CD \cong AD \cong AC$.
(c)   Since $\angle A \cong \angle 1 \cong \angle 4$, $AB \cong BD \cong AD$; and since $\angle 2 \cong \angle C$,
      $BD \cong CD$.

9.   So far we have "applied" our four Principles regarding isosceles and
     equilateral triangles only in the sense of having recognized the congruent
     angles opposite congruent sides (or the reverse). Now we need to use
     what we have learned to prove congruency.

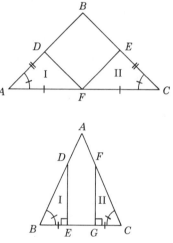

   Example: Prove $\triangle I \cong \triangle II$ and state
   the congruency principle involved.
      Given: $AB \cong BC$
              $AD \cong EC$
              F is the midpoint of AC.
      Prove: $\triangle I \cong \triangle II$
      Solution: Since $AB \cong BC$, then
   $\angle A \cong \angle C$. Therefore $\triangle I \cong \triangle II$, SAS.

   Now it's your turn.
      Given: $AB \cong AC$
              BC trisected at E and G
              $DE \perp BC$, and $FG \perp BC$
      Prove: $\triangle I \cong \triangle II$

   Your reasoning: _____

   - - - - - - - - - - - - - -

   Since $AB \cong AC$, $\angle B \cong \angle C$. Therefore, $\triangle I \cong \triangle II$, ASA.

10. Before leaving the subject of isosceles triangles we should see how the
     principles relating to them are used in a formal proof.

Example:

Given: $AB \cong BC$
        AC is trisected at D and E

Prove: $\angle 1 \cong \angle 2$

Plan: Prove $\triangle\,\text{I} \cong \triangle\,\text{II}$ to obtain
        $BD \cong BE.$

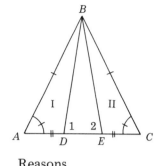

PROOF: Statements

Statements	Reasons
1. AC is trisected at D and E.	1. Given
2. $AD \cong EC$	2. To trisect is to divide into three congruent parts.
3. $AB \cong BC$	3. Given
4. $\angle A \cong \angle C$	4. In a $\triangle$, $\angle$ opposite $\cong$ sides are $\cong$.
5. $\triangle\,\text{I} \cong \triangle\,\text{II}$	5. SAS
6. $BD \cong BE$	6. Corresponding parts of congruent $\triangle$ are $\cong$.
7. $\angle 1 \cong \angle 2$	7. Same as 4.

Since logical proof is the essence of geometry, and because there is no other way to gain skill in proving geometric theorems but by *doing* it, see if you can work out the proof for the following theorem.

THEOREM: *The bisector of the vertex angle of an isosceles triangle is a median to the base.*

Given: Isosceles $\triangle\,ABC$ $(AB \cong BC)$
              BD bisects $\angle B$

Prove: BD is a median to $AC.$
              (Refer to frame 15 of Chapter 1 if you have forgotten what a median is.)

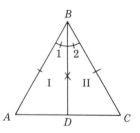

Plan: Prove $\triangle\,\text{I} \cong \triangle\,\text{II}$ to obtain
              $AD \cong DC$

PROOF:

- - - - - - - - - - - - - - - -

PROOF: Statements	Reasons
1. $AB \cong BC$	1. Given
2. BD bisects $\angle B$	2. Given
3. $\angle 1 \cong \angle 2$	3. To bisect is to divide into two congruent parts.
4. $BD \cong BD$	4. Identity
5. $\triangle I \cong \triangle II$	5. SAS
6. $AD \cong DC$	6. Corresponding parts of congruent $\triangle$ are $\cong$.
7. BD is a median to AC	7. A line from a vertex of a $\triangle$ to the midpoint of the opposite side is a median.

Stop now and take the Self-Test which follows, before you go on to the next section, on parallel lines.

SELF-TEST

1. From the following group of four triangles, identify those that are congruent and state the congruency principle involved.

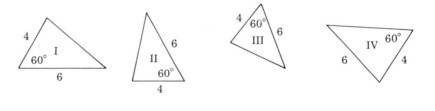

2. Prove (as simply as possible) $\triangle I \cong \triangle II$ in the following problem and state the congruency principle involved.

Given: $BE \perp AD$
    $CF \perp AD$
    $BE \cong CF$
    AD is trisected
Prove: $\triangle I \cong \triangle II$

Proof:

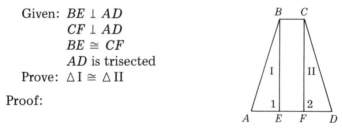

3. State the additional parts needed to prove
    $\triangle$ I $\cong$ $\triangle$ II by the SSS congruency principle.

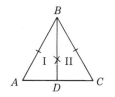

4. Use the two-column format and the congruency method to prove the
    following.

    Given: $\angle 1 \cong \angle 2, BF \cong DE$
                BF bisects $\angle B$
                DE bisects $\angle D$
                $\angle B$ and $\angle D$ are rt $\angle$s

    Prove: $AB \cong CD$

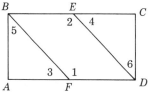

    Plan: _____

    PROOF:

5. Prove: If the legs of one right triangle are congruent respectively to
    the legs of another, their hypotenuses are congruent. (Hint:
    First show congruence.)

6. In the figure at the right, show which
    congruent angles are opposite
    congruent sides of the triangles.

7. In the figure at the right, show
    which congruent sides are opposite
    congruent angles of the triangles.

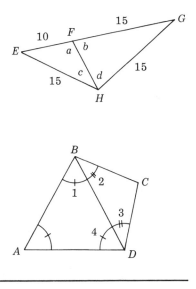

8. Given: $AD \cong CE$
$\qquad \angle 1 \cong \angle 2$
Prove: $\triangle ABD \cong \triangle CBE$
(Not a formal proof; just
show your reasoning.)

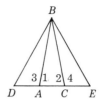

9. Show the formal proof of the following.

Given: $AB \cong AC$
$\qquad F$ is midpoint of BC
$\qquad \angle 1 \cong \angle 2$
Prove: $FD \cong FE$
Plan: Prove $\triangle BDF \cong \triangle CEF$ to
obtain $FD \cong FE$.

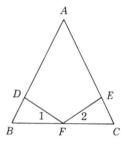

Answers to Self-Test

1. $\triangle I \cong \triangle II \cong \triangle III$, SAS.
Since BE and $CF \perp AD$, $\angle$ 1 and 2 are right $\angle$s, hence congruent. And
since AD is trisected, $AE \cong FD$, also $BE \cong CF$ (given), therefore,
$\triangle I \cong \triangle II$, SAS.
3. $AD \cong DC$
4. Plan: Prove $\triangle ABF \cong \triangle CDE$, hence $AB \cong CD$.

PROOF: Statements	Reasons
1. $\angle 3 \cong \angle 4$	1. Supplements of $\cong$ $\angle$s are $\cong$.
2. $BF \cong DE$	2. Given
3. $\angle B \cong \angle D$	3. Rt. $\angle$s are $\cong$.
4. $\angle 5 \cong \angle 6$	4. Bisected angles are $\cong$.
5. $\triangle ABF \cong \triangle CDE$	5. ASA
6. $AB \cong CD$	6. Corresponding sides of $\cong$ $\triangle$.

5. Given: $AB \cong DE$
$\qquad AC \cong DF$
$\qquad \angle A$ and $\angle D$ are rt. $\angle$s
Prove: $BC \cong EF$
Plan: Prove $\triangle ABC \cong \triangle DEF$, hence
$\qquad BC \cong EF$

PROOF: Statements	Reasons
1. $AB \cong DE$ and $AC \cong DF$	1. Given
2. $\angle A \cong \angle D$	2. Rt. angles are congruent
3. $\triangle ABC \cong \triangle DEF$	3. SAS
4. $BC \cong EF$	4. Corresponding parts of congruent triangles

6. $\angle b \cong \angle d$, $\angle E \cong \angle G$

7. $AB \cong BD \cong AD$, $BC \cong CD$

8. Since $\angle 1 \cong \angle 2$, then $AB \cong BC$ (sides opposite congruent angles). Also, $\angle 3 \cong \angle 4$, since they are supplements of congruent angles. Finally, since $AD \cong CE$ (given), $\triangle ABD \cong \triangle CBE$, SAS.

9. PROOF: Statements

Statements	Reasons
1. $\angle B \cong \angle C$	1. Angles opposite $\cong$ sides are $\cong$.
2. $BF \cong FC$	2. Definition of a midpoint.
3. $\angle 1 \cong \angle 2$	3. Given
4. $\triangle BDF \cong \triangle CEF$	4. ASA
5. $FD \cong FE$	5. Corresponding parts of $\cong \triangle$.

Don't be too concerned if your proofs are not quite as concise as those shown above. This sort of thing takes a lot of practice.

PARALLEL LINES

11. *Parallel lines* are straight lines that lie in the same plane and do not intersect however far they are extended. This concept derives from Euclid's conviction that there is only one line parallel to a given line through a given point. However, because he was unable to prove this, he included it as a postulate. That is, he *assumed* it. Since his day many mathematicians have tried to prove or disprove this postulate by means of other postulates and axioms, but all have failed. As a result, mathematicians have considered what kind of geometry would result if this property were assumed *not* true. Thus, several types of *non-Euclidean* geometry have been developed over a period of a century and a half. All of these have found usefulness in special applications and, granted the assumptions on which they are based, are perfectly valid.

However, since the space available in this Self-Teaching Guide does not permit us to go into any detail regarding these geometries, we will stick to the subject of Euclidean plane geometry for the purposes of our discussion and of this book. If you are interested in learning more about non-Euclidean geometries you will have no trouble finding more information in any standard textbook on geometry or in the separate texts describing the geometries of Lobachevsky, Bolyai, and Riemann.

Now back to Euclid and parallel lines. The symbol for parallel is ||. Thus *AB* || *CD* is read "*AB* is parallel to *CD*."

A *transversal* is a line that cuts across two or more lines. Thus *EF* is a transversal of *AB* and *CD*.

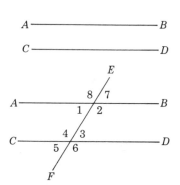

Interior angles are the angles *between* the two lines (1, 2, 3, and 4), while *exterior angles* are those on the outside (5, 6, 7, and 8).

Corresponding angles of two lines cut by a transversal are angles on the *same* side of the transversal line and on the *same* side of the lines. Thus, there are four pairs of corresponding angles: 3 and 7, 4 and 8, 2 and 6, and 1 and 5.

Alternate interior angles are non-adjacent angles *between* the two lines and on opposite sides of the transversal. Thus, 1 and 3, 2 and 4 are alternate interior angles.

Alternate exterior angles are non-adjacent angles *outside* the two lines and on *opposite* sides of the transversal. Thus 5 and 7, 6 and 8 are alternate exterior angles.

Interior angles on the same side of the transversal are 2 and 3, 1 and 4.

All of this may seem like a great deal of attention to give to identifying pairs of angles formed by two lines cut by a transversal. But the fact is that there are a great many principles and properties relating to this situation. It is important, therefore, that we be able to identify the relationships between the various angles as an aid to proving theorems about parallel lines.

See if you can identify the angles shown at the right.

(a)   Interior angles _____

(b)   Exterior angles _____

(c)   Alternate interior angles _____

(d)   Alternate exterior angles _____

(e)   Corresponding angles _____

(f)   Interior angles on the same side of the transversal _____

– – – – – – – – – – – – – – –

(a)   *a*, *b*, *c*, and *d*
(b)   *e*, *f*, *g*, and *h*
(c)   *a* and *c*, *b* and *d*
(d)   *e* and *g*, *f* and *h*
(e)   *h* and *d*, *a* and *e*, *g* and *c*, *b* and *f*
(f)   *b* and *c*, *a* and *d*

12. Now we're going to consider some of the principles of parallel lines.

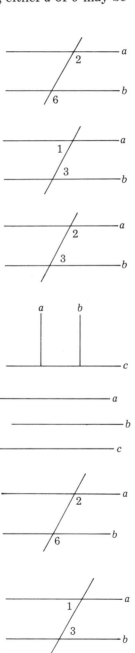

Pr. 1:   Through a given point not on a given
line, one and only one line can be
drawn parallel to a given line.

You will recognize this as the Parallel Line
Postulate we mentioned briefly in frame 11. Thus, either *a* or *b* may be
parallel to *c*, but not both.

Pr. 2:   Two lines are parallel if a pair of
corresponding angles are congruent.

Thus, *a* ∥ *b* if ∠2 ≅ ∠6.

Pr. 3:   Two lines are parallel if a pair of
alternate interior angles are
congruent.

Thus, *a* ∥ *b* if ∠1 ≅ ∠3.

Pr. 4:   Two lines are parallel if a pair of
interior angles on the same side of
a transversal are supplementary.

Thus, *a* ∥ *b* if ∠2 and ∠3 are supplementary.

Pr. 5:   Lines are parallel if they are perpendicular to the same line. (Perpendiculars
to the same line are ∥.)

Thus, *a* ∥ *b* if *a* and *b* are ⊥*c*.

Pr. 6:   Lines are parallel if they are
parallel to the same line. (Parallels
to the same line are parallel.)

Thus, *a* ∥ *b* if *a* and *b* each are ∥ *c*.

Pr. 7:   If two lines are parallel, each pair
of corresponding angles are
congruent. (Corresponding angles
of parallel lines are congruent.)

Thus, if *a* ∥ *b*, then ∠2 ≅ ∠6.

Pr. 8:   If two lines are parallel, each pair of
alternate interior angles are congruent.
(Alternate interior angles of parallel
lines are congruent.)

Thus, if *a* ∥ *b*, then ∠1 ≅ ∠3.

Pr. 9:   If two lines are parallel, each pair of interior angles on the same side of the transversal are supplementary.

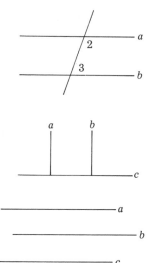

Thus, if $a \parallel b$, then $\angle 2$ and $\angle 3$ are supplementary.

Pr. 10:   If lines are parallel, a line perpendicular to one of them is perpendicular to the other also.

Thus, if $a \parallel b$ and $c \perp a$, then $c \perp b$.

Pr. 11:   If lines are parallel, a line parallel to one of them is parallel to the others also.

Thus, if $a \parallel b$ and $c \parallel a$, then $c \parallel b$.

Pr. 12:   If the sides of two angles are respectively parallel to each other, the angles are either congruent or supplementary.

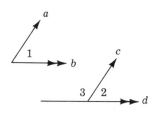

Thus, if $a \parallel c$ and $b \parallel d$, then $\angle 1 \cong \angle 2$ and $\angle 1$ and $\angle 3$ are supplementary.

Use the principles and properties given above to find the missing angular values in the following problems. These represent a numerical application of the principles of parallel lines. (The arrows indicate given pairs of parallel lines.)

Example: Find the values of x and y.
    $x = 130°$ (Pr. 8)
    $y = 180 - 130 = 50°$ (Pr. 9)

Problems:

(a)   Find the values of x and y.

    $x =$ _____

    $y =$ _____

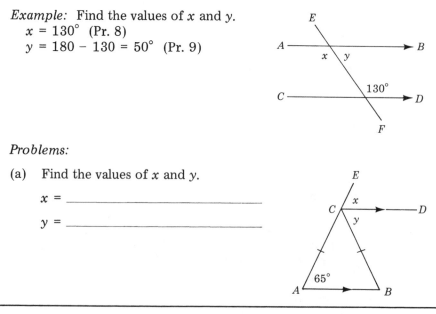

(b)   Find the values of x and y.

$x =$ _____

$y =$ _____

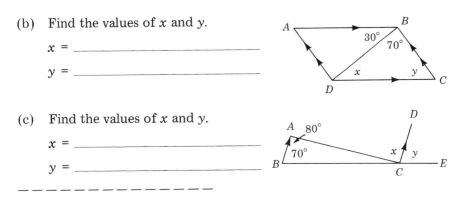

(c)   Find the values of x and y.

$x =$ _____

$y =$ _____

- - - - - - - - - - - - - - - -

(a) $x = 65°$ (Pr. 7). And since $\angle B \cong \angle A$, $\angle B = 65°$. Hence $y = 65°$
(Pr. 8).
(b) $x = 30°$ (Pr. 8). $y = 180 - (30 + 70)$ (Pr. 9), hence $y = 80°$.
(c) $x = 80°$ (Pr. 8). $y = 70°$ (Pr. 7).

13.   Perhaps you noticed that Principles 7 through 11 are simply the converses of Principles 2 through 6 as stated and illustrated in the preceding
frame. Now let's try *applying* these parallel line principles and their
converses.

Example: Given: $\angle 1 \cong \angle 2$
State the parallel line principle needed as
the reason for each of the remaining
statements.

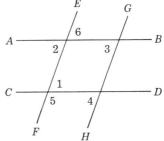

1. $\angle 1 \cong \angle 2$	1. Given
2. $AB \parallel CD$	2. Pr. 3
3. $\angle 3 \cong \angle 4$	3. Pr. 7

Follow this same procedure in the problems below, stating the principle
needed as the reason for each of the remaining statements. (Angle and
line references refer to the diagram above.)

(a)   1. $\angle 2 \cong \angle 3$ 1. Given

2. $EF \parallel GH$ 2. _____

3. $\angle 4$ sup. $\angle 5$ 3. _____

(b)   1. $\angle 5$ sup. $\angle 4$ 1. Given

2. $EF \parallel GH$ 2. _____

3. $\angle 3 \cong \angle 6$ 3. _____

- - - - - - - - - - - - - - - -

(a) Pr. 2, Pr. 9; (b) Pr. 4, Pr. 8

14. It's time now to consider how we use what we have learned so far to establish the formal proof in a parallel line problem.

Example: Given: $AB \cong AC$
     $AE \parallel BC$
  Prove: AE bisects $\angle DAC$
  Plan: Show that $\angle 1$ and $\angle 2$
     are congruent to the
     congruent angles B and C

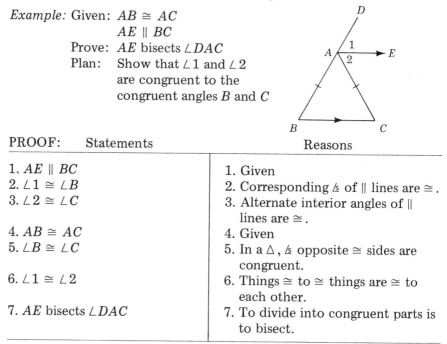

PROOF: Statements   Reasons

Statements	Reasons
1. $AE \parallel BC$	1. Given
2. $\angle 1 \cong \angle B$	2. Corresponding ∡ of $\parallel$ lines are $\cong$.
3. $\angle 2 \cong \angle C$	3. Alternate interior angles of $\parallel$ lines are $\cong$.
4. $AB \cong AC$	4. Given
5. $\angle B \cong \angle C$	5. In a $\triangle$, ∡ opposite $\cong$ sides are congruent.
6. $\angle 1 \cong \angle 2$	6. Things $\cong$ to $\cong$ things are $\cong$ to each other.
7. AE bisects $\angle DAC$	7. To divide into congruent parts is to bisect.

Now together we're going to prove that if the diagonals of a quadrilateral bisect each other, the opposite sides are parallel.

Given: Quad. $ABCD$
   AC and BD bisect each
   other
Prove: $AB \parallel CD$, and $AD \parallel BC$
Plan: Prove $\angle 1 \cong \angle 4$ by
   showing $\triangle$ I $\cong \triangle$ II.
   Prove $\angle 2 \cong \angle 3$ by
   showing $\triangle$ III $\cong \triangle$ IV

PROOF: Statements   Reasons

Statements	Reasons
1. AC and BD bisect each other	1.
2. $BE \cong ED$, $AE \cong EC$	2.
3. $\angle 5 \cong \angle 6, \angle 7 \cong \angle 8$	3.
4. $\triangle$ I $\cong \triangle$ II, $\triangle$ III $\cong \triangle$ IV	4.
5. $\angle 1 \cong \angle 4, \angle 2 \cong \angle 3$	5.
6. $AB \parallel CD$, $BC \parallel AD$	6.

Your job is to fill in the missing Reasons.

– – – – – – – – – – – – – – – –

Reasons: 1. Given
          2. To bisect is to divide into two congruent parts.
          3. Vertical ∡ are congruent.
          4. SAS
          5. Corresponding parts of congruent ▲ are congruent.
          6. Lines cut by a transversal are ‖ if alternate interior angles
          are congruent (Pr. 3).

DISTANCES

15. It is important now that we talk a little about geometric distances and distance principles. Each of the following situations involves the distance between two geometric figures. In each case the distance is measured along a straight line segment that is the *shortest* line between the figures.

A. For the distance *between two points,* such as L and M, use the line segment LM.

B. For the distance *between a point and a line,* such as P and AB, use the line segment PQ, the perpendicular from the point to the line.

C. For the distance *between two parallels,* such as AB and CD, use a line segment like PQ, a perpendicular between the two parallels.

D. For the distance *between a point and a circle,* such as P and circle O, use PQ, the line segment of OP between the point and the circle.

E. For the distance *between two concentric circles,* such as the two circles whose center is O, use PQ, the line segment of the large radius between the two circles.

Following are some of the important distance principles.

Pr. 1:   If a point is on the perpendicular
bisector of a line segment, then it
is equidistant from the ends of the
segment it bisects.

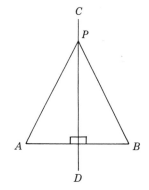

Thus, if *P* is on *CD*, the ⊥ bisector of *AB*,
then *PA* ≅ *PB*.

Pr. 2:   If a point is equidistant from the
ends of a line segment, then it is
on the perpendicular bisector of
the line segment.

Thus, if *PA* ≅ *PB*, then *P* is on *CD*, the ⊥ bisector of *AB*.

Pr. 3:   If a point is on the bisector of an
angle, then it is equidistant from
the sides of the angle.

Thus, if *P* is on *AB*, the bisector of ∠*A*, then
PQ ≅ *PR* where *PQ* and *PR* are the distances
of *P* from the sides of the angle.

Pr. 4:   If a point is equidistant from the
sides of an angle, then it is on the
bisector of the angle.

Thus, if *PQ* ≅ *PR* where *PQ* and *PR* are the distances of *P* from the
sides of ∠*A*, then *P* is on *AB*, the bisector of ∠*A*.

Pr. 5:   Two points each equidistant from the
ends of a line segment determine the
perpendicular bisector of the line
segment. (The line joining the vertices
of two isosceles triangles having a
common base is the perpendicular
bisector of the base.)

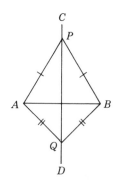

Thus, if *PA* ≅ *PB* and *QA* ≅ *QB*, then *P* and
Q determine *CD*, the ⊥ bisector of *AB*.

Pr. 6: The perpendicular bisectors of the
sides of a triangle meet in a point
that is equidistant from the vertices
of the triangle.

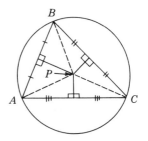

Thus, if *P* is the intersection of the ⊥ bisectors
of the sides of △ *ABC*, then *PA* ≅ *PB* ≅ *PC*.
Also, *P* is the center of the circumscribed
circle and is the *circumcenter* (center of a
circumscribed circle) of △ *ABC*.

Pr. 7: The bisectors of the angles of a
triangle meet in a point that is
equidistant from the sides of the
triangle.

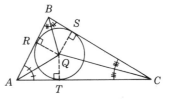

Thus, if *Q* is the intersection of the
bisectors of the angles of △ *ABC*, then
QR ≅ *QS* ≅ *QT* where these are the distances from *Q* to the sides of
△ *ABC*. Also, *Q* is the center of the inscribed circle and is the *incenter*
(center of an inscribed circle) of △ *ABC*.

In order for these distance concepts and
principles to become meaningful you will
need, of course, to apply them. Consider
the following example.

Example: Find the distance and indicate
the *kind* of distance involved:

1. from *P* to *A*. (*PA* = 7, distance between two points.)
2. from *P* to *CD*. (*PG* = 4, distance from a point to a line.)
3. from *A* to *BC*. (*AE* = 10, distance from a point to a line.)
4. from *AB* to *CD*. (*FG* = 6, distance between two ∥ lines.)

Continue to apply what we have
covered in this frame by finding the
distances called for below and indi-
cating the *kind* of distance.

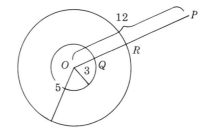

(a) from *P* to the inner circle *O*

_____

(b) from *P* to the outer circle *O*

_____

(c) between the concentric circles

_____

– – – – – – – – – – – – – – –

(a) $PQ = 12 - 3 = 9$, distance from a point to a circle.
(b) $PR = 12 - 5 = 7$, distance from a point to a circle.
(c) $QR = 5 - 3 = 2$, distance between two concentric circles.

16. Now let's practice locating a point in such
     a way as to satisfy some given condition(s).

     Example: Locate *P*, a point on *AB* and
     equidistant from *A* and *C*.
     Solution: Using Pr. 1, we erect a
     perpendicular bisector of *AC*. The point
     at which this line intersects *AB*, at *P*, is the
     point we are seeking. Since *PQ* is a bisector
     of *AC* it is, of course, equidistant from
     A and *C*.

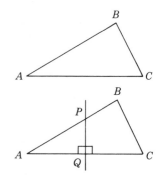

Follow the same general approach, using the appropriate principles, to
locate a point that will satisfy the conditions given in each of the problems below. Draw the point on the figure and write down the principle
you used to locate them.

(a) Locate *Q*, a point on *AB* and
     equidistant from *BC* and *AC*.

     _____

(b) Locate *R*, the center of the
     circumscribed circle of △*ABC*.

     _____

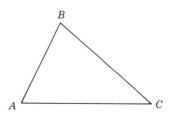

(c) Locate *S*, the center of the inscribed circle of △*ABC*. _____

- - - - - - - - - - - - - - -

(a) Using Pr. 3.   (b) Using Pr. 6.   (c) Using Pr. 7.

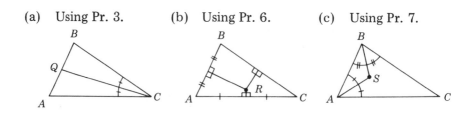

SUM OF THE ANGLES OF A TRIANGLE

17. The sum of the values of the angles of any triangle equals 180° (or, if
     you will, form a *straight angle*).
         This is one of the most widely used theorems of plane geometry,
     and its proof is made possible by the parallel postulate we discussed in

frame 11. Thus, in the figure at the right, if we draw a line through one vertex of the triangle (at B in this case) parallel to the side opposite the vertex, you can see that the straight

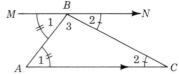

angle B equals the sum of the angles of the triangle. That is, $\angle 1 + \angle 2 + \angle 3 = 180°$. This is true because both $\angle$ 1 and 2 are alternate interior angles of the parallel lines AC and MN and therefore are congruent. $\angle 3$ is, of course, the remaining angle of the triangle and the completing angle at B.

We will be discussing several angle sum principles that derive from or relate to this theorem, but first let's take a moment to consider another concept that we are going to need. This is the interior and exterior angles of a polygon.

In frame 14, Chapter 1, we learned that a *polygon* is a closed figure bounded by straight line segments as sides.

An *exterior angle* of a polygon is formed whenever one of its sides is extended through a vertex. If *each* of the sides of a polygon is extended (as shown at the right), there will be an exterior angle formed at each vertex.

Each of these exterior angles is the *supplement* of its adjacent *interior angle*. In the case of the pentagon (five-sided polygon) $ABCDE$, there will be five exterior angles, one at each vertex. *Notice that each exterior angle is the supplement of an adjacent interior angle.* For example, $\angle a + \angle a' = 180°$.

Now back to our angle sum principles.

Pr. 1:   The sum of the angles of a triangle equals a straight angle or $180°$.

Thus, in $\triangle ABC$, $\angle A + \angle B + \angle C = 180°$.

Pr. 2:   If two angles of one triangle are congruent respectively to two angles of another triangle, the remaining angles are congruent.

Thus, in $\triangle ABC$ and $\triangle A'B'C'$, if $\angle A \cong \angle A'$ and $\angle B \cong \angle B'$, then $\angle C \cong \angle C'$.

Pr. 3:   The sum of the values of the angles of a quadrilateral equals $360°$.

Thus, in the quadrilateral $ABCD$, $\angle A + \angle B + \angle C + \angle D = 360°$.

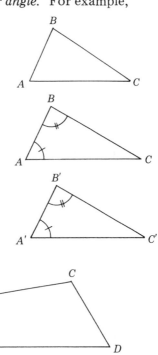

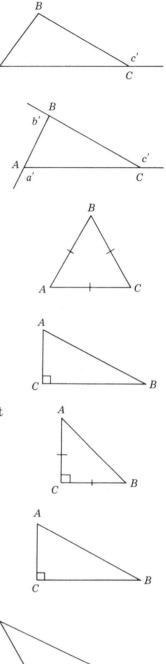

Pr. 4:   The value of each exterior angle
of a triangle equals the sum of
its two non-adjacent (opposite)
interior angles.

Thus, in $\triangle ABC$, $\angle c' = \angle A + \angle B$.

Pr. 5:   The sum of the exterior angles
of a triangle equals $360°$.

Thus, in $\triangle ABC$, $\angle a' + \angle b' + \angle c' = 360°$.

Pr. 6:   Each angle of an equilateral
triangle equals $60°$.

Thus, if $\triangle ABC$ is equilateral, then $\angle A = 60°$,
$\angle B = 60°$, and $\angle C = 60°$.

Pr. 7:   The acute angles of a right triangle
are complementary.

Thus, in the rt. $\triangle ABC$, if $\angle C = 90°$, then
$\angle A + \angle B = 90°$.

Pr. 8:   Each acute angle of an isosceles right
triangle equals $45°$.

Thus, in isos. rt. $\triangle ABC$, if $\angle C = 90°$, then
$\angle A = 45°$ and $\angle B = 45°$.

Pr. 9:   A triangle can have no more than
one right angle.

Thus, in rt. $\triangle ABC$, if $\angle C = 90°$, then $\angle A$ and
$\angle B$ cannot be rt. $\angle$s .

Pr. 10:   A triangle can have no more
than one obtuse angle.

Thus, in obtuse $\triangle ABC$, if $\angle C$ is obtuse,
then $\angle A$ and $\angle B$ cannot be obtuse
angles.

Pr. 11: Two angles are congruent or supplementary if their sides are respectively perpendicular to each other.

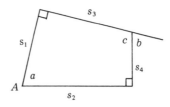

Thus, if $s_1 \perp s_3$ and $s_2 \perp s_4$, then $\angle a \cong \angle b$ and $\angle a$ and $\angle b$ are supplementary.

These principles may all seem a bit abstract to you at this point. But don't worry; we'll find plenty of use for them later. Now it's time to apply some of these principles.

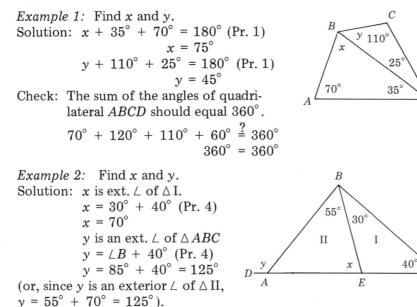

Example 1: Find x and y.
Solution: $x + 35° + 70° = 180°$ (Pr. 1)
$$x = 75°$$
$$y + 110° + 25° = 180° \text{ (Pr. 1)}$$
$$y = 45°$$
Check: The sum of the angles of quadrilateral $ABCD$ should equal $360°$.
$$70° + 120° + 110° + 60° \overset{?}{=} 360°$$
$$360° = 360°$$

Example 2: Find x and y.
Solution: x is ext. $\angle$ of $\triangle$ I.
$$x = 30° + 40° \text{ (Pr. 4)}$$
$$x = 70°$$
y is an ext. $\angle$ of $\triangle ABC$
$$y = \angle B + 40° \text{ (Pr. 4)}$$
$$y = 85° + 40° = 125°$$
(or, since y is an exterior $\angle$ of $\triangle$ II,
$y = 55° + 70° = 125°$).

Find x and y in the following.

(a) _____

$x =$ _____

_____

$y =$ _____

(b) _____

$x =$ _____

_____

$y =$ _____

_ _ _ _ _ _ _ _ _ _ _ _ _ _ _

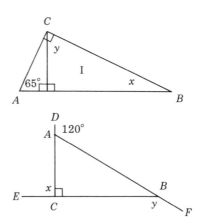

(a)   In $\triangle ABC$, $x + 65° = 90°$ (Pr. 7)

$$x = 25°$$

In $\triangle I$, $x + y = 90°$ (Pr. 7)

$$25° + y = 90°$$

$$y = 65°$$

(b)   Since $DC \perp EB$, $x = 90°$

$$x + y + 120° = 360°\ \text{(Pr. 5)}$$

$$90° + y + 120° = 360°$$

$$y = 150°$$

18.   So you may become a little more familiar with their special properties we're going to apply some of our angle sum principles to isosceles and equilateral triangles.

Example: Find x and y.
Solution: By Pr. 8, $x = 45°$.
  Since $\angle ABC = 60°$ (Pr. 6)
  and $\angle CBD = 45°$ (Pr. 8)
  $y = 60° + 45° = 105°$.

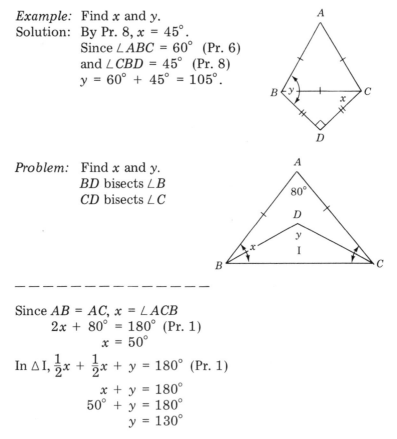

Problem: Find x and y.
  BD bisects $\angle B$
  CD bisects $\angle C$

– – – – – – – – – – – – – – –

Since $AB = AC$, $x = \angle ACB$

$$2x + 80° = 180°\ \text{(Pr. 1)}$$

$$x = 50°$$

In $\triangle I$, $\frac{1}{2}x + \frac{1}{2}x + y = 180°$ (Pr. 1)

$$x + y = 180°$$

$$50° + y = 180°$$

$$y = 130°$$

19.   Some of the solutions to the preceding problems involving the application of angle sum principles required the use of a little basic algebra. Does solving for x and y and working with simple linear equations seem

familiar to you? It certainly should from your study of algebra. Try to use what you have learned about algebra wherever you get a chance. It will simplify your work and sharpen your skills. With this thought in mind let's try using algebra to prove an angle sum problem. Don't be alarmed, it's not going to be anything very difficult.

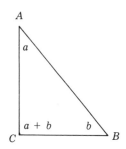

Prove: If one angle of a triangle equals the sum of the other two, then the triangle is a right triangle.

   Given: $\triangle ABC$

             $\angle C = \angle A + \angle B$

   Prove: $\triangle ABC$ is a right triangle

   Plan: Prove $\angle C = 90°$

Algebraic Proof:

      Let a = number of degrees in $\angle A$

        b = number of degrees in $\angle B$

      Then $a + b$ = number of degrees in $\angle C$

$$a + b + (a + b) = 180° \text{ (Pr. 1)}$$
$$2a + 2b = 180°$$
$$a + b = 90°$$

   Since $\angle C = 90°$, $\triangle ABC$ is a right triangle.

Not that hard, was it? Ready to try one?

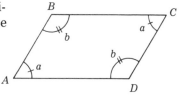

Prove: If the opposite angles of a quadrilateral are equal, then its opposite sides are parallel.

   Given: Quadrilateral $ABCD$

             $\angle A \cong \angle C, \angle B \cong \angle D$

   Prove: $AB \parallel CD$ and $BC \parallel AD$

   Plan: Prove int. $\angle$s on same side of transversal are supplementary.

Algebraic Proof:

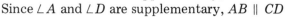

Let a = number of degrees in $\angle A$ and $\angle C$

   b = number of degrees in $\angle B$ and $\angle D$

$$2a + 2b = 360° \text{ (Pr. 3)}$$
$$a + b = 180°$$

Since $\angle A$ and $\angle B$ are supplementary, $BC \parallel AD$

Since $\angle A$ and $\angle D$ are supplementary, $AB \parallel CD$

SUM OF THE ANGLES OF A POLYGON

20. Once again (see frame 14, Chapter 1, frame 17, Chapter 2), a polygon is a closed figure in a plane bounded by straight line segments as sides.

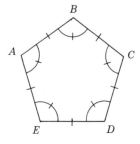

   An *n-gon* is a polygon of *n* sides. Thus a polygon of 20 sides is a 20-gon.

   A *regular polygon* is an equilateral and equiangular polygon. Thus, a regular *pentagon* (you learned in frame 17) is a polygon having 5 equal angles and 5 equal sides. A *square* is a regular polygon of 4 sides.

   Since it frequently is convenient to identify polygons according to the number of sides, below is a chart that may be helpful to you in learning their names.

Sides	Polygon	Sides	Polygon
3	Triangle	8	Octagon
4	Quadrilateral	9	Nonagon
5	Pentagon	10	Decagon
6	Hexagon	12	Dodecagon
7	Heptagon	*n*	*n*-gon

Without looking back see how many of the following polygons you can name correctly.

2 sides _____   12 sides _____

7 sides _____   3 sides _____

9 sides _____   14 sides _____

4 sides _____

— — — — — — — — — — — — — —

2 sides: no such thing   12 sides: Dodecagon
7 sides: Heptagon   3 sides: Triangle
9 sides: Nonagon   14 sides: 14-gon
4 sides: Quadrilateral

21. There are some very interesting things to learn about polygons. For example, by drawing diagonals from any vertex to each of the other vertices, a polygon of 7 sides is divisible into 5 triangles. Each of these triangles contains one side of the polygon except the first and last triangles, which contain two such sides.

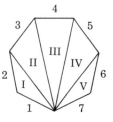

In general, this process will divide a polygon of n sides into $(n - 2)$ triangles. That is, the number of such triangles is always two less than the number of sides of the polygon.

The sum of the interior angles of the polygon equals the sum of the interior angles of the triangles. Hence: Sum of interior angles of a polygon of n sides = $(n - 2)180°$.

Another important fact concerns the *exterior* angles of a polygon. In the figure at the right notice the one-to-one correspondence between each vertex of the polygon and each pair of angles marked. Angles a, b, c, d, and e are *interior* angles whereas their supplements — angles a', b', c', d', and e' — are *exterior* angles.

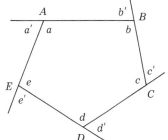

The sum of the angles at the five vertices is, of course, 5 times $180°$ or $900°$. On the other hand, the sum of the *interior* angles is, as we have just learned, $(n - 2)180°$ or $3(180)° = 540°$. Thus the sum of the *exterior* angles is $900° - 540° = 360°$.

Hence we can conclude that: The sum of the exterior angles of a polygon of n sides = $360°$.

What we have learned above about the interior and exterior angles of a polygon is summed up in the following polygon angle principles.

Pr. 1: If S is the sum of the interior angles of a polygon of n sides, then S (measured in st. ∡) = $n - 2$, or S (in degrees) = $(n - 2)180°$.
(Note: by st. (straight) angles we mean angles of $180°$.)

Thus, the sum of the interior angles of a polygon of 10 sides (decagon) equals: $(n - 2)180° = 8(180°) = 1440°$.

Pr. 2: The sum of the exterior angles of any polygon equals $360°$.

Thus, the sum of the exterior angles of a polygon of 23 sides equals $360°$.

Pr. 3: If a *regular polygon* of n sides has an interior angle i and an exterior angle e (in degrees), then $i = \dfrac{180°(n - 2)}{n}$,

$e = \dfrac{360°}{n}$, and $i + e = 180°$.

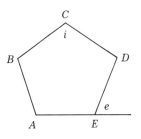

Thus, for a regular polygon of 20 sides,

$i = \dfrac{180°(20 - 2)}{20} = 162°$, $e = \dfrac{360°}{20} = 18°$,

and $i + e = 162° + 18° = 180°$.

Now for a little practice in using these principles.

(a)   Find the sum of the interior angles of a polygon of 9 sides and express your answer in straight angles and in degrees.

(b)   Find the number of *sides* of a polygon if the sum of the interior angles is $3600°$.

(c)   Is it possible to have a polygon the sum of whose interior angles is $1890°$?

————————————————

(a) S (in straight angles) $= n - 2 = 9 - 2 = 7$ st. $\measuredangle$.
       S (in degrees) $= (n - 2)180° = 7(180°) = 1260°$.
(b) S (in degrees) $= (n - 2)180°$. Then $3600° = (n - 2)180°$, from which $n = 22°$.
(c)   Since $1890° = (n - 2)180°$, then $n = 12\frac{1}{2}$. A polygon cannot have $12\frac{1}{2}$ sides, hence the answer is No.

22.   In the problems that follow you are to apply the angle principles to a *regular* polygon. Don't forget (frame 20) that the angles and sides of a regular polygon are all *equal.*

(a)   Find the exterior angle (the word "angle" here is singular because there is just *one* value for the exterior angles, that is, they are all the same size) of a regular polygon having 9 sides. (Hint: Apply Pr. 2.)

(b)   Find each interior angle (don't let the "each" fool you; they're all the same size) of a regular polygon having 9 sides. (Hint: Use Pr. 1.)

(c) Find the number of sides of a regular polygon if each exterior angle is 5°. (Hint: Substitute 5 for e in your exterior angle formula and solve for n.)

(d) Find the number of sides of a regular polygon if each interior angle is 165°. (Hint: Substitute 165° for i in your $i + e$ formula from Pr. 3 to find the value of e, then substitute this value for e in the exterior angles formula to solve for the number of sides.)

- - - - - - - - - - - - -

(a) Since $n = 9$, $e = \dfrac{360°}{9} = 40°$.

(b) Since $n = 9$, $i = \dfrac{(n - 2)180°}{n} = \dfrac{(9 - 2)180°}{9} = 140°$.

Or, since $i + e = 180°$, $i = 180° - e = 180° - 40° = 140°$.

(c) Substituting $e = 5°$ in $e = \dfrac{360°}{n}$, gives us $5° = \dfrac{360°}{n}$, from which

$5n = 360$, or $n = 72$ sides.

(d) Substituting $i = 165°$ in $i + e = 180°$, we get $e = 15°$.

Using $e = 15°$ in the formula $e = \dfrac{360°}{n}$ gives us $n = 24$ sides.

TWO NEW CONGRUENCY THEOREMS

23. So far we have studied three methods of proving triangles congruent. These are (from frame 2), side-angle-side (SAS), angle-side-angle (ASA), and side-side-side (SSS). Now we are going to discuss two additional ways in which to prove that triangles are congruent.

Pr. 1:   If two angles and a side opposite one of them of one triangle are congruent to the corresponding parts of another, the triangles are congruent. (side-angle-angle or SAA theorem)

Thus, if $\angle A \cong \angle A'$, $\angle B \cong \angle B'$, and $BC \cong B'C'$, then $\triangle ABC \cong \triangle A'B'C'$.

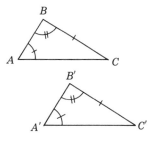

Pr. 2:   If the hypotenuse (H) and a leg (L) of one right triangle are congruent to the corresponding parts of another right triangle, the triangles are congruent.

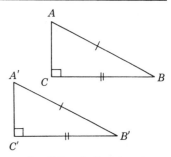

Thus, if hyp. $AB \cong$ hyp. $A'B'$, and leg $BC \cong$ leg $B'C'$, then rt. $\triangle ABC \cong$ rt. $\triangle A'B'C'$.

Use these new theorems to prove $\triangle I \cong \triangle II$ in each of the following two problems. This will not require formal proof. Just show the equal parts in the diagrams containing the two triangles and state the reason for the congruency.

(a)   Given: $BD \perp BC$, $BD \perp AD$, and
         $AB \cong CD$
      Prove: $\triangle I \cong \triangle II$

(b)   Given: $AB \cong BC$
         $FD \perp AB$
         $FE \perp BC$
         F is the midpoint of AC
      Prove: $\triangle I \cong \triangle II$

––––––––––––––––––––

(a)   Since triangles I and II are right triangles, contain a common side, and their hypotenuses are congruent, $\triangle I \cong \triangle II$, by HL.

(b)   Since FE and FD are perpendicular respectively to sides BC and AB, triangles I and II are right triangles. And since $AB \cong BC$, their opposite angles, A and C, are congruent. Also, since F is a midpoint, then $AF \cong CF$. Therefore, $\triangle I \cong \triangle II$, by SAA.

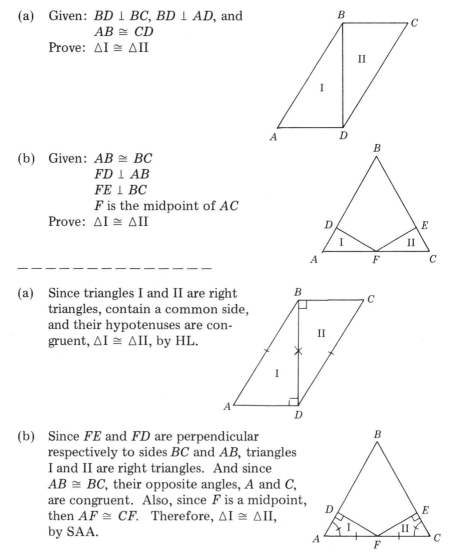

24. Now we're going to collaborate by developing a formal proof for congruency using one of the theorems you've just learned. You'll have to discover as you go along which one it is. Your part will be to fill in the missing reasons in the following proof.

Given: Quadrilateral $ABCD$
       $DF \perp AC$, $BE \perp AC$
       $AE \cong FC$, $BC \cong AD$

Prove: $BE \cong FD$

Plan: Prove $\triangle I \cong \triangle II$

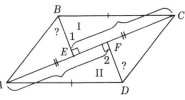

PROOF: Statements	Reasons
1. $BC \cong AD$	1.
2. $DF \perp AC$, $BE \perp AC$	2.
3. $\angle 1 \cong \angle 2$	3.
4. $AE \cong FC$	4.
5. $EF \cong EF$	5.
6. $AF \cong EC$	6.
7. $\triangle I \cong \triangle II$	7.
8. $BE \cong FD$	8.

– – – – – – – – – – – – – – –

Reasons: 1. Given
          2. Given
          3. Perpendiculars form rt. $\angle$, and all rt. $\angle$ are congruent.
          4. Given
          5. Identity
          6. If equals are added to equals, the sums are equal.
          7. HL
          8. Corresponding parts of congruent $\triangle$ are congruent.

SELF-TEST

1. Find the values of x and y in the figure at the right.

(frame 12)

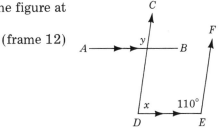

2. State the parallel line principle needed as the reason for each of the remaining statements.

    1. $EF \perp AB$, $GH \perp AB$,
        $EF \perp CD$ 1. Given

    2. $EF \parallel GH$ 2. _____

    3. $CD \perp GH$ 3. _____

                 (frame 13)

3. Prove the following.

    Given: Quadrilateral $ABCD$
           $AB \cong CD$, $BC \cong AD$
    Prove: $AB \parallel CD$, $BC \parallel AD$
               (frame 14)

4. Given: $AC \cong BC$
          $\angle B \cong \angle E$
    Prove: $AB \parallel DE$
         (frame 14)

5. Find the following distances.

    (a) From A to B. _____

    (b) From E to AC. _____

    (c) From A to BC. _____

    (d) From ED to BC. _____
             (frame 15)

6. Find the following distances.

    (a) From P to the outer circle. _____

    (b) From P to the inner circle. _____

    (c) Between the concentric circles. _____

    (d) From P to O. _____

7. (a)   Locate P, a point on AD, equidistant
          from B and C.

    (b)   Locate Q, a point on AD, equidistant
          from AB and BC.
                              (frame 16)

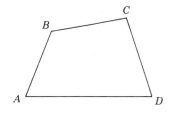

8. Find x and y. Given: CD bisects $\angle C$.

    $x =$ _____

    $y =$ _____
                              (frame 17)

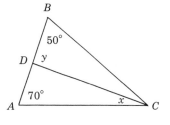

9. Find x and y. Given: AD bisects $\angle A$.

    $x =$ _____

    $y =$ _____
                              (frame 18)

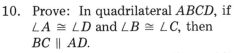

10. Prove: In quadrilateral $ABCD$, if
    $\angle A \cong \angle D$ and $\angle B \cong \angle C$, then
    $BC \parallel AD$.
                              (frame 19)

11. Show that a triangle is equilateral
    if its angles are represented by
    $x + 15$, $3x - 75$, and $2x - 30$.
                              (frame 19)

12. (a) Find the sum of the interior angles (in straight angles) of a polygon of 9 sides.

    (b) 32 sides.   (frame 21)

13. (a) Find each exterior angle of a regular polygon having 18 sides.

    (b) 20 sides.

    (c) 40 sides.   (frame 22)

14. Prove triangles I and II are congruent. On the diagram show the congruent parts of both triangles and state the reason for congruency.

    Given: Isosceles triangle ABC $(AB \cong BC)$
            BD is altitude to AC
    Prove: $\triangle I \cong \triangle II$ (Use informal reasoning.)
                                              (frame 23)

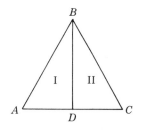

15. Given: $\angle B \cong \angle D$
            $BC \parallel AD$
    Prove: $BC \cong AD$ (By formal proof.)
                          (frame 24)

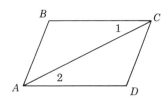

Answers to Self-Test

1. $x = 180° - 110° = 70°$ (Pr. 9)
   $y = 110°$ (Pr. 12)
2. 2. Pr. 5; 3. Pr. 10
3. PROOF:

Statements	Reasons
1. $AB \cong CD$, $BC \cong AD$	1. Given
2. $BD \cong BD$	2. Identity
3. $\triangle I \cong \triangle II$	3. SSS
4. $\angle ADB \cong \angle DBC$	4. Corresponding parts of $\cong$ $\triangle$.
5. $BC \parallel AD$	5. Lines cut by a transversal are $\parallel$ if the alternate interior $\angle$ are $\cong$.
6. $\angle ABD \cong \angle BDC$	6. Same as 4.
7. $AB \parallel CD$	7. Same as 5.

4. PROOF:

Statements	Reasons
1. $AC \cong BC$	1. Given
2. $\angle A \cong \angle B$	2. Angles opposite the $\cong$ sides.
3. $\angle B \cong \angle E$	3. Given
4. $\angle A \cong \angle E$	4. Things $\cong$ to the same thing are $\cong$ to each other.
5. $AB \parallel DE$	5. Lines cut by a transversal are $\parallel$ if the alt. interior $\angle$ are $\cong$.

5. (a) 25; (b) 9; (c) 20; (d) 8
6. (a) 8; (b) 10; (c) 2; (d) 14
7.

8. $x = 30°$, $y = 100°$
9. $x = 56°$, $y = 68°$
10. Given: Quadrilateral $ABCD$
    $\angle A \cong \angle D$, $\angle B \cong \angle C$
    Prove: $BC \parallel AD$
    Plan: Prove int. $\angle$ on same side of transversal (of $\parallel$ lines) are supplementary.
    Proof: Let a = number of degrees in $\angle A$ and $\angle D$.
    b = number of degrees in $\angle B$ and $\angle C$.
    $2a + 2b = 360°$ (Pr. 3)
    $a + b = 180°$
    Since $\angle A$ and $\angle B$ are supplementary, $BC \parallel AD$.

11. $(x + 15) + (3x - 75) + (2x - 30) = 180$ (Pr. 1)
     Hence $6x = 270$
     or $x = 45$, therefore each angle = $60°$.

12. (a) $S = n - 2$, or $9 - 2 = 7$ st. ∠
     (b) $32 - 2 = 30$ st. ∠.

13. (a) $e = \dfrac{360}{n}$, or $e = \dfrac{360}{18} = 20°$

     (b) $\dfrac{360}{20} = 18°$

     (c) $\dfrac{360}{40} = 9°$

14. Since BD is an altitude to AC, it is perpendicular
     to AC (by definition), hence triangles I and II
     are rt. ∠. And because $\triangle ABC$ is isosceles,
     $AB \cong BC$ and $\angle A \cong \angle C$. Therefore
     $\triangle I \cong \triangle II$, HL.

15. PROOF: Statements

Statements	Reasons
1. $\angle B \cong \angle D$	1. Given
2. $BC \parallel AD$	2. Given
3. $\angle 1 \cong \angle 2$	3. Alternate interior angles.
4. $AC \cong AC$	4. Identity
5. $\triangle I \cong \triangle II$	5. SAA
6. $BC \cong AD$	6. Corresponding parts.

PARALLELOGRAMS, TRAPEZOIDS, MEDIANS, AND MIDPOINTS

25. Beginning with frame 11 we considered some of the properties of
     parallel lines and the relationships between the angles produced when
     two parallel lines were cut by a transversal. We also discussed quadri-
     laterals (four-sided figures) and polygons in general. Now we are going
     to give some attention to particular *kinds* of quadrilaterals.
          The first figure we will consider is the
     trapezoid. A *trapezoid* is a quadrilateral
     having two — and only two — parallel sides.
     The *bases* of a trapezoid are its parallel
     sides. The *legs* are its non-parallel sides.
     The *median* of a trapezoid is the line
     segment joining the midpoints of its legs.

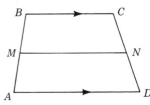

     Thus, in trapezoid $ABCD$, the bases are AD and BC, and the legs are AB
     and CD. If M and N are midpoints, then MN is the median of the
     trapezoid.

An *isosceles trapezoid* is a trapezoid whose legs are congruent. Thus, in the isosceles trapezoid *ABCD*, *AB* ≅ *CD*. The base angles of a trapezoid are the angles at the ends of its longer base. Thus, ∠*A* and ∠*D* are the base angles of the isosceles trapezoid *ABCD*.

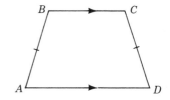

The following principles apply to trapezoids.

Pr. 1:   The base angles of an isosceles trapezoid are congruent.

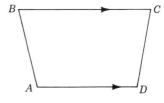

Thus, in trapezoid *ABCD*, if *AB* ≅ *CD*, then ∠*A* ≅ ∠*D* (and ∠*B* ≅ ∠*C*).

Pr. 2:   If the base angles of a trapezoid are congruent, the trapezoid is isosceles.

Thus, in trapezoid *ABCD*, if ∠*A* ≅ ∠*D*, then *AB* ≅ *CD*.

Answer (or complete) the following questions and statements about trapezoids.

(a)   How many parallel sides does a trapezoid have? ⎯⎯⎯⎯⎯⎯

(b)   The bases of a trapezoid are its ⎯⎯⎯⎯⎯⎯⎯⎯⎯⎯ .

(c)   The legs of a trapezoid are its ⎯⎯⎯⎯⎯⎯⎯⎯⎯⎯ sides.

(d)   The median of a trapezoid is a line joining which two points?

⎯⎯⎯⎯⎯⎯⎯⎯⎯⎯⎯⎯⎯⎯⎯⎯⎯⎯⎯⎯⎯⎯⎯⎯

(e)   How does an isosceles trapezoid differ from other trapezoids?

⎯⎯⎯⎯⎯⎯⎯⎯⎯⎯⎯⎯⎯⎯⎯⎯⎯⎯⎯⎯⎯⎯⎯⎯

(f)   What is unique about the base angles of an isosceles trapezoid?

⎯⎯⎯⎯⎯⎯⎯⎯⎯⎯⎯⎯⎯⎯⎯⎯⎯⎯⎯⎯⎯⎯⎯⎯

⎯ ⎯ ⎯ ⎯ ⎯ ⎯ ⎯ ⎯ ⎯ ⎯ ⎯ ⎯ ⎯ ⎯ ⎯

(a) 2; (b) parallel sides; (c) non-parallel; (d) the midpoints of its legs; (e) its legs are congruent; (f) they are congruent

26.   Now that you know a little something about trapezoids, you will want to apply your knowledge. A good way to exercise what you know is by using it algebraically to find the missing values of certain angles in a trapezoid.

Example: *ABCD* is a trapezoid.
         Find x and y.

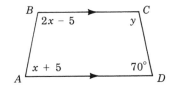

Solution: Since $AD \parallel BC$, $(2x - 5) +$
$(x + 5) = 180°$, $3x = 180°$, or $x = 60°$.
$y + 70° = 180°$, or $y = 110°$.

Follow this procedure in solving the following.

(a)   *ABCD* is an isosceles trapezoid.
      Find x and y.

(b)   *ABCD* is an isosceles trapezoid.
      $\angle B : \angle A = 3 : 2$
      Find x and y.

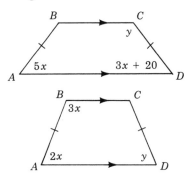

- - - - - - - - - - - - - - - -

(a)   Since $\angle A \cong \angle D$, $5x = 3x + 20$
$$2x = 20$$
$$x = 10$$
      Since $BC \parallel AD$, $y + (3x + 20) = 180$
$$y + 50 = 180$$
$$y = 130$$

(b)   Let the number of degrees in $\angle B$ and $\angle A$ be $3x$ and $2x$ respectively.
      Since $BC \parallel AD$, $3x + 2x = 180$, or $x = 36$.
      Since $\angle D = \angle A$, $y = 2x$, or $y = 72$.

27. A *parallelogram* is a quadrilateral whose
     opposite sides are parallel. The symbol
     for parallelogram is $\square$.   Thus, in
     $\square ABCD$, $AB \parallel CD$ and $AD \parallel BC$.

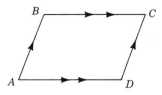

     If the opposite sides of a quadri-
     lateral are parallel, then it is a
     parallelogram. (This is simply the converse of the above definition.)
     Thus, if $AB \parallel CD$ and $AD \parallel BC$, then *ABCD* is a $\square$. Now let's con-
     sider some of the properties of parallelograms.

     Pr. 1:   The opposite sides of a parallelo-
               gram are parallel. (Definition)

Pr. 2:   A diagonal of a parallelogram divides it into two congruent triangles.

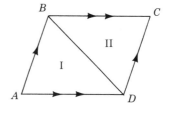

Thus, if *BD* is a diagonal of ▱*ABCD*, then △I ≅ △II.

Pr. 3:   The opposite sides of a parallelogram are congruent.

Thus, in ▱*ABCD*, *AB* ≅ *CD* and *AD* ≅ *BC*.

Pr. 4:   The opposite angles of a parallelogram are congruent.

Thus, in ▱*ABCD*, ∠*A* ≅ ∠*C* and ∠*B* ≅ ∠*D*.

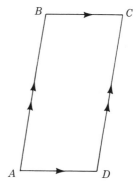

Pr. 5:   Any two adjacent angles of a parallelogram are supplementary.

Thus, in ▱*ABCD*, ∠*A* is the supplement of ∠*B* or ∠*D*, ∠*B* is the supplement of ∡ *A* and *C*, and so on.

Pr. 6:   The diagonals of a parallelogram bisect each other.

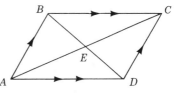

Thus, in ▱ *ABCD*, *AE* ≅ *EC* and *BE* ≅ *ED*.

Take a few minutes to review what you have read above and then answer the following questions.

(a)   Is the figure at the right a parallelo-

gram? _____ Why? _____

_____

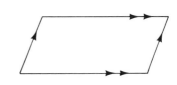

(b)   What does a diagonal of a parallelo-

gram divide it into? _____

(c)   What relationship do the opposite sides of a parallelogram bear to one another (other than the fact they are parallel, of course)?

_____

(d)   What relationship exists between the opposite angles of a

parallelogram? _____

(e)   What is the relationship between angles C and D in the figure at the right? _____

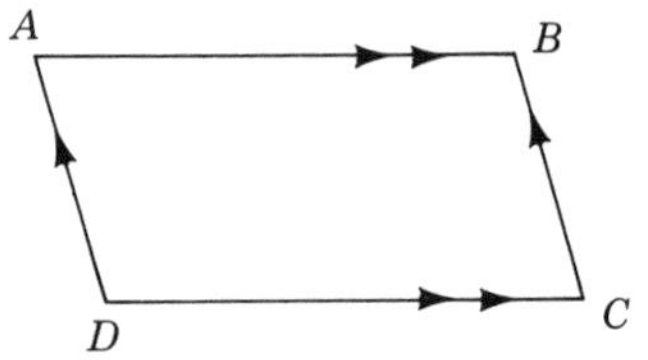

(f)   What is the relationship between line segments BE and DE in the figure at the right? _____

- - - - - - - - - - - - - - - - -

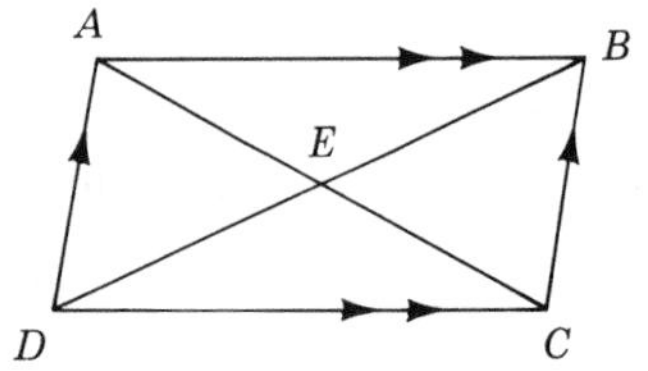

(a) Yes. Because the opposite sides are parallel.   (b) Two congruent triangles.   (c) They are congruent.   (d) They are congruent.
(e) They are supplementary.   (f) They are congruent.

28.   The foregoing principles provide us with a number of ways of *proving* a quadrilateral is a parallelogram. Here are some of them.

Pr. 7:   A quadrilateral is a parallelogram if its opposite sides are parallel.

Thus, if $AB \parallel CD$ and $AD \parallel BC$, then $ABCD$ is a $\square$.

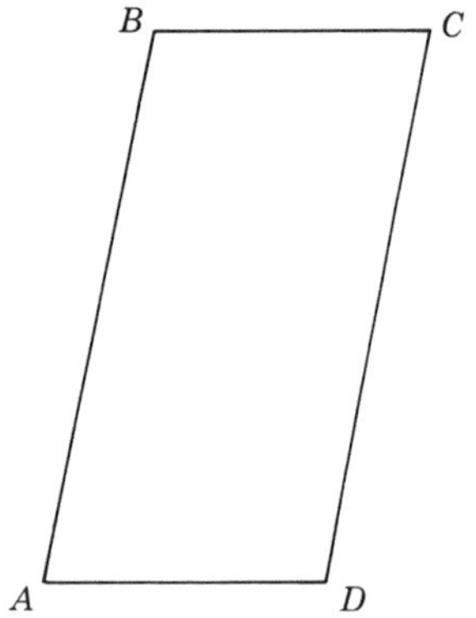

Pr. 8:   A quadrilateral is a parallelogram if its opposite sides are congruent.

Thus, if $AB \cong CD$ and $AD \cong BC$, then $ABCD$ is a $\square$.

Pr. 9:   A quadrilateral is a parallelogram if two sides are congruent and parallel.

Thus, if $BC \cong AD$ and $BC \parallel AD$, then $ABCD$ is a $\square$.

Pr. 10:   A quadrilateral is a parallelogram if its opposite angles are congruent.

Thus, if $\angle A \cong \angle C$ and $\angle B \cong \angle D$, then $ABCD$ is a $\square$.

Pr. 11:   A quadrilateral is a parallelogram if its diagonals bisect each other.

Thus, if $AE \cong EC$ and $BE \cong ED$, then $ABCD$ is a $\square$.

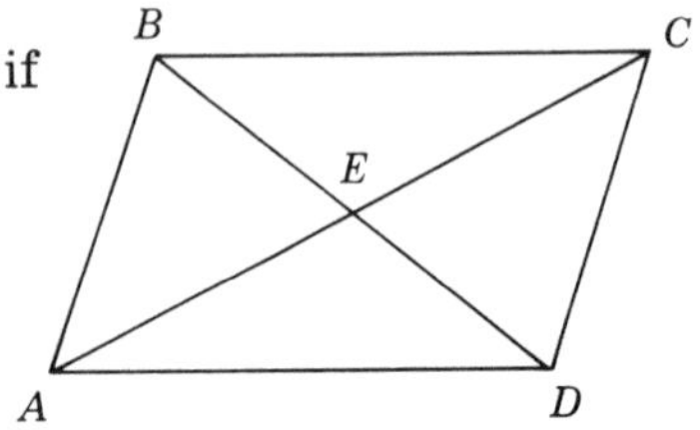

Study Principles 7 through 11 and then see if you can write down, briefly, the five characteristics that will prove that a quadrilateral is a parallelogram.

(a) _____

(b) _____

(c) _____

(d) _____

(e) _____

- - - - - - - - - - - - - - -

(a) opposite sides are parallel
(b) opposite sides are congruent
(c) two sides are congruent and parallel
(d) opposite angles are congruent
(e) diagonals bisect each other

29. Now we're going to apply the properties of parallelograms.

Example: Find x and y in the parallelogram *ABCD*. Perimeter = 40.

Solution: Since, by Pr. 3, $BC = AD = 3x$, and $CD = 2x$, then $2(2x + 3x) = 40$ (the perimeter, or distance around the figure). Therefore, $10x = 40$, and $x = 4$. Also by Pr. 3, $2y - 2 = 3x$, hence $2y - 2 = 3(4)$, $2y = 14$, and $y = 7$.

The following two problems will require the application of some of the *other* parallelogram principles.

(a)   Find x and y in the figure at the right.

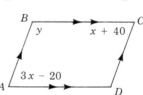

(b)   Find x and y in the figure at the right.

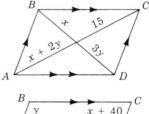

- - - - - - - - - - - - - - -

(a)   By Pr. 6, $x + 2y = 15$ and $x = 3y$. Hence, substituting $3y$ for x in the first equation, $3y + 2y = 15$, or $y = 3$. Therefore, $x = 3y = 9$.

(b)   By Pr. 4, $3x - 20 = x + 40$, $2x = 60$, and $x = 30$. By Pr. 5,
$y + (x + 40) = 180$, or $y + (30 + 40) = 180$, and $y = 110$.

30.   Apply Pr. 7 to determine which quadrilaterals in the following problems are parallelograms. State the parallelograms in each.

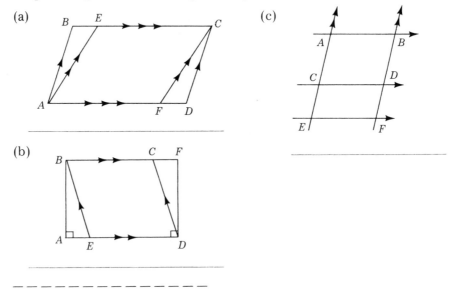

(a) *ABCD, AECF;*   (b) *ABFD, BCDE;*   (c) *ABDC, CDFE, ABFE*

31.   Apply Pr. 9, 10, and 11 to help you state *why ABCD* is a parallelogram in each of the following examples.

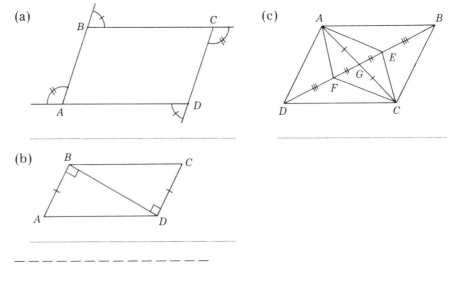

(a) Since supplements of congruent angles are congruent, opposite angles *A*, *B*, *C*, and *D* are congruent. Thus, by Pr. 10, *ABCD* is a ▱.

(b) Since perpendiculars to the same line are parallel, *AB* ∥ *CD*. Hence by Pr. 9, *ABCD* is a ▱.

(c) Using the addition axiom, *DG* ≅ *GB*. Hence by Pr. 11, *ABCD* is a ▱.

SOME SPECIAL PARALLELOGRAMS: RECTANGLE, RHOMBUS, SQUARE

32. Since we have talked about quite a few figures bounded by straight lines, this seems an appropriate place to try to organize them a bit in the form of a diagram showing their relationships to one another. The diagram below should help summarize some of the things we have covered thus far and also give you an indication of where the next three figures we're going to discuss fit into the larger scheme of things.

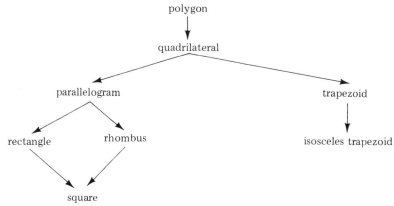

What do the terms *polygon*, *quadrilateral*, and *parallelogram* tell you already about the characteristics we are going to find in rectangles, rhombuses, and squares? _____

_____

— — — — — — — — — — — — — — — —

They are bounded by straight lines and are four-sided figures, the opposite sides of which are parallel.

33. Now let's define these new terms (not new to *you*, perhaps, but new to our discussion).

A *rectangle* is an *equiangular* parallelogram. A *rhombus* is an *equilateral* parallelogram. A *square* is an equilateral *and* equiangular parallelogram, hence it is both a rectangle and a rhombus.

Here are the properties of the special parallelograms.

Pr. 1: A rectangle, rhombus, or square has all the properties of a parallelogram.

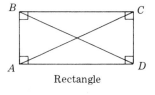

Pr. 2: Each angle of a rectangle is a right angle.

Pr. 3: The diagonals of a rectangle are congruent.

Thus, in rectangle $ABCD$, $AC \cong BD$.

Pr. 4: All the sides of a rhombus are congruent.

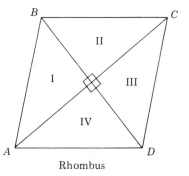

Pr. 5: The diagonals of a rhombus are perpendicular bisectors of each other.

Thus, in rhombus $ABCD$, AC and BD are ⊥ bisectors of each other.

Pr. 6: The diagonals of a rhombus bisect the vertex angles.

Thus, in rhombus $ABCD$, AC bisects $\angle A$ and $\angle C$, and BD bisects $\angle B$ and $\angle D$.

Pr. 7: The diagonals of a rhombus form four congruent triangles.

Thus, in rhombus $ABCD$, $\triangle I \cong \triangle II \cong \triangle III \cong \triangle IV$.

Pr. 8: A square has all the properties of both the rhombus and the rectangle.

By definition, a square is both a rectangle and a rhombus.

Since by now you may be having trouble keeping track of the characteristics or properties of the diagonals in the various figures, following is a chart that should help sort them out for you. A checkmark indicates a diagonal property of the figure.

Diagonal Properties	Parallel-ogram	Rectangle	Rhombus	Square
Diagonals bisect each other.	✓	✓	✓	✓
Diagonals are congruent.		✓		✓
Diagonals are perpendicular.			✓	✓
Diagonals bisect vertex angles.			✓	✓
Diagonals form 2 pairs of congruent triangles.	✓	✓	✓	✓
Diagonals form 4 congruent triangles.			✓	✓

Draw diagrams of the following figures and mark the congruent sides and angles. Assume opposite sides parallel.

(a) A rhombus   (b) A rectangle   (c) A square   (d) A parallelogram

— — — — — — — — — — — — — — —

(a)   (b)   (c)   (d)

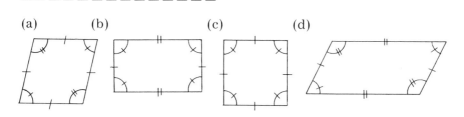

34. The basic or minimum definition of a rectangle is: *A rectangle is a parallelogram having one right angle.* Since the consecutive angles of a parallelogram are supplementary, if *one* angle is a right angle then the *remaining* angles must be right angles.

   The converse of this definition of a rectangle provides a useful method of proving that a parallelogram is a rectangle.

Pr. 9:   If a parallelogram has one right angle, then it is a rectangle.

Thus, if *ABCD* is a ▱ and *A* = 90°, then *ABCD* is a rectangle.

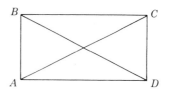

Pr. 10:   If a parallelogram has congruent diagonals, then it is a rectangle.

Thus, if *ABCD* is a ▱ and $AC \cong BD$, then *ABCD* is a rectangle.

(a)   Add the minimum markings to the parallelogram at the right to show that it is a rectangle, in accordance with Pr. 9.

(b)   Add the minimum markings to the parallelogram at the right to show that it is a rectangle, in accordance with Pr. 10.

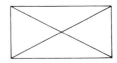

- - - - - - - - - - - - - - - -

(a)   (b)

35.   The basic or minimum definition of a rhombus is: *A rhombus is a parallelogram having two equal adjacent sides.*

The converse of this definition of a rhombus furnishes a useful method of proving that a parallelogram is a rhombus.

Pr. 11:   If a parallelogram has congruent adjacent sides, then it is a rhombus.

Thus, if *ABCD* is a ▱ and $AB \cong BC$, then *ABCD* is a rhombus.

And in the case of a square:

Pr. 12:   If a parallelogram has a right angle and two equal adjacent sides, then it is a square.

This follows from the fact that a square is both a rectangle *and* a rhombus.

Now let's apply what we have learned about the rhombus to solve some problems.

Example: If *ABCD* is a rhombus, find *x* and *y*.
Solution: Since $AB \cong AD$, $3x - 7 = 20$, or $x = 9$. (Pr. 4) And since $\triangle ABD$ is equilateral, $y = 20$. (Pr. 6)

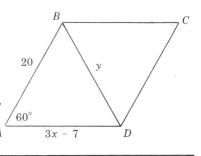

Solve the following rhombuses similarly for x and y.

(a) 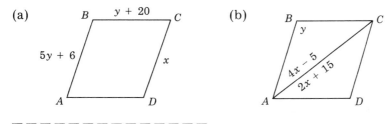 (b)

(a) Since $BC \cong AB$ (Pr. 4), $5y + 6 = y + 20$, or $y = 3\frac{1}{2}$.
And since $CD \cong BC$ (Pr. 4), $x = y + 20$, or $x = 23\frac{1}{2}$.

(b) Since AC bisects $\angle A$ (Pr. 6), $4x - 5 = 2x + 15$, or $x = 10$. Also, since $\angle B$ and $\angle A$ are supplementary (Pr. 5 of a parallelogram), $y + 70 = 180$, or $y = 110°$.

36. Now we come to the case where we have three or more parallels. This will lead us to a further consideration of *medians* and *midpoints*.

Pr. 1: If three or more parallels cut off congruent segments on one transversal, then they cut off congruent segments on any other transversal.

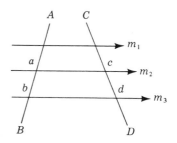

Thus, if $m_1 \parallel m_2 \parallel m_3$ and segments a and b of transversal AB are congruent, then segments c and d of transversal CD are congruent.

Pr. 2: If a line is drawn from the midpoint of one side of a triangle and parallel to a second side, then it passes through the midpoint of the third side.

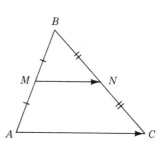

Thus, in $\triangle ABC$, if M is the midpoint of AB, and $MN \parallel AC$, then N is the midpoint of BC.

Pr. 3: If a line joins the midpoints of two sides of a triangle, then it is parallel to the third side and equal to one-half of it.

Thus, in $\triangle ABC$, of M and N are the midpoints of AB and BC, then $MN \parallel AC$ and $MN = \frac{1}{2}AC$.

Pr. 4:   The median of a trapezoid is
parallel to its bases and equal
to one-half of their sum.

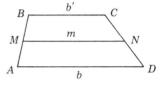

Thus, if m is the median of a trapezoid
$ABCD$, then $m \parallel b$, $m \parallel b'$, and $m = \frac{1}{2}(b + b')$.

Pr. 5:   The median to the hypotenuse of a
right triangle equals one-half of the
hypotenuse.

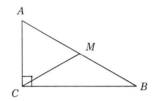

Thus, in rt. $\triangle ABC$, if CM is the median to
hypotenuse AB, then $CM = \frac{1}{2}AB$; that is,
$CM \cong AM \cong MB$.

Pr. 6:   The medians of a triangle meet in a
point which is two-thirds of the
distance from any vertex to the
midpoint of the opposite side.

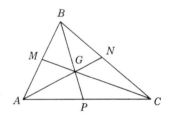

Thus, if AN, BP, and CM are medians of
$\triangle ABC$, then they meet in a point, G,
which is two-thirds of the distance from
A to N, B to P, and C to M. (Note: G is the center of gravity or the
centroid of $\triangle ABC$. If the triangle is made of a firm substance, such as
cardboard, it can be balanced at the centroid on the point of a pin.)

To help you gain familiarity with these principles and see how they can
be applied we're going to work with them a bit in solving some problems.   Let's start with Pr. 1.

Example: Use Pr. 1 to find x and y.
Solution: Since $BF \cong FA$, we know that
$BE \cong ED$ and $CG = \frac{1}{2}CD$ (Pr. 1), then
$x = 8$ and $y = 7\frac{1}{2}$.

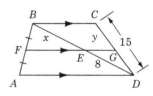

Apply Pr. 1 to the following to find x and y.

(a)   (b)

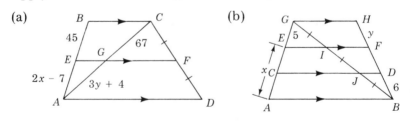

(a) Since $BE \cong EA$ and $CG \cong AG$, $2x - 7 = 45$ and $3y + 4 = 67$, hence $x = 26$ and $y = 21$.

(b) Since $AC \cong CE \cong EG$ and $HF \cong FD \cong DB$, $x = 10$ and $y = 6$.

37. Next we will apply Principles 2 and 3.

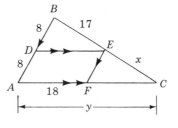

Example: Use Principles 2 and 3 as necessary to find x and y.
Solution: Since, by Pr. 2, E is the midpoint of BC and F is the midpoint of AC, then $x = 17$ and $y = 36$.

With a little courage you can do the same thing. Try it. Use Principles 2 and 3 to find x and y in the following problems.

(a)   (b)

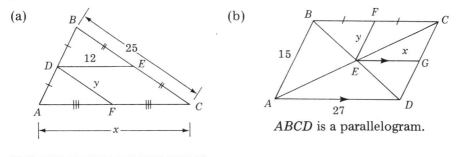

ABCD is a parallelogram.

- - - - - - - - - - - - - - - -

(a) Since, by Pr. 3, $DE = \frac{1}{2}AC$ and $DF = \frac{1}{2}BC$, then $x = 24$ and $y = 12\frac{1}{2}$.

(b) Since $ABCD$ is a parallelogram, E is the midpoint of AC. Also by Pr. 2, G is the midpoint of CD. Therefore (by Pr. 3), $x = \frac{1}{2}(27) = 13\frac{1}{2}$; $y = \frac{1}{2}(15) = 7\frac{1}{2}$.

38. Now let's turn our attention once more to the trapezoid, applying Pr. 4. Apply the formula $m = \frac{1}{2}(b + b')$ to solve the following.

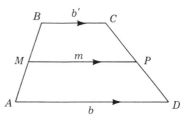

If MP is the median of trapezoid $ABCD$,

(a) Find m if $b = 20$ and $b' = 28$.

_____

(b) Find b' if $b = 30$ and $m = 26$.

_____

(c)   Find b if $b' = 35$ and $m = 40$.

_____

_ _ _ _ _ _ _ _ _ _ _ _ _ _

(a) $m = \frac{1}{2}(20 + 28)$, $m = 24$.
(b) $26 = \frac{1}{2}(30 + b')$, $b' = 22$.
(c) $40 = \frac{1}{2}(b + 35)$, $b = 45$.

39.   Principles 5 and 6 apply, as you may recall, to the medians of a triangle. We will use these principles to find x and y in the two problems below; but first an example.

Example: Find x and y.
Solution: Since $AM \cong MB$, CM is the median to hypotenuse AB. Hence (by Pr. 5), $3x = 20$ and $\frac{1}{3}y = 20$. Thus, $x = 6\frac{2}{3}$ and $y = 60$.

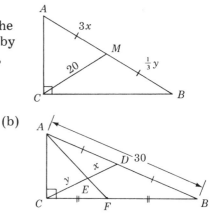

Now it's your turn. Find x and y.

(a) 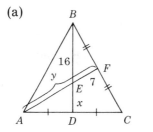   (b)

_ _ _ _ _ _ _ _ _ _ _ _ _ _ _

(a) BD and AF are medians of $\triangle ABC$. Hence, by Pr. 6, $x = \frac{1}{2}(16) = 8$, and $y = 3(7) = 21$.
(b) CD is the median to hypotenuse AB hence, by Pr. 5, $CD = 15$. CD and AF are medians of $\triangle ABC$, hence, by Pr. 6, $x = \frac{1}{3}(15) = 5$, $y = \frac{2}{3}(15) = 10$.

The following Self-Test should assist your review of what we have covered concerning parallelograms, trapezoids, medians, and midpoints. If you find you are a little weak on any of the principles discussed, by sure to re-read the appropriate frames before going on.

SELF-TEST

1. *ABCD* is a trapezoid. Find *x* and *y*.
    (frame 26)

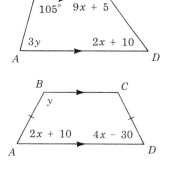

2. *ABCD* is an isosceles trapezoid. Find *x*
    and *y*.   (frame 26)

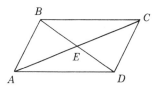

3. If *ABCD* is a parallelogram, find *x* and *y*.
    AD = 5*x*, *AB* = 2*x*, *CD* = *y*,
    perimeter = 84.   (frame 29)

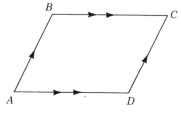

4. *ABCD* is a parallelogram. If $\angle A$ = 4*y* −
    60, $\angle C$ = 2*y*, and $\angle D$ = *x*, find *x* and *y*.
    (frame 29)

5. If *ABCD* is a parallelogram, find *x* and *y*
    when *AE* = *x*, *EC* = 4*y*, *BE* = *x* − 2*y*,
    and *ED* = 9.   (frame 29)

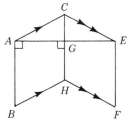

6. Name the parallelograms in the figure at
    the right.   (frame 30)

7. State why *ABCD* is a parallelogram.
    (frame 30)

8. Name the following figures. (Opposite sides are parallel.)   (frame 32)

    (a)   (b)   (c)   (d)

    _____   _____   _____   _____

9. State the minimum requirement for a parallelogram to be a rectangle.
    (frame 33)

_____

10. *ABCD* is a rhombus. Find *x* and *y* if *AB* = 7*x*,
    AD = 3*x* + 10, and *BC* = *y*.   (frame 34)

    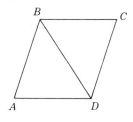

11. Find *x* and *y* in the figure at the right.
    (frame 35)

    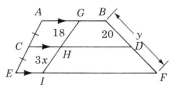

12. Find *x* and *y* in the triangle at the right.
    (frame 36)

    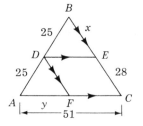

13. If *MP* is the median of trapezoid *ABCD*,
    find *m* if *b* = 23 and *b'* = 15.
    (frame 37)

    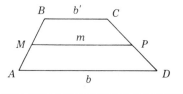

14. In a right triangle, find the length of the median to a hypotenuse whose length is 45.   (frame 39)

Answers to Self-Test

1. $x = 15$, $y = 25$
2. $x = 20$, $y = 130$
3. $x = 6$, $y = 12$
4. $x = 120$, $y = 30$
5. $x = 18$, $y = 4\frac{1}{2}$
6. $ACHB$, $CEFH$
7. Opposite sides are congruent.
8. (a)   rhombus (Note: There is nothing to indicate that the interior angles are right angles, hence it is not a square.)
   (b)   rhombus
   (c)   rectangle (a parallelogram with at least one right angle)
   (d)   parallelogram (Again, no right angle is indicated and we cannot assume the interior angles are such unless so indicated.)
9. Pr. 9: It must have one right angle.
10. $x = 2\frac{1}{2}$, $y = 17\frac{1}{2}$
11. $x = 6$, $y = 40$
12. $x = 28$, $y = 25\frac{1}{2}$
13. $m = 19$
14. $22\frac{1}{2}$

CHAPTER THREE

Plane Geometry:
Circles and Similarity

We are going to be studying similar figures in this chapter. But not just that. We also are going to be studying the circle, ratios and proportions, finding the areas of various geometric figures, regular polygons, determining a locus, and, finally, constructions — which you should find a lot of fun, since it is nice to draw geometric figures as well as to reason about them.

Specifically, when you complete this chapter you will be able to recognize and use the basic principles relating to:

- the circle—including its various elements such as the radius, diameter, circumference, arcs, chords, central angles, tangents, and secants, as well as inscribed and circumscribed figures;

- tangents to a circle—including the length of a tangent from a point to a circle, internally and externally tangent circles, and applying tangent principles to solve geometric problems;

- measuring arcs and angles in a circle—including angle measurement principles, inscribed angles, and using angle measurement principles to find the values of unknown angles and arcs;

- similarity—including the concepts of geometric ratio and proportion, proportional lines, similar triangles and polygons, mean proportionals in right triangles, the Law of Pythagoras, and special right triangles.

CIRCLES

1. We are going to talk first about the circle and circle relationships. And in order to do so it is important that you be familiar with the important terms associated with the circle. Some of these you will recognize because we have mentioned them earlier. However, we will repeat them here so that they will all be together in one spot for ready reference.

A *circle* is a closed curve, all of whose points lie in the same plane and are at the same distance from a point within called the *center*. The symbol for circle is ⊙, and for circles, ⊚.

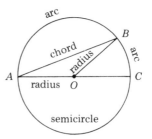

The *circumference* of a circle is the distance around the circle. It contains $360°$.

A *radius* of a circle is a line joining the center to a point on the circumference. Thus, *AO, BO,* and *CO* are radii (plural of radius).

A *central angle* is an angle formed by two radii. Thus, $\angle AOB$ and $\angle BOC$ are central angles.

An *arc* is a part of the circumference of a circle. The symbol for arc is ⌒. Thus, $\overset{\frown}{AB}$ stands for arc *AB*.

A *semicircle* is an arc equal to one-half of the circumference of a circle. Thus, *ABC* is a semicircle.

A *minor arc* is an arc less than a semicircle. A *major arc* is an arc greater than a semicircle. Thus, in the figure above, $\overset{\frown}{BC}$ is a minor arc and $\overset{\frown}{BAC}$ is a major arc. Three letters are required to indicate a major arc.

To *intercept* an arc is to cut off the arc. Thus, in the figure above, $\angle BAC$ and $\angle BOC$ intercept $\overset{\frown}{BC}$.

A *chord* of a circle is a line segment joining two points of the circumference. Thus, in the figure at the right, *AB* is a chord.

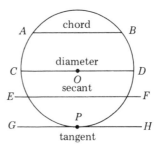

A *diameter* of a circle is a chord through the center. Thus, *CD* is a diameter of circle *O*.

A *secant* of a circle is a line that intersects the circle at two points. Thus, *EF* is a secant.

A *tangent* of a circle is a line that touches the circle at one and only one point, no matter how far extended. Thus, *GH* is a tangent to the circle at *P*. *P* is the point of tangency, or point of contact.

An *inscribed polygon* is a polygon all of whose sides are chords of a circle. Thus, $\triangle ABD$, $\triangle BCD$, and quadrilateral *ABCD* are inscribed polygons of circle *O*.

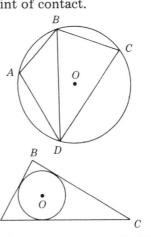

A *circumscribed circle* is a circle passing through each vertex of a polygon. Thus, circle *O* is a circumscribed circle of quadrilateral *ABCD*.

A *circumscribed polygon* is a polygon all of whose sides are tangents to a circle. Thus, $\triangle ABC$ is a circumscribed polygon of circle *O*.

An *inscribed circle* is a circle to which all the sides of a polygon are tangents. Thus, circle O is an inscribed circle of $\triangle ABC$ (on page 109).

Concentric circles are circles that have the same center. Thus, the two circles shown are concentric circles because they have the common center O. AB is a tangent of the inner circle and a chord of the outer one. CD is a secant of the inner circle and a chord of the outer one.

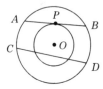

Study the above terms until you are quite certain about their meanings. When you think you are ready, give yourself the following little quiz.

Quiz on Circle Definitions

In the figure at the right identify at least one each of the following.

Example: radius OE

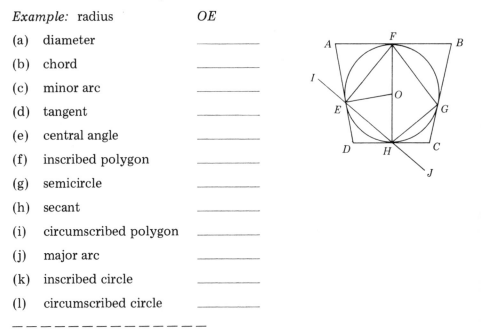

(a) diameter   _____

(b) chord   _____

(c) minor arc   _____

(d) tangent   _____

(e) central angle   _____

(f) inscribed polygon   _____

(g) semicircle   _____

(h) secant   _____

(i) circumscribed polygon   _____

(j) major arc   _____

(k) inscribed circle   _____

(l) circumscribed circle   _____

(a) FH; (b) FG, EF, GH, HE; (c) $\overset{\frown}{EF}$, $\overset{\frown}{FG}$, $\overset{\frown}{GH}$, $\overset{\frown}{HE}$; (d) CD, AB, BC, DA; (e) $\angle EOF$, $\angle EOH$, $\angle FOH$; (f) quadrilateral $EFGH$, $\triangle EFH$, $\triangle FGH$; (g) $\overset{\frown}{FEH}$ or $\overset{\frown}{FGH}$; (h) IJ; (i) quadrilateral $ABCD$; (j) $\overset{\frown}{FEG}$, $\overset{\frown}{GHF}$, $\overset{\frown}{HEG}$; (k) circle O in $ABCD$; (l) circle O about $EFGH$

2. As with most geometric figures—the basic ones, at least—there are a
    number of important principles associated with circles. And since much
    of our further work in this section depends upon your being familiar
    with these principles, we will go on to them next. As you might sus-
    pect, they relate mainly to the terms you have just learned.

> *Pr. 1:*   A diameter divides a circle into two
>   congruent parts.

Thus, diameter AB divides circle O into two
congruent semicircles, $\overparen{ACB}$ and $\overparen{ADB}$.

> *Pr. 2:*   If a chord divides a circle into two
>   congruent parts, then it is a diameter.
>   (This is the converse of Pr. 1.)

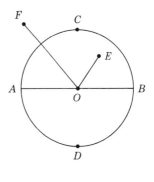

Thus, if $\overparen{ACB} \cong \overparen{ADB}$, then AB is a diameter.

> *Pr. 3:*   A point is outside, on, or inside a
>   circle according to whether its distance
>   from the center is greater than, equal
>   to, or less than the radius.

Thus, F is outside circle O since FO is greater than a radius. E is inside
circle O since EO is less than a radius. And A is on circle O since AO is
a radius.

> *Pr. 4:*   Radii of the same or congruent circles
>   are congruent.

Thus, in circle O, $OA \cong OC$.

> *Pr. 5:*   Diameters of the same or congruent
>   circles are congruent.

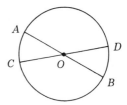

Thus, in circle O, $AB \cong CD$.

> *Pr. 6:*   In the same or congruent circles,
>   congruent central angles have congruent
>   arcs.

Thus, in circle O, if $\angle 1 \cong \angle 2$, then $\overparen{AC} \cong \overparen{CB}$.

> *Pr. 7:*   In the same or congruent circles,
>   congruent arcs have congruent central
>   angles.

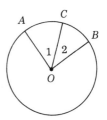

Thus, in circle O, if $\overparen{AC} \cong \overparen{CB}$, then $\angle 1 \cong \angle 2$.
(Principles 6 and 7 are converses.)

Pr. 8:   In the same or congruent circles, congruent
            chords have congruent arcs.

Thus, in circle *O*, if $AB \cong AC$, then $\overparen{AB} \cong \overparen{AC}$.

Pr. 9:   In the same or congruent circles, congruent
            arcs have congruent chords.

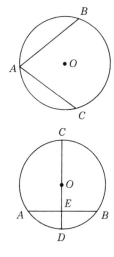

Thus, in circle *O*, if $\overparen{AB} \cong \overparen{AC}$, then $AB \cong AC$.
(Principles 8 and 9 are converses.)

Pr. 10:   A diameter perpendicular to a chord
            bisects the chord and its arcs.

Thus, in circle *O*, if $CD \perp AB$, then *CD* bisects *AB*,
$\overparen{AB}$, and $\overparen{ACB}$.

Pr. 11:   A perpendicular bisector of a chord
            passes through the center of the
            circle.

Thus, in circle *O*, if *PD* is the perpendicular
bisector of *AB*, then *PD* passes through center *O*.

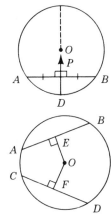

Pr. 12:   In the same or congruent circles, congruent
            chords/are equally distant from the center.

Thus, in circle *O*, if $AB \cong CD$, $OE \perp AB$ and
$OF \perp CD$, then $OE \cong OF$.

Pr. 13:   In the same or congruent circles, chords that are equally distant
            from the center are congruent.

Thus, in circle *O* above, if $OE \cong OF$, $OE \perp AB$ and $OF \perp CD$, then
$AB \cong CD$.

Apply Principles 4 and 5 in solving the following problems.

(a)   What kind of triangle is $\triangle OCD$?

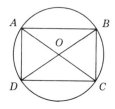

_____

(b)   What kind of quadrilateral is *ABCD*?

_____

(c) If circle $O \cong$ circle Q, what kind of quadrilateral is $OAQB$?

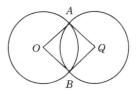

_____

— — — — — — — — — — — — — — —

(a) Since radii or diameters of the same or congruent circles are congruent, $OC \cong OD$, hence $\triangle OCD$ is isosceles.

(b) Since diagonals AC and BD are congruent and bisect each other, $ABCD$ is a rectangle.

(c) Since the circles are congruent, $OA \cong AQ \cong QB \cong BO$, hence $OAQB$ is a rhombus.

3. Now let's see how some of our principles apply in proving a circle problem.

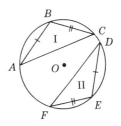

Given: $AB \cong DE$, $BC \cong EF$
Prove: $\angle B \cong \angle E$
Plan: Prove $\triangle I \cong \triangle II$

PROOF: Statements

Statements	Reasons
1. $AB \cong DE$, $BC \cong EF$	1. Given
2. $\overset{\frown}{AB} \cong \overset{\frown}{DE}$, $\overset{\frown}{BC} \cong \overset{\frown}{EF}$	2. In a circle, $\cong$ chords have $\cong$ arcs.
3. $\overset{\frown}{ABC} \cong \overset{\frown}{DEF}$	3. Equals added to equals are equal.
4. $AC \cong DF$	4. In a circle, $\cong$ arcs have $\cong$ chords.
5. $\triangle I \cong \triangle II$	5. SSS
6. $\angle B \cong \angle E$	6. Corresponding parts

Practice proving a circle problem by completing the proof (reasons) missing below. Prove the following statement: _If a radius bisects a chord, then it is perpendicular to the chord._

Given: Circle O
        OC bisects AB
Prove: $OC \perp AB$
Plan: Prove $\triangle AOD \cong \triangle BOD$, hence $\angle 1 \cong \angle 2$. Also, $\angle 1$ and $\angle 2$ are supplementary.

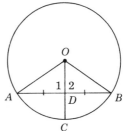

PROOF: Statements	Reasons
1. Draw *OA* and *OB*.	1. A straight line may be drawn between two points. (Definition.)
2. *OA* ≅ *OB*	2.
3. *OC* bisects *AB*	3.
4. *AD* ≅ *DB*	4.
5. *OD* ≅ *OD*	5.
6. △*AOD* ≅ △*BOD*	6.
7. ∠1 ≅ ∠2	7.
8. ∠1 is the supplement of ∠2.	8.
9. ∠1 and ∠2 are rt. angles.	9.
10. *OC* ⊥ *AB*	10.

– – – – – – – – – – – – – – – –

2.   Radii of a circle are congruent.
3.   Given
4.   To bisect is to divide into two congruent parts.
5.   Identity
6.   SSS
7.   Corresponding parts of congruent ▲ are congruent.
8.   Adjacent ∡ are supplementary if exterior sides lie in a straight line.
9.   Equal supplementary angles are right angles.
10.   Rt. ∡ are formed by perpendiculars.

TANGENTS

4.   The *length of a tangent* from a point to a circle is the length of the line segment from the given point to the point of tangency. Thus, *PA* is the length of the tangent from *P* to circle *O*.

Following are some tangent principles.

Pr. 1:   A tangent is perpendicular to the radius drawn to the point of contact.

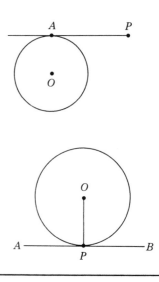

Thus, if AB is a tangent to circle O at P, and OP is drawn, then $AB \perp OP$.

Pr. 2:   A line is tangent to a circle if it is perpendicular to the outer end of a radius.

Thus, if $AB \perp$ radius OP at P, then AB is tangent to circle O.

Pr. 3:   A line passes through the center of a circle if it is perpendicular to a tangent at its point of contact.

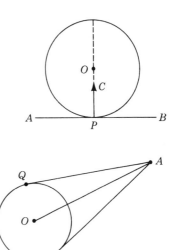

Thus, if AB is tangent to circle O at P, and $CP \perp AB$ at P, then CP extended will pass through the center O.

Pr. 4:   Tangents to a circle from an outside point are congruent.

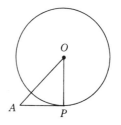

Thus, if AP and AQ are tangent to circle O at P and Q, then $AP \cong AQ$.

Pr. 5:   The line from the center of a circle to an outside point bisects the angle between the two tangents from the point to the circle.

Thus, OA (in the figure above) bisects $\angle PAQ$ if AP and AQ are tangents to circle O.

Here are some examples of the ways in which we can apply the above tangent principles.

Example 1:   In the figure at the right AP is a tangent. If $AP \cong OP$, what kind of triangle is OPA?

Solution: AP is tangent to the circle at P; then by Pr. 1, $\angle OPA$ is a right angle. Also $AP \cong OP$ (given). Therefore, $\triangle OPA$ is an isosceles right triangle.

Example 2: AP is a tangent. If $\angle A : \angle O =$ 2 : 3, what is the value of $\angle A$?
Solution: By Pr. 1, $\angle P = 90°$, hence $\angle A + \angle O = 90°$. If we let $\angle A = 2x$ and $\angle O = 3x$ (to set up the proportionality 2 : 3), then $2x + 3x = 5x$, and $5x = 90$, hence $x = 18$. Therefore, $\angle A = 36°$.

Example 3: AP, BQ, and AB are tangents. Find y.
Solution: By Pr. 4, $AR = 6$, and $RB = y$. Then $RB = AB - AR = 14 - 6 = 8$. Hence $y = RB = 8$.

Use the foregoing examples as a general guide in solving the following problems.

(a) AP and AQ are tangents. If $AP \cong PQ$, what kind of triangle is APQ?

_____

(b)   (Also in the figure at the right), if $AP \cong OP$, what kind of quadrilateral is $OPAQ$. _____

_____

(c) AP, AB, and BR are tangents. If $OQ \perp PR$, what kind of quadrilateral is $PABR$?

_____

(d)   If AP and AQ are tangents, find $\angle 1$ if $\angle O = 140°$. _____

_____

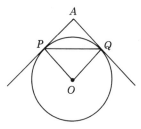

(e) $\triangle ABC$ is circumscribed. Find x.

_____

– – – – – – – – – – – – – –

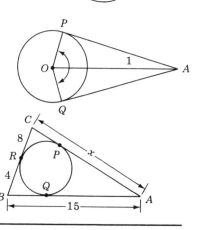

(a) AP and AQ are tangents from a point to the circle, hence, by Pr. 4, $AP \cong AQ$. Also, $AP \cong PQ$, therefore $\triangle APQ$ is an equilateral triangle.

(b)   By Pr. 4, $AP \cong AQ$. Also, OP and OQ are congruent radii and $AP \cong OP$. Thus, $AP \cong AQ \cong OP \cong OQ$. Also $AP \perp OP$ (Pr. 1). Hence $OPAQ$ is a rhombus with a right angle, or a square.

(c)   By Pr. 1, $AP \perp PR$ and $BR \perp PR$, hence $AP \parallel BR$ since both are $\perp$ to PR. By Pr. 1, $AB \perp OQ$, also $PR \perp OQ$ (given), hence $AB \parallel PR$ since both are $\perp$ to OQ. Therefore, $PABR$ is a parallelogram with a right angle, or a rectangle.

(d)   By Pr. 1, $\angle P = \angle Q = 90°$. Since $\angle P + \angle Q + \angle O + \angle A = 360°$, $\angle A + \angle O = 180°$. And since $\angle O = 140°, \angle A = 40°$. Then by Pr. 5, $\angle 1 = \frac{1}{2}\angle A = 20°$.

(e)   By Pr. 4, $PC = 8$, $QB = 4$, and $AP = AQ$. Then $AQ = AB - QB = 11$. Hence $x = AP + PC = 11 + 8 = 19$.

5.   The *line of centers of two circles* is the line joining their centers. Thus, OO' is the line of centers of circles O and O'.

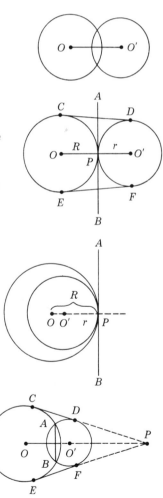

In the figure at the right, circles O and O' are tangent *externally* at P. AB is the common internal tangent of both circles. The line of centers OO' passes through P, is perpendicular to AB, and equals the sum of the radii, $R + r$. Also, AB bisects each of the common external tangents, CD and EF.

As shown at the right, circles O and O' are tangent *internally* at P. AB is the common external tangent of both circles. The line of centers OO', if extended, passes through P, is perpendicular to AB, and equals the difference of the radii, $R - r$.

The figure at the right represents overlapping circles since, as shown, circles O and O' overlap. Their common chord is AB. If the circles are not congruent, their (congruent) common external tangents CD and EF meet at P. The line of centers OO' is the perpendicular bisector of AB and, if extended, passes through P.

As shown, circles O and O' are entirely outside of each other. The common internal tangents, AB and CD, meet at P. If the circles are not congruent, their common external tangents, EF and GH, if extended, meet at P'. The line of centers OO' passes through P and P'. Also, $AB \cong CD$ and $EF \cong GH$.

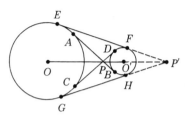

Apply the above relationships between two circles in varying positions to answer the following questions. If two circles have radii of 9 and 4 respectively, find their line of centers:

(a)   If the circles are tangent externally.

_____

(b)   If the circles are tangent internally.

_____

(c)   If the circles are concentric.

_____

(d)   If the circles are 5 units apart ($d = 5$).

_____

– – – – – – – – – – – – – – – –

Let R = radius of larger circle, r = radius of smaller circle.
(a)   Since $R = 9$ and $r = 4$, $OO' = R + r = 9 + 4 = 13$.
(b)   Since $R = 9$ and $r = 4$, $OO' = R - r = 9 - 4 = 5$.
(c)   Since the circles have the same center, their line of centers has zero length.
(d)   Since $R = 9$ and $r = 4$ and $d = 5$, $OO' = R + d + r = 9 + 5 + 4 = 18$.

MEASUREMENT OF ANGLES AND ARCS IN A CIRCLE

6. A *central angle* has the same number of
    degrees as the arc it intercepts. Thus, a
    central angle which is a right angle inter-
    cepts a 90° arc; a 40° central angle
    intercepts a 40° arc, and a central angle
    which is a straight line intercepts a semi-
    circle of 180°.

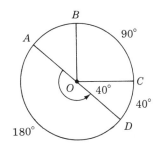

Since the numerical measures in
degrees of both the central angle and its
intercepted arc are the same, we can
restate the above principle as follows: *A central angle is measured by
its intercepted arc.*

The symbol $\stackrel{\circ}{=}$ frequently is used to mean "is measured by." (Be
careful not to say that a central angle *equals* its intercepted arc; an
angle cannot equal an arc since they are different things.)

An *inscribed angle* is an angle formed by
two chords drawn from the same point on a
circle. An inscribed angle is said to *intercept*
the arc between its sides. Also, it is said to be
inscribed in an arc if its vertex is on the arc
and its sides terminate in the ends of the arc.
Thus, $\angle A$ is an inscribed angle whose sides
are the chords AB and AC. Note that $\angle A$
intercepts $\overset{\frown}{BC}$ and is *inscribed* in $\overset{\frown}{BAC}$.

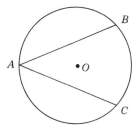

Now let's consider some angle measurement principles.

Pr. 1:   A central angle is measured by its intercepted arc.

Pr. 2:   An inscribed angle is measured by one-half of its *intercepted* arc.

Pr. 3:   In the same or congruent circles,
           congruent inscribed angles have
           congruent intercepted arcs.

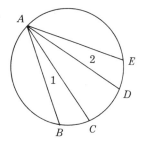

Thus, if $\angle 1 \cong \angle 2$, then $\overset{\frown}{BC} \cong \overset{\frown}{DE}$.

Pr. 4:   In the same or congruent circles,
           inscribed angles having congruent
           intercepted arcs are equal.

Thus, if $\overset{\frown}{BC} \cong \overset{\frown}{DE}$, then $\angle 1 \cong \angle 2$. (This is
the converse of Pr. 3.)

Pr. 5:   Angles inscribed in the same or
            congruent arcs are congruent.

Thus, if $\angle C$ and $\angle D$ are inscribed in $\overset{\frown}{ACB}$,
then $\angle C \cong \angle D$.

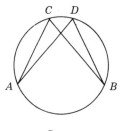

Pr. 6:   An angle inscribed in a semicircle
            is a right angle.

Thus, since $\angle C$ is inscribed in semicircle
ACD, $\angle C = 90°$.

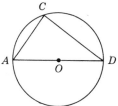

Pr. 7:   Opposite angles of an inscribed quadri-
            lateral are supplementary.

Thus, if $ABCD$ is an inscribed quadrilateral,
$\angle A$ is the supplement of $\angle C$, and $\angle B$ is the
supplement of $\angle D$.

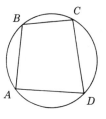

Pr. 8:   Parallel lines intercept congruent arcs
            on a circle.

Thus, if $AB \parallel CD$, then $\overset{\frown}{AC} \cong \overset{\frown}{BD}$. If tangent
FG is parallel to CD, then $\overset{\frown}{PC} \cong \overset{\frown}{PD}$.

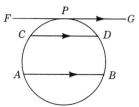

Pr. 9:   An angle formed by a tangent and a
            chord is measured by one-half of its
            intercepted arc.

Thus, $\angle A = \frac{1}{2}a°$.

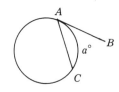

Pr. 10:   An angle formed by two intersecting
            chords is measured by one-half the
            sum of the intercepted arcs.

Thus, $\angle 1 = \frac{1}{2}(a° + b°)$.

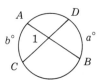

Pr. 11:   An angle formed by two secants intersecting outside a circle is measured by one-half the difference of the intercepted arcs.

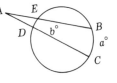

Thus, $\angle A = \frac{1}{2}(a° - b°)$.

Pr. 12:   An angle formed by a tangent and a secant intersecting outside a circle is measured by one-half the difference of the intercepted arcs.

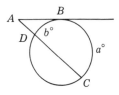

Thus, $\angle A = \frac{1}{2}(a° - b°)$.

Pr. 13:   An angle formed by two tangents intersecting outside a circle is measured by one-half the difference of the intercepted arcs.

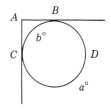

Thus, $\angle A = \frac{1}{2}(a° - b°)$.

Now it is time we applied some of these principles. Let's start with Principles 1 and 2—measuring central and inscribed angles.

Example: If $\angle y = 46°$, find $\angle x$.
Solution: $\angle y \overset{\circ}{=} \overarc{BC}$, therefore $\overarc{BC} = 46°$.
$\angle x \overset{\circ}{=} \frac{1}{2}\overarc{BC} = \frac{1}{2}(46°) = 23°$.

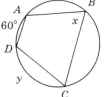

Solve these similarly:

(a)   If $\angle y = 112°$, find $\angle x$.

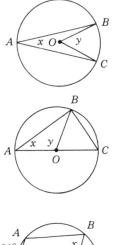

(b)   If $\angle x = 75°$, find $\hat{y}$.

(a) $y \stackrel{\circ}{=} \overset{\frown}{AB}$, hence $\overset{\frown}{AB} = 112°$. $\overset{\frown}{BC} = \overset{\frown}{ABC} - \overset{\frown}{AB} = 180° - 112°$
$= 68°$. $x \stackrel{\circ}{=} \frac{1}{2}\overset{\frown}{BC} = \frac{1}{2}(68°) = 34°$.
(b) $x \stackrel{\circ}{=} \frac{1}{2}\overset{\frown}{ADC}$, hence $\overset{\frown}{ADC} = 150°$. $y = \overset{\frown}{ADC} - \overset{\frown}{AD} = 150° - 60°$
$= 90°$.

7. Apply Principles 3 through 8 (measuring angles and arcs) to find x and y in each of the following.

Example: Find arc x and angle y.
Solution: Since $\angle 1 \cong \angle 2$, $\overset{\frown}{x} = \overset{\frown}{AB} = 50°$. (Pr. 3)
Since $\overset{\frown}{AD} \cong \overset{\frown}{CD}$, $\angle y = \angle ABD = 65°$. (Pr. 5)

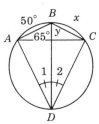

(a) Find angles x and y.

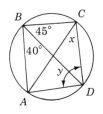

(b) Find angles x and arc $\overset{\frown}{y}$.

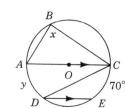

(a) $\angle ABD$ and $\angle x$ are inscribed in $\overset{\frown}{ABD}$, hence $\angle x = \angle ABD = 40°$.
$ABCD$ is an inscribed quadrilateral, hence $\angle y = 180° - \angle B = 95°$.
(b) Since $\angle x$ is inscribed in a semicircle, $\angle x = 90°$. And since
$AC \parallel DE$, $\overset{\frown}{y} = \overset{\frown}{CE} = 70°$.

8. Now let's try applying Pr. 9 (measuring an angle formed by a tangent and a chord). In the example and problems that follow, CD is a tangent at P.

Example: Find $\angle x$ if $\overset{\frown}{y} = 220°$.
Solution: $\angle z \stackrel{\circ}{=} \frac{1}{2}(220°) = 110°$.
$\angle x = 180° - 110° = 70°$.

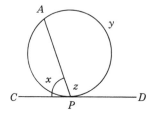

(a)   Find $\angle x$ if $\hat{y} = 140°$.

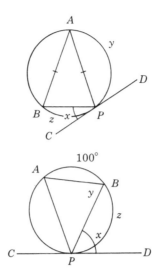

(b)   Find $\angle x$ if $\angle y = 75°$.

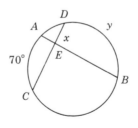

- - - - - - - - - - - - - - - -

(a)   Since $AB \cong AP$, $\overset{\frown}{AB} = \hat{y} = 140°$, and $\hat{z} = 360° - 140° - 140°$
      $= 80°$. $x \overset{\circ}{=} \frac{1}{2}\hat{z} = 40°$.
(b) $\angle y \overset{\circ}{=} \frac{1}{2}\overset{\frown}{AP}$, hence $\overset{\frown}{AP} \overset{\circ}{=} 2y$ or $150°$. $\hat{z} = 360° - 100° - 150°$
      $= 110°$. Therefore, $\angle x \overset{\circ}{=} \frac{1}{2}\hat{z} = 55°$.

9. Pr. 10 states that an angle formed by two intersecting chords is measured
    by one-half the sum of the intercepted arcs. Apply this in the example
    and problems that follow. (Remember, intercepted arcs are the arcs
    lying *between* the sides of the angle.)

Example: Find $\hat{y}$ if $\angle x = 95°$.
Solution: $\angle x \overset{\circ}{=} \frac{1}{2}(\overset{\frown}{AC} + \hat{y})$, hence
$95° = \frac{1}{2}(70° + \hat{y})$, or $\hat{y} = 120°$.

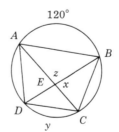

(a)   Find $\angle x$ if $\hat{y} = 80°$.

(b)   Find $\angle y$ if $\hat{x} = 78°$.

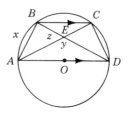

- - - - - - - - - - - - - -

(a) $\angle z \overset{\circ}{=} \frac{1}{2}(\hat{y} + \overset{\frown}{AB}) = \frac{1}{2}(80° + 120°) = 100°$. $\angle x = 180° - \angle z = 80°$.
(b) $BC \parallel AD$, hence $\overset{\frown}{CD} = x = 78°$. $\angle z \overset{\circ}{=} \frac{1}{2}(x + \overset{\frown}{CD}) = 78°$.
      $\angle y = 180° - \angle z = 102°$.

10.   And finally, Principles 11 to 13 tell us that an angle formed by two
      secants, by a secant and a tangent, or by two tangents is measured by
      one-half the difference of the intercepted arcs. Apply this in the example
      and problems below.

Example: Find $\hat{y}$ if $\angle x = 40°$.
Solution: $\angle x \overset{\circ}{=} \frac{1}{2}(\overset{\frown}{BC} - \hat{y})$, or $40° = \frac{1}{2}(200° - \hat{y})$, $\hat{y} = 120°$.

(a)   Find $\hat{y}$ if $\angle x = 67°$.

(b)   Find $\hat{y}$ if $\angle x = 61°$.

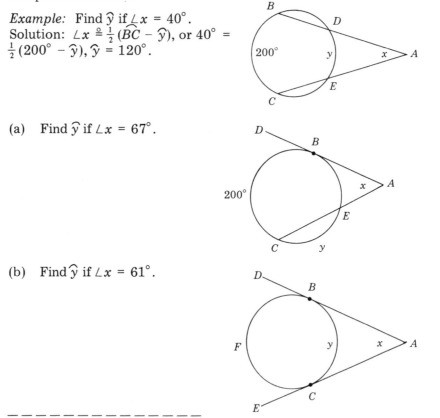

- - - - - - - - - - - - - -

(a) $\angle x \overset{\underline{o}}{=} \frac{1}{2}(\widehat{BC} - \widehat{BE})$, hence $67° = \frac{1}{2}(200° - \widehat{BE})$, or $\widehat{BE} = 66°$.
$\widehat{y} = 360° - 266° = 94°$.

(b) $\angle x \overset{\underline{o}}{=} \frac{1}{2}(\widehat{BFC} - \widehat{y})$, hence $61° = \frac{1}{2}[(360° - \widehat{y}) - \widehat{y}]$, or $\widehat{y} = 119°$.

11.   Now let's try using these principles to solve some slightly more general problems involving the measurement of angles and arcs.

Example: Find x and y in the figure at the right.

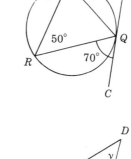

Solution: $50° = \frac{1}{2}\widehat{PQ}$ or $\widehat{PQ} = 100°$. (Pr. 2)
$70° = \frac{1}{2}\widehat{QR}$ or $\widehat{QR} = 140°$. (Pr. 9)
Then $\widehat{PR} = 360° - \widehat{PQ} - \widehat{QR} = 120°$.
$\angle x \overset{\underline{o}}{=} \frac{1}{2}\widehat{PR} = 60°$. (Pr. 9)
$\angle y \overset{\underline{o}}{=} \frac{1}{2}(\widehat{PRQ} - \widehat{PQ})$ (Pr. 13)
$= \frac{1}{2}(260° - 100°) = 80°$

Find x and y.

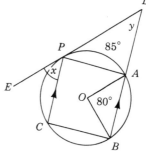

- - - - - - - - - - - - - - - - - -

$\widehat{AB} = 80°$ (Pr. 1)
$\widehat{BC} = \widehat{PA} = 85°$ (Pr. 8)
Then $\widehat{PC} = 360° - \widehat{PA} - \widehat{AB} - \widehat{BC} = 110°$.
$\angle x \overset{\underline{o}}{=} \frac{1}{2}\widehat{PC} = 55°$ (Pr. 9)
$\angle y \overset{\underline{o}}{=} \frac{1}{2}(\widehat{PCB} - \widehat{PA})$ (Pr. 12)
$= \frac{1}{2}(195° - 85°) = 55°$.

It's time you checked up on yourself to see how much you have learned about circles, tangents, and the measurement of angles and arcs. The following Self-Test should help you do so.

SELF-TEST

1. Given: $AB \cong DE$, $AC \cong DF$; Prove: $\angle B \cong \angle E$.
    (frame 2)

2. Prove formally the following: If a radius bisects a chord, then it bisects its arcs. (frame 3)

3. DP and CQ are tangents. Find $\angle 2$ and $\angle 3$ if $\angle OPD$ is trisected and PQ is a diameter. (frame 4)

4. Quadrilateral $ABCD$ is circumscribed. Find x. (frame 4)

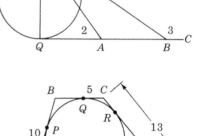

5. If two circles have radii of 20 and 13 respectively, find their line of centers:

    (a) If the circles are concentric. _____

    (b) If the circles are 7 units apart. _____

    (c) If the circles are tangent externally. _____

    (d) If the circles are tangent internally. _____

    (frame 5)

6. If the line of centers of two circles is 30, what is the relation between the circles:

 (a) If their radii are 25 and 5. _____

 (b) If their radii are 35 and 5. _____

 (c) If their radii are 20 and 5. _____

 (d) If their radii are 25 and 10. _____

 (frame 6)

7. Find the number of degrees in a central angle which intercepts an arc of:

 (a) 40° _____   (e) $2x°$ _____

 (b) 90° _____   (f) $(180 - x)°$ _____

 (c) 170° _____   (g) $(2x - 2y)°$ _____

 (d) 180° _____   (frame 6)

8. Find the number of degrees in an inscribed angle that intercepts an arc of:

 (a) 40° _____   (f) 348° _____

 (b) 90° _____   (g) $2x°$ _____

 (c) 170° _____   (h) $(180 - x)°$ _____

 (d) 180° _____   (i) $(2x - 2y)°$ _____

 (e) 260° _____   (frame 6)

9. If quadrilateral $ABCD$ is inscribed in a circle, find $\angle A$ if $\angle C = 45°$. (frame 7)

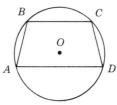

10. If BC and AD are the parallel sides of inscribed trapezoid $ABCD$, as shown, find $\overset{\frown}{AB}$ if $\overset{\frown}{CD} = 85°$. (frame 7)

11. Find the number of degrees in the angle formed by a tangent and a chord drawn to the point of tangency if the intercepted arc is $38°$.

(frame 8)

12. Find the number of degrees in the arc intercepted by an angle formed by a tangent and a chord drawn to the point of tangency if the angle equals $55°$.

(frame 8)

13. Find the values of $\hat{x}$ and $\angle y$ in the figure at the right.   (frame 9)

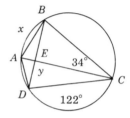

14. If AB and AC are intersecting secants as shown, find $\angle A$ if $\widehat{c} = 100°$ and $\widehat{a} = 40°$.

(frame 10)

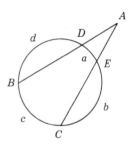

Answers to Self-Test

1. PROOF: Statements

Statements	Reasons
1. $AB \cong DE$, $AC \cong DF$	1. Given
2. $\widehat{AB} \cong \widehat{DE}$, $\widehat{ABC} \cong \widehat{DEF}$	2. In a circle, $\cong$ chords have $\cong$ arcs.
3. $\widehat{BC} \cong \widehat{EF}$	3. If equals are subtracted from equals, the differences are equal.
4. $BC \cong EF$	4. Same as 2.
5. $\Delta\text{I} \cong \Delta\text{II}$	5. SSS
6. $\angle B \cong \angle E$	6. Corresponding parts of $\cong$ $\triangle$.

2. Given: Circle O
   OC bisects AB
   Prove: $\overset{\frown}{AC} \cong \overset{\frown}{CB}$
   Plan: Prove $\triangle AOD \cong \triangle BOD$, $\angle 1 \cong \angle 2$,
   hence $\overset{\frown}{AC} \cong \overset{\frown}{CB}$.

PROOF: Statements	Reasons
1. Draw OA and OB	1. A straight line may be drawn between any two points.
2. $OA \cong OB$	2. Radii of a circle are congruent.
3. OC bisects AB	3. Given
4. $AD \cong DB$	4. To bisect is to divide into two congruent parts.
5. $OD \cong OD$	5. Identity
6. $\triangle AOD \cong \triangle BOD$	6. SSS
7. $\angle 1 \cong \angle 2$	7. Corresponding parts of congruent triangles are congruent.
8. $\overset{\frown}{AC} \cong \overset{\frown}{CB}$	8. In the same or congruent circles, congruent central angles have congruent arcs.

3. By Pr. 1, $\angle DPQ = \angle PQC = 90°$. Since $\angle 1 = 30°$, then $\angle 2 = 60°$. And since $\angle 3$ is an exterior angle of $\triangle PQB$, $\angle 3 = 90° + 60° = 150°$.

4. By Pr. 4, $AS = 10$, $CR = 5$, and $RD = SD$. Then $RD = CD - CR = 8$. Hence, $x = AS + SD = 10 + 8 = 18$.

5. (a) 0; (b) 40; (c) 33; (d) 7

6. (a) tangent externally; (b) tangent internally; (c) the circles are 5 units apart; (d) overlapping

7. (a) $40°$; (b) $90°$; (c) $170°$; (d) $180°$; (e) $2x°$; (f) $(180 - x)°$; (g) $(2x - 2y)°$

8. (a) $20°$; (b) $45°$; (c) $85°$; (d) $90°$; (e) $130°$; (f) $174°$; (g) $x°$; (h) $(90 - \frac{1}{2}x)°$; (i) $(x - y)°$

9. $135°$

10. $85°$

11. $19°$

12. $110°$

13. $\overset{\frown}{x} = 68°$, $\angle y = 95°$

14. $30°$

RATIOS AND PROPORTION

12. From your study of algebra you will recall that *ratios* are used to compare quantities by division. Thus, the ratio of two quantities is the first divided by the second. A ratio is an abstract number, that is, a number

without a unit of measure. The ratio of 10 inches to 2 inches is 10 ÷ 2, or 5.

You will also recall that a ratio can be expressed in several ways:

(1) by use of a colon, as 3:5;
(2) as a common fraction, $\frac{3}{5}$;
(3) as a decimal, .60;
(4) as a percent, 60%; or
(5) by use of the word "to," as 3 to 5.

To find ratios, the quantities involved must have the same unit. A ratio should be simplified by reducing it to lowest terms and eliminating any fractions contained in the ratio. Thus, to find the ratio of 1 foot to 3 inches, first change the foot to 12 inches, then take the ratio of 12 inches to 3 inches. The result is a ratio of 4 to 1, or 4. Also, the ratio of $2\frac{1}{2}:\frac{1}{2}$ equals 5:1, or 5.

The ratio of three or more quantities may be expressed as a *continued ratio*. Thus the ratio of $3 to $4 to $5 is the continued ratio 3:4:5. This enlarged ratio is a combination of three separate ratios, namely, 3:4, 3:5, and 4:5.

Express each of the following ratios in lowest terms.

(a) 20° to 5° _____   (h) 30 to 50 _____

(b) $1.50 to $6.00 _____   (i) 5.6 to .7 _____

(c) $3\frac{1}{2}$ yrs to $1\frac{1}{2}$ yrs _____   (j) 12 to $\frac{3}{8}$ _____

(d) 2 yrs to 6 mos _____   (k) $3x$ to $5x$ _____

(e) 60¢ to $3.00 _____   (l) $3a^2$ to a^3 _____

(f) 1 gal to 2 qt to 2 pt _____   (m) p to $5p$ to $7p$ _____

(g) 1 ton to 1 lb to 8 oz _____

- - - - - - - - - - - - - - - -

(a) $\frac{20}{5} = 4$; (b) $\frac{1.50}{6.00} = \frac{1}{4}$; (c) $\frac{3\frac{1}{2}}{1\frac{1}{2}} = \frac{7}{3}$; (d) 24 mos to 6 mos = $\frac{24}{6} = 4$; (e) 60¢ to 300¢ = $\frac{60}{300} = \frac{1}{5}$; (f) 4 qt to 2 qt to 1 qt = 4:2:1;

(g) 2000 lb to 1 lb to $\frac{1}{2}$ lb = $2000:1:\frac{1}{2}$ = 4000:2:1; (h) $\frac{30}{50} = \frac{3}{5}$;

(i) $\frac{5.6}{.7} = 8$; (j) $12 \div \frac{3}{8} = 12\left(\frac{8}{3}\right) = 32$; (k) $\frac{3x}{5x} = \frac{3}{5}$; (l) $\frac{3a^2}{a^3} = \frac{3}{a}$;

(m) $p:5p:7p$ = 1:5:7

13. No doubt you noticed that the above problems included the ratio of two quantities with the *same* unit, the ratio of two quantities with *different* units, the *continued ratio* of three quantities, and several

numerical and *algebraic ratios.* The intent was to provide you with a general review of various kinds of ratios so that you will recognize them when you see them again.

Now let us consider the use of ratios in angle problems.

Example: If two angles are in the ratio of 3:2, find the angles if they are adjacent and form an angle of $40°$.
Solution: Since the ratio between the angles is 3:2, let $3x$ and $2x$ represent the number of degrees in the angles. Then, $3x + 2x = 40$, $5x = 40$, or $x = 8$. Hence the angles are $24°$ and $16°$.

Assuming again that two angles are in the ratio of 3:2, find the angles if:

(a)   they are the acute angles of a right triangle.

(b)   they are two angles of a triangle whose third angle is $70°$.

(a) $3x + 2x = 90$, $5x = 90$, $x = 18$. Hence the angles are $54°$
      and $36°$.
(b) $3x + 2x + 70 = 180$, $5x = 110$, $x = 22$. Hence the angles are
      $66°$ and $44°$.

14.   Now consider a situation where three angles are in the ratio of $4:3:2$.

      Example: Find the angles if the first and the third are supplementary.
      Solution: Let $4x$, $3x$, and $2x$ represent the number of degrees in the angles. Then $4x + 2x = 180$, $6x = 180$, $x = 30$. Hence the angles are $120°$, $90°$, and $60°$.

      Now assuming the same ratio between the three angles, find their values if the angles are the three angles of a triangle.

      $4x + 3x + 2x = 180$, $9x = 180$, $x = 20$. Therefore, the angles are $80°$, $60°$, and $40°$.

15.   Again from your study of algebra, you know that a *proportion* is an equality of two ratios. Thus 2:5 = 4:10, or $\frac{2}{5} = \frac{4}{10}$ is a proportion.

      The fourth term of a proportion is the *fourth proportional* to the other three, taken in order. Thus, in 2:3 = 4:x, x is the fourth proportional to 2, 3, and 4.

The *means* of a proportion are its *middle* terms (that is, the second and third terms). The *extremes* of a proportion are its *outside* terms (that is, its first and fourth terms). Thus, in the proportion $a:b = c:d$, b and c are the means, and a and d are the extremes.

If the two means of a proportion are the same, either mean is the *mean proportional* between the first and fourth terms. Thus, in $9:3 = 3:1$, 3 is the mean proportional between 9 and 1.

Now we need to consider some proportion principles.

Pr. 1:   In any proportion, the product of the means equals the product of the extremes.

Thus, if $a:b = c:d$, then $ad = bc$.

Pr. 2:   If the product of two numbers equals the product of two other numbers, either pair may be made the means of a proportion and the other pair may be made the extremes.

Thus, if $3x = 5y$, then $x:y = 5:3$, or $y:x = 3:5$, or $3:y = 5:x$, or $5:x = 3:y$.

The next four principles have to do with methods of changing a proportion into a new proportion.

Pr. 3:   *Inversion Method.*   A proportion may be changed into a new (equal) proportion by inverting each ratio.

Thus, if $\dfrac{1}{x} = \dfrac{4}{5}$, then $\dfrac{x}{1} = \dfrac{5}{4}$.

Pr. 4:   *Alternation Method.*   A proportion may be changed into a new proportion by interchanging the means or by interchanging the extremes.

Thus, if $\dfrac{x}{3} = \dfrac{y}{2}$, then $\dfrac{x}{y} = \dfrac{3}{2}$, or $\dfrac{2}{3} = \dfrac{y}{x}$.

Pr. 5:   *Addition Method.*   A proportion may be changed into a new proportion by adding the terms of each ratio to obtain new first and third terms.

Thus, if $\dfrac{a}{b} = \dfrac{c}{d}$, $\dfrac{a}{b}$ is the first ratio. Adding its terms (a and b) gives us the new first term (of the proportion) $a + b$. And adding the terms of the second ratio, $\dfrac{c}{d}$, gives us the new third term, namely, $c + d$. Therefore our proportion now becomes $\dfrac{a + b}{b} = \dfrac{c + d}{d}$. Similarly, the proportion $\dfrac{x - 2}{2} = \dfrac{9}{1}$ becomes $\dfrac{(x - 2) + 2}{2} = \dfrac{9 + 1}{1}$ or simply $\dfrac{x}{2} = \dfrac{10}{1}$.

Pr. 6: *Subtraction Method.* A proportion may be changed into a new proportion by subtracting the terms of each ratio to obtain new first and third terms.

Thus, if $\frac{a}{b} = \frac{c}{d}$, then $\frac{a - b}{b} = \frac{c - d}{d}$. Or if $\frac{x + 3}{3} = \frac{9}{1}$, then $\frac{(x + 3) - 3}{3} = \frac{9 - 1}{1}$, or $\frac{x}{3} = \frac{8}{1}$.

Here are two other proportion principles.

Pr. 7: If any three terms of one proportion equal the corresponding three terms of another proportion, the remaining terms are equal.

Thus, if $\frac{x}{y} = \frac{3}{5}$ and $\frac{x}{4} = \frac{3}{5}$, then $y = 4$.

Pr. 8: In a series of equal ratios, the sum of the numerators is to the sum of the denominators as any one numerator is to its denominator.

Thus, if $\frac{a}{b} = \frac{c}{d} = \frac{e}{f}$, then $\frac{a + c + e}{b + d + f} = \frac{a}{b}$. Or if $\frac{x - y}{4} = \frac{y - 3}{5} = \frac{3}{1}$, then $\frac{x - y + y - 3 + 3}{4 + 5 + 1} = \frac{3}{1}$ or $\frac{x}{10} = \frac{3}{1}$.

Now let's practice using these principles. Solve for x in the following proportions.

(a) $x:4 = 6:8$ _____

(b) $3:x = x:27$ _____

(c) $x:5 = 2x:(x + 3)$ _____

(d) $\frac{3}{x} = \frac{2}{5}$ _____

(e) $\frac{x}{2x - 3} = \frac{3}{5}$ _____

(f) $\frac{x - 2}{4} = \frac{7}{x + 2}$ _____

- - - - - - - - - - - - - -

(a) $4(6) = 8x$, $8x = 24$, $x = 3$
(b) $x^2 = 3(27)$, $x^2 = 81$, $x = \pm 9$
(c) $5(2x) = x(x + 3)$, $10x = x^2 + 3x$, $x^2 - 7x = 0$, $x = 0$ or 7
(d) $2x = 3(5)$, $2x = 15$, $x = 7\frac{1}{2}$
(e) $3(2x - 3) = 5x$, $6x - 9 = 5x$, $x = 9$
(f) $4(7) = (x - 2)$, $28 = x^2 - 4$, $x^2 = 32$, $x = \pm 4\sqrt{2}$

16. The next few problems involve finding the fourth proportional to three given numbers.

Example: Find the fourth proportional to 2, 4, 6.
Solution: $2:4 = 6:x$, $2x = 24$, $x = 12$

Follow this same procedures to find the fourth proportionals in the following problems.

(a)   4, 2, 6 _____

(b) $\frac{1}{2}$, 3, 4 _____

(c) b, d, c _____

– – – – – – – – – – – – – –

(a) $4:2 = 6:x$, $4x = 12$, $x = 3$
(b) $\frac{1}{2}:3 = 4:x$, $\frac{1}{2}x = 12$, $x = 24$
(c) $b:d = c:x$, $bx = cd$, $x = \dfrac{cd}{b}$

17. Now let's try finding the mean proportional to two given numbers. (Remember, from frame 15, this is a case where the second and third terms are equal; either is the mean proportional.)

Example: Find the positive mean proportional (x) between 5 and 20.
Solution: $5:x = x:20$, $x^2 = 100$, $x = 10$

Find the positive mean proportional between $\frac{1}{2}$ and $\frac{8}{9}$.

_____

– – – – – – – – – – – – – –

$\frac{1}{2}:x = x:\frac{8}{9}$, $x^2 = \frac{4}{9}$, $x = \frac{2}{3}$

18. Occasionally you will find equal products and need to change these into proportions. The procedure for doing so is essentially contained in Pr. 2. Thus if we have the equal products $ad = bc$, we can use Pr. 2 to form proportion $a:b = c:d$. Or suppose we had the product $ay = bx$ and wished to find the ratio of x to y. Using Pr. 1 and Pr. 2 we can easily form the proportion $x:y = a:b$.

In each of the following, form a proportion whose fourth term is x.

(a) $cx = bd$ _____

(b) $pq = ax$ _____

(c) $2bx = 3s^2$ _____

– – – – – – – – – – – – – –

(a) $\frac{c}{b} = \frac{d}{x}$,   (b) $\frac{a}{p} = \frac{q}{x}$,   (c) $\frac{2b}{3s} = \frac{s}{x}$ or $\frac{2b}{3} = \frac{s^2}{x}$

19. Try selecting the correct method (Pr. 3, 4, 5, or 6) and change the proportions shown below into new proportions.

Example: Starting with the proportion $\frac{15}{x} = \frac{3}{4}$, form a new proportion whose *first* term is x.

Solution: By Pr. 3, $\frac{x}{15} = \frac{4}{3}$.

Form new proportions whose first terms are x.

(a) $\frac{x - 6}{6} = \frac{5}{3}$ _____

(b) $\frac{x + 8}{8} = \frac{4}{3}$ _____

(c) $\frac{5}{2} = \frac{15}{x}$ _____

- - - - - - - - - - - - - -

(a) By Pr. 5, $\frac{x}{6} = \frac{8}{3}$

(b) By Pr. 6, $\frac{x}{8} = \frac{1}{3}$

(c) By Pr. 4, $\frac{x}{2} = \frac{15}{5}$

20. Use Pr. 8 to find x in the following problems.

Example: $\frac{x - 2}{9} = \frac{2}{3}$ or, by Pr. 8, $\frac{x - 2 + 2}{9 + 3} = \frac{2}{3}, \frac{x}{12} = \frac{2}{3}, x = 8$.

(a) $\frac{x + y}{8} = \frac{x - y}{4} = \frac{2}{3}$ _____

(b) $\frac{3x - y}{15} = \frac{y - 3}{10} = \frac{3}{5}$ _____

- - - - - - - - - - - - - -

(a) $\frac{(x + y) + (x - y)}{8 + 4} = \frac{2}{3}, \frac{2x}{12} = \frac{2}{3}, x = 4$

(b) $\frac{(3x - y) + (y - 3) + 3}{15 + 10 + 5} = \frac{3}{5}, \frac{3x}{30} = \frac{3}{5}, x = 6$ (adding in the third ratio simplified the solution)

21. So far our discussion of ratios and proportions probably has seemed to you a lot more like algebra than geometry. And of course it was. But one of the reasons you studied algebra was so that you could use it to help you solve a variety of problems, and you are about to find another use for it here. Already we have used algebra to help solve a number of simple equations that we have encountered in our study of geometry. The study of proportionality provides still another opportunity. And remember, our overall approach throughout this book is a combined algebraic and geometric view of the mathematical concepts that will prepare you for the study of calculus. Now we are ready to consider the subject of *proportional segments* and see how what we have been learning about proportion can be applied in plane geometry. First, let's examine some of the basic properties of proportional segments.

If two segments are divided proportionately,
(1) the corresponding segments are in proportion, and
(2) the two segments and either pair of corresponding segments are in proportion.

Thus, if AB and AC are divided proportionately by DE, a proportion such as

$\frac{a}{b} = \frac{c}{d}$ may be obtained using the four segments, or a proportion such as

$\frac{a}{AB} = \frac{c}{AC}$ may be obtained using the two lines and two of their segments.

A proportion such as $\frac{a}{b} = \frac{c}{d}$ can be arranged in eight ways. To obtain the eight variations simply let each term in the proportion represent a segment of the above diagram. Each of the possible proportions then is obtained by using the same direction, as follows:

Direction Down

$\frac{a}{b} = \frac{c}{d}$ or $\frac{c}{d} = \frac{a}{b}$

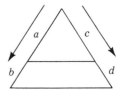

Direction Up

$\frac{b}{a} = \frac{d}{c}$ or $\frac{d}{c} = \frac{b}{a}$

Direction Right

$\frac{a}{c} = \frac{b}{d}$ or $\frac{b}{d} = \frac{a}{c}$

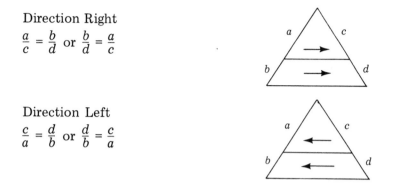

Direction Left

$\frac{c}{a} = \frac{d}{b}$ or $\frac{d}{b} = \frac{c}{a}$

Below are four fundamental principles relating to proportional lines.

Pr. 1: If a line is parallel to one side of a triangle, then it divides the other two sides proportionately.

Thus, in $\triangle ABC$, if $DE \parallel BC$, then $\frac{a}{b} = \frac{c}{d}$.

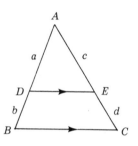

Pr. 2: If a line divides two sides of a triangle proportionately, it is parallel to the third side. (Converse of Pr. 1.)

Thus, in $\triangle ABC$ if $\frac{a}{b} = \frac{c}{d}$, then $DE \parallel BC$.

Example of Pr. 1: Find x in the figure at the right.

Solution: $DE \parallel BC$, hence $\frac{x}{12} = \frac{28}{14}$, or

$x = 24$.

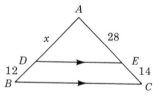

Now try this problem: Find x in the adjacent diagram.

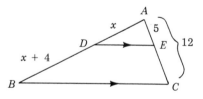

_____

- - - - - - - - - - - - - - -

$EC = 7$, $DE \parallel BC$. Hence $\frac{x}{x + 4} = \frac{5}{7}$, $7x = 5x + 20$, $x = 10$

22. Now let's consider Principle 3.

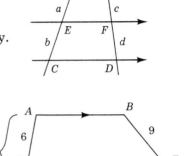

Pr. 3: Three or more parallel lines divide any two transversals proportionately.

Thus, if $AB \parallel EF \parallel CD$, then $\frac{a}{b} = \frac{c}{d}$.

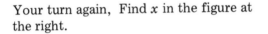

Example: Find x in the figure at the right.
Solution: $EC = 4$, and $AB \parallel EF \parallel CD$. Hence

$\frac{x}{9} = \frac{4}{6}$ or $x = 6$.

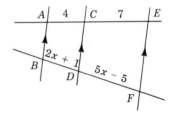

Your turn again, Find x in the figure at the right.

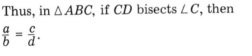

_____

- - - - - - - - - - - - - - - -

$AB \parallel CD \parallel EF$, hence $\frac{5x - 5}{2x + 1} = \frac{7}{4}$, $20x - 20 = 14x + 7$, $6x = 27$,

$x = 4\frac{1}{2}$

23. The fourth proportional line principle is as follows.

Pr. 4: A bisector of an angle of a triangle divides the opposite side into segments which are proportional to the adjacent sides.

Thus, in $\triangle ABC$, if CD bisects $\angle C$, then
$\frac{a}{b} = \frac{c}{d}$.

Example: Find x in the figure at the right.
Solution: BD bisects $\angle B$, hence

$\frac{x}{10} = \frac{18}{15}$, or $x = 12$.

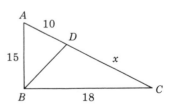

Use this same approach to find x in the adjacent figure.

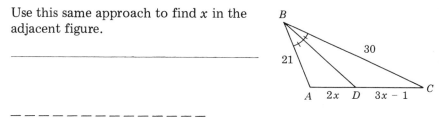

_____

– – – – – – – – – – – – – – – –

BD bisects $\angle B$, hence $\dfrac{3x-1}{2x} = \dfrac{30}{21} = \dfrac{10}{7}$, or $21x - 7 = 20x$, $x = 7$

SIMILARITY

24. We come now to the topic of similar triangles. Because triangles are three-sided polygons we can be a bit more general in our definition if we define similar polygons, since this will include triangles.

   Similar polygons are polygons whose corresponding angles are congruent and whose corresponding sides are in proportion. Thus, similar polygons have the same *shape*, although not necessarily the same *size*. If they have the same shape *and* size, then they will be congruent.)

   The symbol $\sim$ means "similar." Therefore, if we wish to say that two triangles (such as those at the right) are similar, we write this $\triangle ABC \sim A'B'C'$. We read this as "triangle A-prime B-prime C-prime."

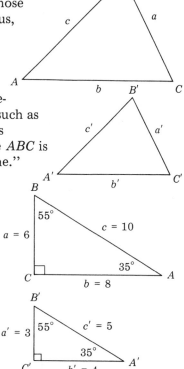

   Like congruent triangles, *corresponding sides of similar triangles are opposite congruent angles*. And for convenience sake, corresponding sides and angles are usually identified by the same letters with primes. Thus, $\triangle ABC \sim \triangle A'B'C'$, since $\angle A = 35° = \angle A', \angle B = 55° = \angle B',$

$\angle C = 90° = \angle C',$ and $\dfrac{a}{a'} = \dfrac{b}{b'} = \dfrac{c}{c'}$, or

$\dfrac{6}{3} = \dfrac{8}{4} = \dfrac{10}{5}$.

Now let's consider some of the principles relating to similar triangles.

Pr. 1:   Corresponding angles of similar triangles are congruent. (By definition.)

Pr. 2:   Corresponding sides of similar triangles are in proportion. (By definition.)

Example: In similar triangles ABC and $A'B'C'$, find x and y if $\angle A \cong \angle A'$ and $\angle B \cong \angle B'$.

Solution: Since $\angle A \cong \angle A'$ and $\angle B \cong \angle B'$, x and y correspond to 32 and 26 respectively. Hence

$\frac{x}{32} = \frac{15}{20}$ and $x = 24$. Similarly,

$\frac{y}{26} = \frac{15}{20}$ or $y = 19\frac{1}{2}$.

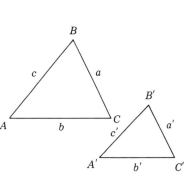

Pr. 3:   Two triangles are similar if two angles of one triangle are congruent respectively to two angles of the other.

Thus, if $\angle A \cong \angle A'$ and $\angle B \cong \angle B'$, then $\triangle ABC \sim \triangle A'B'C'$.

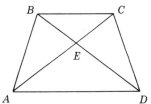

Example: In the figure at the right, two pairs of congruent angles can be used to prove $\triangle BEC \sim \triangle AED$. Indicate which angles are congruent and state the reason. (*ABCD* is a trapezoid.)

Solution: $\angle CBD \cong \angle BDA$ and $\angle BCA \cong \angle CAD$, since alternate interior angles of parallel lines are congruent ($BC \parallel AD$). Also, $\angle BEC$ and $\angle AED$ are congruent vertical angles.

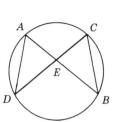

Apply Pr. 3 similarly to solve this problem: Name two pairs of angles that can be used to prove $\triangle AED \sim \triangle CEB$ and state the reason.

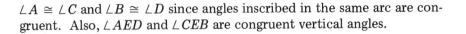

$\angle A \cong \angle C$ and $\angle B \cong \angle D$ since angles inscribed in the same arc are congruent. Also, $\angle AED$ and $\angle CEB$ are congruent vertical angles.

25. *Pr. 4:*   Two triangles are similar if an
angle of one triangle is con-
gruent to an angle of the
other and the sides including
these angles are in proportion.

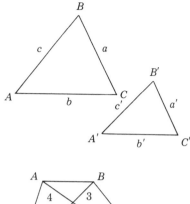

Thus, if $\angle C \cong \angle C'$ and $\frac{a}{a'} = \frac{b}{b'}$, then
$\triangle ABC \sim \triangle A'B'C'$.

Example: Name the pair of congruent
angles and the proportion needed to
prove $\triangle AEB \sim \triangle DEC$.
Solution: $\angle AEB \cong \angle DEC$,
$\frac{3}{9} = \frac{4}{12}$

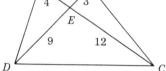

Name the pair of congruent angles and
the proportion needed to prove
$\triangle AED \sim \triangle ABC$.

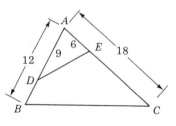

- - - - - - - - - - - - - - - -

$\angle A \cong \angle A, \ \frac{6}{12} = \frac{9}{18}$

26. *Pr. 5:*   Two triangles are similar if
their corresponding sides are
in proportion.

Thus, if $\frac{a}{a'} = \frac{b}{b'} = \frac{c}{c'}$, then
$\triangle ABC \sim \triangle A'B'C'$.

Example: Determine the proportion
needed to prove $\triangle ABC \sim \triangle DEF$.
Solution: $\frac{6}{12} = \frac{8}{16} = \frac{12}{24}$

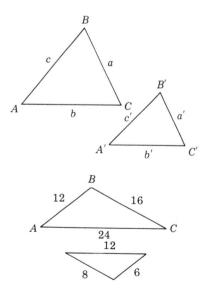

Set up the proportion needed to prove
$\triangle ABD \sim \triangle BDC$.

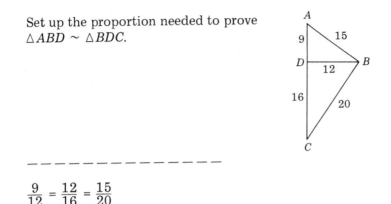

$$\frac{9}{12} = \frac{12}{16} = \frac{15}{20}$$

27. *Pr. 6:*   Two right triangles are similar if an acute angle of one is congruent to an acute angle of the other.   (Corollary of Pr. 3.)

Example: Name the angles that can be
used to prove $\triangle ACD \sim \triangle ACB$.
Solution: $\angle ACB$ and $\angle ADC$ are right
angles; $\angle A \cong \angle A$.

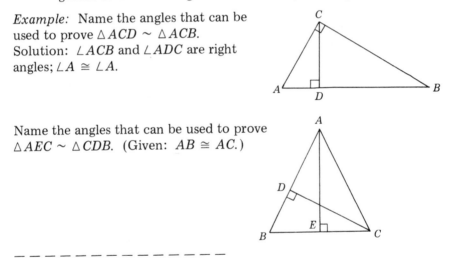

Name the angles that can be used to prove
$\triangle AEC \sim \triangle CDB$.   (Given: $AB \cong AC$.)

$\angle AEC$ and $\angle BDC$ are right angles. Also, $\angle B \cong \angle ACE$ since angles in a
triangle opposite congruent sides are congruent.

28.   There are innumerable applications of the similar triangle and proportionality concepts we have been discussing. They can be applied in the
solution of a great many routine types of problem that occur daily in
engineering design, architectural layout and drafting, shop work, sheet
metal work, machinery design, and so on. Unfortunately, there is
neither time nor space in a relatively brief book such as this — and one
that is intended primarily to provide you with general guidance to
several branches of mathematics leading to calculus — to allow for the
introduction of any large number of applied examples. However, you

will have no difficulty in finding as many of these as you wish in almost any standard textbook on plane geometry.

Nevertheless, we will introduce such examples where we can, and the present subject provides a good opportunity. Consider the following problem.

Example: A tree casts a 15-foot shadow at a time when a nearby upright pole, 6 feet in height, casts a shadow of 2 feet. We wish to find the height of the tree if both the tree and the pole make right angles with the ground.

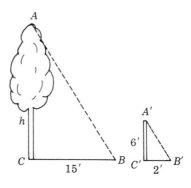

Solution: At the same time, in localities near each other, the rays of the sun strike the ground at congruent angles, hence $\angle B \cong \angle B'$. And since the tree and pole make right angles with the ground, $\angle C \cong \angle C'$. Therefore, $\triangle ABC \sim \triangle A'B'C'$, $\frac{h}{6} = \frac{15}{2}$, and $h = 45$ feet.

Now try this problem (be sure to draw diagrams to assist you): A 7-foot upright pole near a vertical tree casts a 6-foot shadow. At that time,

(a)   find the height of the tree if its shadow is 36 feet.

(b)   find the shadow of the tree if its height is 77 feet.

- - - - - - - - - - -

(a) $\frac{7}{h} = \frac{6}{36}$, or $h = 42$ feet.

(b) $\frac{7}{77} = \frac{6}{s}$, or $s = 66$ feet.

29.   There are two useful mean proportionals in a right triangle with which you should be familiar. They are as follows.

Pr. 1:   The altitude to the hypotenuse of a right triangle is the mean proportional between the segments of the hypotenuse.

Thus, in right $\triangle ABC$, $\frac{BD}{CD} = \frac{CD}{DA}$.

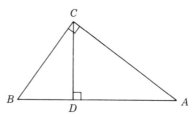

Pr. 2:   In a right triangle, either leg is the mean proportional between the hypotenuse and the projection of that leg on the hypotenuse (i.e., portion of the hypotenuse lying under that leg).

Thus, in right $\triangle ABC$, $\dfrac{AB}{BC} = \dfrac{BC}{BD}$, and $\dfrac{AB}{AC} = \dfrac{AC}{AD}$.

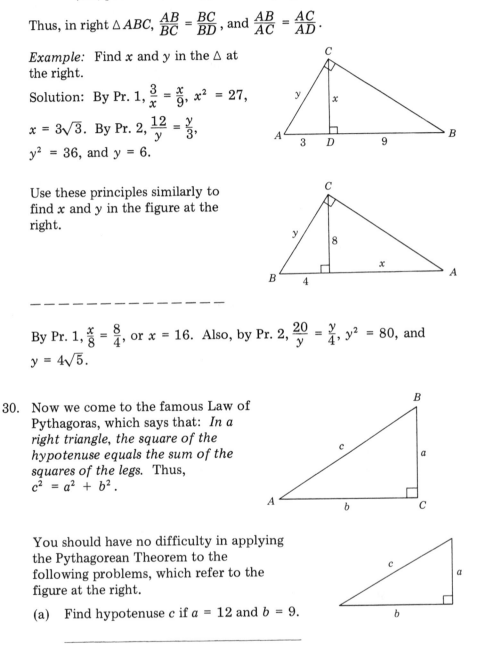

Example: Find x and y in the $\triangle$ at the right.

Solution: By Pr. 1, $\dfrac{3}{x} = \dfrac{x}{9}$, $x^2 = 27$,

$x = 3\sqrt{3}$. By Pr. 2, $\dfrac{12}{y} = \dfrac{y}{3}$,

$y^2 = 36$, and $y = 6$.

Use these principles similarly to find x and y in the figure at the right.

By Pr. 1, $\dfrac{x}{8} = \dfrac{8}{4}$, or $x = 16$. Also, by Pr. 2, $\dfrac{20}{y} = \dfrac{y}{4}$, $y^2 = 80$, and $y = 4\sqrt{5}$.

30. Now we come to the famous Law of Pythagoras, which says that: *In a right triangle, the square of the hypotenuse equals the sum of the squares of the legs.* Thus, $c^2 = a^2 + b^2$.

You should have no difficulty in applying the Pythagorean Theorem to the following problems, which refer to the figure at the right.

(a) Find hypotenuse c if $a = 12$ and $b = 9$.

_____

(b) Find leg a if $b = 6$ and $c = 8$.

_____

(c)   Find leg b if $a = 4\sqrt{3}$ and $c = 8$.

_ _ _ _ _ _ _ _ _ _ _ _ _ _ _

(a) $c^2 = a^2 + b^2$, or $c^2 = 12^2 + 9^2 = 225$, or $c = 15$.
(b) $a^2 = c^2 - b^2 = 8^2 - 6^2 = 28$, or $a = 2\sqrt{7}$.
(c) $b^2 = c^2 = a^2 = 8^2 - (4\sqrt{3})^2$, or $b^2 = 64 - 48$, from which $b = 4$.

31.   Use the Law of Pythagoras to find the
altitude to the base of an isosceles triangle
if the base is 8 and the congruent sides
are 12. (Note: The altitude, h, of an
isosceles triangle bisects the base.)

_ _ _ _ _ _ _ _ _ _ _ _ _ _ _

Since the altitude of an isosceles triangle bisects the base, then
$h^2 = a^2 - (\frac{1}{2}b)^2$, or $h^2 = 12^2 - 4^2 = 128$, from which $h = 8\sqrt{2}$.

32.   The Law of Pythagoras also applies very nicely to the rhombus.

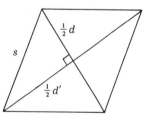

Example: In a rhombus, find side s if the
diagonals are 30 and 40.
Solution: Keeping in mind that the
diagonals of a rhombus are perpendicular
bisectors of each other, we can write
$s^2 = (\frac{1}{2}d)^2 + (\frac{1}{2}d')^2$. Or, substituting
the values for d and d',
$s^2 = 15^2 + 20^2 = 625$, $s = 25$.

Find diagonal d if a side is 26 and the other diagonal is 20.

_ _ _ _ _ _ _ _ _ _ _ _ _ _ _

Since $s = 26$ and $d' = 20$, then $26^2 = (\frac{1}{2}d)^2 + 10^2$, or $576 = (\frac{1}{2}d)^2$,
from which $\frac{1}{2}d = 24$, $d = 48$.

33. Let's see if you can apply the Law of Pythagoras to a trapezoid. It will be good practice for you.

Find x in the isosceles trapezoid $ABCD$ at the right. (Note: The dotted perpendicular line shown in the diagram is an additional line needed only for solution. Observe how a rectangle is formed by this added line.)

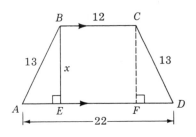

– – – – – – – – – – – – – – –

$EF = BC = 12$, $AE = \frac{1}{2}(22 - 12) = 5$. Then $x^2 = 13^2 - 5^2 = 144$, $x = 12$.

34. Finally, there are two special right triangles having unique properties that we need to talk about. One is the $30°$-$60°$-$90°$ triangle and the other is the $45°$-$45°$-$90°$ triangle. These unique properties will be especially useful when we get to the subject of trigonometry. But let's see what they are.

A $30°$-$60°$-$90°$ triangle is one-half of an equilateral triangle, as you can see from the two figures at the right.

Thus, in right $\triangle ABC$, $b = \frac{1}{2}c$. Therefore, if we let $c = 2$, then $b = 1$ and, applying the Law of Pythagoras, $a^2 = c^2 - b^2 = 2^2 - 1^2 = 3$, or $a = \sqrt{3}$, and the ratio of the sides is $b:c:a = 1:2:\sqrt{3}$.

Here are some important principles relating to $30°$-$60°$-$90°$ triangles and to the equilateral triangle.

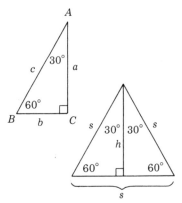

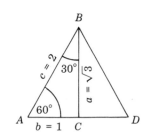

Pr. 1:   The leg opposite the $30°$ angle equals one-half the hypotenuse, i.e., $a = \frac{1}{2}c$.

Pr. 2:   The leg opposite the $60°$ angle equals one-half the hypotenuse times the square root of 3, i.e., $b = \frac{1}{2}c\sqrt{3}$.

Pr. 3:   The leg opposite the 60° angle equals the leg opposite the 30° angle times the square root of 3, i.e., $b = a\sqrt{3}$·

Pr. 4:   The altitude of an equilateral triangle equals one-half a side times the square root of 3, i.e., $h = \frac{1}{2}s\sqrt{3}$. (This is a corollary of Pr. 2.)

Apply these principles in the following problems. Be sure to draw a diagram to assist you.

(a)   If the hypotenuse of a 30°–60°–90° triangle is 12, find its legs.

(b)   Each leg of an isosceles trapezoid is 18. If the base angles are 60° and the upper base is 10, find the altitude and the lower base.

(a)   By Pr. 1, $a = \frac{1}{2}(12) = 6$.
      By Pr. 2, $b = \frac{1}{2}(12)\sqrt{3}$, or $b = 6\sqrt{3}$.

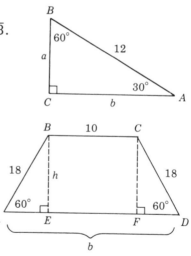

(b)   By Pr. 2, $h = \frac{1}{2}(18)\sqrt{3} = 9\sqrt{3}$.
      By Pr. 1, $AE = FD = \frac{1}{2}(18) = 9$, hence $b = 9 + 10 + 9 = 28$.

35.   A 45°–45°–90° triangle is one-half a square. Thus, in right triangle ABC, $c^2 = a^2 + a^2$, or $c = a\sqrt{2}$, hence the ratio of the sides is $a:a:c = 1:1:\sqrt{2}$.
      Principles of the 45°–45°–90° triangle and of the square are as follows:

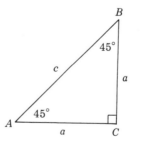

Pr. 5: The leg opposite a $45°$ angle equals one-half the hypotenuse times the square root of 2, i.e., $a = \frac{1}{2}c\sqrt{2}$.

Pr. 6: The hypotenuse equals a side times the square root of 2, i.e., $c = a\sqrt{2}$.

Pr. 7: In a square, a diagonal equals a side times the square root of 2, i.e., $d = s\sqrt{2}$.

Apply Principles 5 and 6 in the following problems (again, be sure to draw diagrams).

(a)   Find the leg of an isosceles right triangle whose hypotenuse is 28.

(b)   An isosceles trapezoid has base angles of $45°$. If the upper base is 12 and the altitude is 3, find the lower base and each leg.

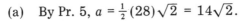

(a)   By Pr. 5, $a = \frac{1}{2}(28)\sqrt{2} = 14\sqrt{2}$.

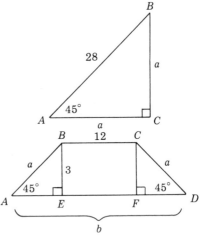

(b)   By Pr. 6, $a = 3\sqrt{2}$.
       $AE = BE = 3$ and $EF = 12$.
       Hence $b = 3 + 12 + 3 = 18$.

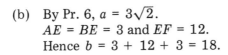

It's time to take a look back now over what we have covered on the subject of similarity. The following Self-Test will help you review the principal concepts and perhaps show you where you need to do some reviewing before going on to the next chapter.

SELF-TEST

1. Express each ratio in lowest terms.   (frame 12)

   (a)   20¢ to 5¢ _____   (e)   \$2.20 to \$3.30 _____

   (b)   30 lb. to 25 lb. _____   (f) $\frac{1}{2}$ lb. to $\frac{1}{4}$ lb. _____

   (c)   27 min. to 21 min. _____   (g)   5 ft. to $\frac{1}{4}$ ft. _____

   (d)   15% to 75% _____   (h) $16\frac{1}{2}$ ft. to $5\frac{1}{2}$ ft. _____

2. If two angles in the ratio of $5:4$ are represented by $5x$ and $4x$, express
   each statement as an equation, then find x and the angles.   (frame 13)

   (a)   The angles are adjacent and form an angle of $45°$.

   (b)   The angles are complementary.

   (c)   The angles are supplementary.

   (d)   The angles are two angles of a triangle whose third angle is their
         difference.

3. If three angles in the ratio of $7:6:5$ are represented by $7x$, $6x$, and $5x$,
   express each statement as an equation and find x and the angles.
                                                                   (frame 14)

   (a)   The first and second are adjacent and form an angle of $91°$.

   (b)   The first and third are supplementary.

   (c)   The angles are the three angles of a triangle.

4. Solve for x.   (frame 15)

   (a) $x:6 = 8:3$ _____

   (b) $5:4 = 20:x$ _____

   (c) $(x + 4):3 = 3:(x - 4)$ _____

   (d) $(2x + 8):(x + 2) = (2x + 5):(x + 1)$ _____

5. Find the fourth proportional to each set of numbers.   (frame 16)

(a) 1, 3, 5 _____

(b) 2, 3, 4 _____

(c) $\frac{1}{3}$, 2, 5 _____

(d) b, $2a$, $3b$ _____

6. Find the positive mean proportional between each pair of numbers.
                                                                 (frame 17)

(a) 4 and 9 _____

(b) $\frac{1}{3}$ and 27 _____

(c) 2 and 5 _____

(d) p and q _____

7. In each, form a proportion whose fourth term is x.   (frame 18)

(a) $hx = a^2$ _____

(b) $3x = 7$ _____

(c) $x = \frac{ab}{c}$ _____

8. In each, form a new proportion whose first term is x, then find x.
                                                                 (frame 19)

(a) $\frac{3}{2} = \frac{9}{x}$ _____

(b) $\frac{a}{x} = \frac{2}{b}$ _____

(c) $\frac{x - 20}{20} = \frac{1}{4}$ _____

9. Find x in each.   (frame 20)

(a) $\frac{x - 7}{8} = \frac{7}{4}$ _____

(b) $\frac{x + y}{6} = \frac{x - y}{3} = \frac{1}{3}$ _____

10. Find x in the figure at the right.
    (frame 21)

11. Find x in the figure at the right.
    (frame 22)

12. Find x in the figure at the right.
    (frame 23)

13. In the figure shown opposite, two pairs of angles can be used to prove triangles ABC and DEF are similar. Determine the congruent angles.   (frame 24)

14. What pair of congruent angles and what proportion are needed to prove triangles ADE and ABC similar?
    (frame 25)

15. Indicate the proportion needed to prove triangles ADE and ABC are similar.
    (frame 26)

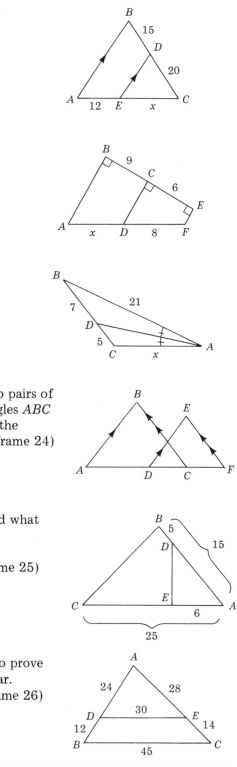

16. What angles can be used to prove
    $\triangle AED \sim FGB$; ($ABCD$ is a
    parallelogram.)   (frame 27)

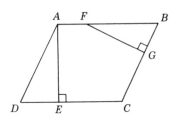

17. A 10 ft. upright pole near a vertical tree casts a 12 ft. shadow. At that
    time,

    (a) find the height of the tree if its shadow is 30 feet.

    (b) find the shadow of the tree if its height is 30 feet. (Draw yourself
         a diagram.)   (frame 28)

18. CD is the altitude to the hypotenuse AB.

    (a) If $p = 2$ and $q = 6$, find a and h.

    _____

    (b) If $p = 4$ and $a = 6$, find c and h.

    _____

    (c) If $p = 16$ and $h = 8$, find q and b.

    _____

    (d) If $b = 12$ and $q = 6$, find p and h.

    _____

                    (frame 29)

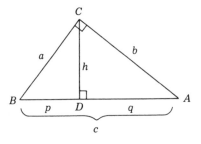

19. In a right triangle whose legs are a and b, find
    the hypotenuse c when

    (a) $a = 15$, $b = 20$ _____

    (b) $a = 5$, $b = 4$ _____

    (c) $a = 7$, $b = 7$ _____
                              (frame 30)

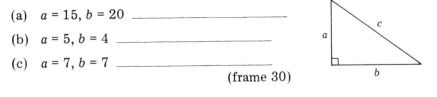

20. In isosceles trapezoid $ABCD$,

    (a)   Find a if $b = 32$, $b' = 20$, and $h = 8$.

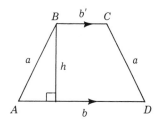

    _____

    (b)   Find h if $b = 24$, $b' = 14$, and $a = 13$.

    _____

    (c)   Find b if $a = 15$, $b' = 10$, and $h = 12$.

    _____

    (d)   Find b' if $a = 6$, $b = 21$, and $h = 3\sqrt{3}$.

    _____

                        (frame 33)

21. In a $30°-60°-90°$ triangle, find: (frame 34)

    (a)   the legs if the hypotenuse is 20.

    (b)   the other leg and hypotenuse if the leg opposite $30°$ is 7.

    (c)   the other leg and hypotenuse if the leg opposite $60°$ is $5\sqrt{3}$.

22. In an isosceles right triangle, find: (frame 35)

    (a)   each leg if the hypotenuse is 34.

    (b)   the hypotenuse if each leg is $15\sqrt{2}$.

Answers to Self-Test

1. (a) 4; (b) $\frac{6}{5}$; (c) $\frac{9}{7}$; (d) $\frac{1}{5}$; (e) $\frac{2}{3}$; (f) 2; (g) 20; (h) 3

2. (a) $5x + 4x = 45$, $x = 5$, 25° and 20°
    (b) $5x + 4x = 90$, $x = 10$, 50° and 40°
    (c) $5x + 4x = 180$, $x = 20$, 100° and 80°
    (d) $5x + 4x + x = 180$, $x = 18$, 90° and 72°

3. (a) $7x + 6x = 91$, $x = 7$, 49°, 42°, and 35°
    (b) $7x + 5x = 180$, $x = 15$, 105°, 90°, and 75°
    (c) $7x + 6x + 5x = 180$, $x = 10$, 70°, 60°, and 50°

4. (a) 16; (b) 16; (c) ±5; (d) 2

5. (a) 15; (b) 6; (c) 30; (d) $6a$

6. (a) 6; (b) 3; (c) $\sqrt{10}$; (d) $\sqrt{pq}$

7. (a) $\frac{h}{a} = \frac{a}{x}$; (b) $\frac{3}{7} = \frac{1}{x}$; (c) $\frac{c}{a} = \frac{b}{x}$

8. (a) $\frac{x}{2} = \frac{9}{3}$; $x = 6$; (b) $\frac{x}{a} = \frac{b}{2}$, $x = \frac{ab}{2}$; (c) $\frac{x}{20} = \frac{5}{4}$, $x = 25$

9. (a) 21; (b) $\frac{3}{2}$

10. 16

11. 12

12. 15

13. $\angle A \cong \angle EDF$, $\angle F \cong \angle BCA$

14. $\angle A \cong \angle A$, $\dfrac{10}{25} = \dfrac{6}{15}$

15. $\dfrac{24}{36} = \dfrac{28}{42} = \dfrac{30}{45}$

16. $\angle D \cong \angle B$, $\angle AED \cong \angle FGB$

17. (a) 25 feet; (b) 36 feet

18. (a) $a = 4$, $h = \sqrt{12}$ or $2\sqrt{3}$
    (b) $c = 9$, $h = \sqrt{20}$ or $2\sqrt{5}$
    (c) $q = 4$ and $b = \sqrt{80}$ or $4\sqrt{5}$
    (d) $p = 18$, $h = \sqrt{108}$ or $6\sqrt{3}$

19. (a) 25; (b) $\sqrt{41}$; (c) $7\sqrt{2}$

20. (a) 10; (b) 12; (c) 28; (d) 15

21. (a) 10 and $10\sqrt{3}$; (b) $7\sqrt{3}$ and 14; (c) 5 and 10

22. (a) $17\sqrt{2}$; (b) 30

CHAPTER FOUR

Plane Geometry:
Areas, Polygons, and Locus

Having learned something about circles, tangents, similarity, and the methods of measuring angles and arcs, we are going to turn our attention now to learning some formulas for *area* measurement and how to apply these in a variety of problems. We also are going to investigate the properties of regular polygons and how to find the area of a circle as well as of a segment and sector of a circle. We will then discuss the concept of the locus of a point — something that will come in very handy when we get to the subject of analytic geometry. Finally, we will have some fun with geometric constructions.

When we get to the end of this chapter you will have learned about:

- finding the area of such geometric figures as rectangles, squares, parallelograms, triangles, trapezoids, and the rhombus;

- the regular polygon, including such elements as its radius, apothem, central angles, calculating its area, and its relation to the circle;

- the ratio π, finding the areas and circumferences of inscribed and circumscribed circles, and the areas of segments and sectors;

- determining the locus of a point equidistant from two given points, from two parallel lines, from the sides of a given angle, from intersecting lines, and from a point and a circle;

- a number of basic constructions made with the use of a straight edge and compass only.

AREAS

1. No doubt you have a general familiarity with areas and some of the methods of computing them. Figuring the number of square yards of

carpeting you need for your living room or how many "yards" (this is a little trickier because of the differing widths of materials) you need for a dress are common enough calculations. But now we need to be a bit more precise as we consider methods of calculating the areas of a wider variety of geometric shapes. As usual, we will begin by defining a few terms.

A *square unit* is the surface enclosed by a square whose side is 1 unit.

1 in.

1 in. | 1 square inch

The *area of a closed plane figure*, such as a polygon, is the number of square units contained in its surface. Since a rectangle 5 units long and 4 units wide can be divided into 20 unit squares, its area is 20 square units.

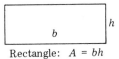

4

5

The *area of a rectangle* equals the product of its base and altitude. Thus, if $b = 8$ in. and $h = 3$ in., then $A = 24$ sq. in.

Rectangle: $A = bh$

The *area of a square* equals the square of a side. Thus, if $s = 6$, then $A = s^2 = 36$. It follows, therefore, that the area of a square also equals one-half the square of a diagonal. Since $A = s^2$ and $s = d/\sqrt{2}$, $A = \frac{1}{2}d^2$.

Square: (1) $A = s^2$
          (2) $A = \frac{1}{2}d^2$

Here are a few practice exercises for you.

(a) Find the area of a rectangle if the base is 15 and the perimeter (distance around) is 50.

$A = $ _____

(b) Find the area of a rectangle if the altitude is 10 and the diagonal is 26.

$A = $ _____

(c) Find the base and altitude of a rectangle if its area is 70 and its perimeter is 34.

$b = $ _____ $h = $ _____

— — — — — — — — — — — — — —

(a) $p = 50$, $b = 15$. Since $p = 2b + 2h$, $50 = 2(15) + 2h$, or $h = 10$. Therefore $A = bh = 15(10) = 150$.

(b) $d = 26$, $h = 10$. In right $\triangle ACD$, $d^2 = b^2 + h^2$, or $26^2 = b^2 + 10^2$, from which $b = 24$. Hence $A = bh = 24(10) = 240$.

(c) $A = 70$, $p = 34$. Since $p = 2b + 2h$, $34 = 2(b + h)$ and $h = 17 - b$. Then $A = bh$, or $70 = b(17 - b)$, $b^2 - 17b + 70 = 0$, and $b = 7$ or 10. Since $h = 17 - b$, we obtain $h = 10$ or 7.

2. The above problems involved working with rectangles. The problems below will provide you with a little practice working with squares. (Use the diagrams to assist you.)

(a) Find the area of a square if the perimeter is 30.

$A = $ _____

(b) Find the area of a square if the radius of the circumscribed circle is 10.

$A = $ _____

(c) Find the side and the perimeter of a square whose area is 20.

$s = $ _____ $p = $ _____

(d) Find the number of square inches in a square foot.

_____

— — — — — — — — — — — — —

(a) $p = 30$. Since $p = 4s$, $30 = 4s$ and $s = 7\frac{1}{2}$. Then $A = s^2 = (7\frac{1}{2})^2 = 56\frac{1}{4}$.

(b) Since $r = 10$, $d = 2r = 20$. Then $A = \frac{1}{2}d^2 = \frac{1}{2}(20)^2 = 200$.

(c) $A = 20$ and $A = s^2$, hence $s^2 = 20$, $s = 2\sqrt{5}$. Perimeter $= 4s = 8\sqrt{5}$.

(d) $A = s^2$. Since 1 ft. $= 12$ in., $A = 12^2 = 144$. Therefore, 1 sq. ft. $= 144$ sq. in.

3. Having considered the rectangle and the square, let's turn our attention now to the parallelogaram. Here is a very useful area theorem relating to the parallelogram.

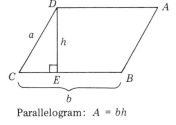

Parallelogram: $A = bh$

The area of a parallelogram equals the product of a side and the altitude to that side.

Thus, in $\square ABCD$, if $b = 10$ and $h = 2.7$, then $A = 10(2.7) = 27$.

Apply this in the following two problems.

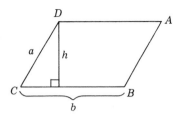

(a) Find the area of a parallelogram if
the area is represented by $x^2 - 4$,
a side by $x + 4$, and the altitude to
that side by $x - 3$.

(b) In a parallelogram, find the altitude
if the area is 54 and the ratio of the
altitude to the base is 2:3.

(a) $A = x^2 - 4$, $b = x + 4$, and $h = x - 3$.
Then $A = bh$, or $x^2 - 4 = (x + 4)(x - 3)$, $x^2 - 4 = x^2 + x - 12$,
and $x = 8$. Hence $A = x^2 - 4 = 64 - 4 = 60$.

(b) Let $h = 2x$, $b = 3x$. Then $A = bh$, or $54 = (3x)(2x)$, $54 = 6x^2$,
$9 = x^2$, and $x = 3$. Hence $h = 2x = 2(3) = 6$.

4. Next we come to the area of a triangle.

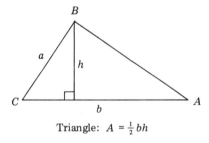

*The area of a triangle equals
one-half the product of a side
and the altitude to that side.*

Thus, $A = \frac{1}{2}bh$. This just involves a
little straightforward arithmetic which
you should have no trouble applying
in the following problem.

Triangle: $A = \frac{1}{2}bh$

Find the area of a triangle if two adjacent
sides of 15 and 8 include an angle of $150°$.

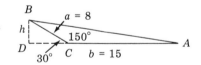

$b = 15$, $a = 8$. Since $\angle BCA = 150°$, $\angle BCD = 180° - 150° = 30°$.
In $\triangle BCD$, h is opposite $\angle BCD$, hence $h = \frac{1}{2}a = 4$. Then
$A = \frac{1}{2}bh = \frac{1}{2}bh = \frac{1}{2}(15)(4) = 30$.

5. The trapezoid is equally easy to work with. Here are the relevant
theorems.

The area of a trapezoid equals one-half the product of its altitude and the sum of its bases.

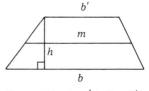

Thus, if $h = 20$, $b = 27$, and $b' = 23$, then $A = \frac{1}{2}(20)(27 + 23) = 500$.

Trapezoid: $A = \frac{1}{2}h\,(b + b')$

The area of a trapezoid equals the product of its altitude and median.

Since (in the figure above) $A = \frac{1}{2}h(b + b')$ and $m = \frac{1}{2}(b + b')$, then $A = hm$.

Use the above relationships in the following problems.

(a) Find the area of a trapezoid if the bases are 7.3 and 2.7, and the altitude is 3.8.

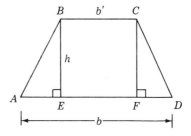

(b) Find the area of an isosceles trapezoid if the bases are 22 and 10, and the legs are each 10.

(c) Find the bases of an isosceles trapezoid if the area is $52\sqrt{3}$, the altitude is $4\sqrt{3}$, and each leg is 8.

- - - - - - - - - - - - - - - -

(a) $b = 7.3$, $b' = 2.7$, $h = 3.8$ Therefore $A = \frac{1}{2}h(b + b') = \frac{1}{2}(3.8)(7.3 + 2.7) = 19$.

(b) $b = 22$, $b' = 10$, $AB = 10$. $EF = b' = 10$ and $AE = \frac{1}{2}(22 - 10) = 6$. In $\triangle BEA$, $h^2 = 10^2 - 6^2 = 64$, or $h = 8$.
Then $A = \frac{1}{2}h(b + b') = \frac{1}{2}(8)(22 + 10) = 128$.

(c) $AE = \sqrt{(AB)^2 - h^2} = \sqrt{64 - 48} = 4$, $FD = AE = 4$,
$b' = b - (AE + FD) = b - 8$. Then $A = \frac{1}{2}h(b + b') = \frac{1}{2}h(2b - 8)$
or $52\sqrt{3} = \frac{1}{2}(4\sqrt{3})(2b - 8)$, from which $26 = 2b - 8$ or $b = 17$.
(This is good practice both in algebra and in reasoning.) Thus, $b = 17$, $b' = 9$.

6. Finally, among the quadrilaterals, we have
    this theorem giving us a means of finding
    the area of the rhombus.

    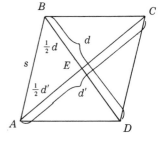

    *The area of a rhombus equals one-half
    the product of its diagonals.*

    Since we know (from frame 33,
    Chapter 2) that each diagonal is the perpen-
    dicular bisector of the other, the area of $\triangle I$

    Rhombus: $A = \frac{1}{2} dd'$

    is $\frac{1}{2} (\frac{1}{2} d)(\frac{1}{2} d') = \frac{1}{8} dd'$. Thus the rhombus,
    which consists of 4 triangles congruent to $\triangle I$, has an area of $4(\frac{1}{8} dd')$
    or $\frac{1}{2} dd'$.

    Use this information to solve the following problems.

    (a) Find the area of a rhombus if one
         diagonal is 30 and a side is 17.

    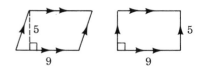

    (b) Find a diagonal of a rhombus if the
         other diagonal is 8 and the area of
         the rhombus is 52.

    (a) $d' = 30$, $s = 17$. In right $\triangle AEB$, $s^2 = (\frac{1}{2} d)^2 + (\frac{1}{2} d')^2$.
         $17^2 = (\frac{1}{2} d)^2 + 15^2$, $\frac{1}{2} d = 8$, or $d = 16$. Then
         $A = \frac{1}{2} dd' = \frac{1}{2}(16)(30) = 240$.
    (b) $d' = 8$, $A = 52$. Then $A = \frac{1}{2} dd'$, or $52 = \frac{1}{2}(d)(8)$ and $d = 13$.

7. We will conclude our discussion of areas by stating the following four
    principles and giving you an illustration of each.

    Pr. 1: Parallelograms have equal areas
             if they have congruent bases
             and congruent altitudes. (This
             is a corollary of the theorem in
             frame 3.)

    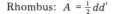

    Thus, the parallelograms shown at the
    right have equal areas.

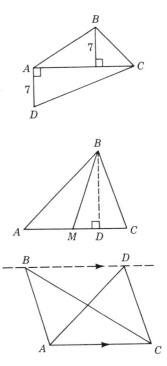

Pr. 2: Triangles have equal areas if they have congruent bases and congruent altitudes. (This is a corollary of the theorem in frame 4.)

Thus, in the figure at the right, $\triangle CAB = \triangle CAD$ in area.

Pr. 3: A median divides a triangle into two triangles of equal area.

Thus, in the figure at the right where BM is a median, $\triangle AMB = \triangle BMC$ since they have congruent bases ($AM \cong MC$) and common altitude BD.

Pr. 4: Triangles have equal areas if they have a common base and their vertices lie on a line parallel to the base.

Thus, $\triangle ABC = \triangle ADC$ in the figure at the right.

Now it's time for a review of the main facts we have discussed about areas.

SELF-TEST

1. Find the area of a rectangle if:

    (a) the base is 11 in. and the altitude is 9 in.

    _____

    (b) the base is 25 and the perimeter is 90

    _____

    (c) the diagonal is 12 and the angle between the diagonal and the base is 60°.

    _____
    (frame 1)

2. Find the area of a rectangle inscribed in a circle if:

    (a) the radius of the circle is 5 and the base is 6

    _____

(b)   the radius and the altitude are both 5.

_____

(frame 1)

3.   Find the base and altitude of a rectangle if:

(a)   its area is 28 and the base is 3 more than the altitude

_____

(b)   its area is 72 and the base is twice the altitude

_____

(c)   its area is 12 and the perimeter is 16.

_____

(frame 1)

4.   Find the area of:

(a)   a square yard in square inches

_____

(b)   a square meter in square decimeters (1 meter = 10 decimeters).

_____

(frame 2)

5.   Find the area of a square if:

(a)   a side is 15

_____

(b)   the perimeter is 44

_____

(c)   the diagonal is 8.

_____

(frame 2)

6.   Find the area of a square if:

(a)   the radius of the circumscribed circle is 8

_____

(b)   the diameter of the circumscribed circle is $10\sqrt{2}$.

(frame 2)

7. Find the area of a parallelogram if the base and altitude are, respectively:

(a)   3 ft. and $5\frac{1}{3}$ ft.

(b)   4 ft. and 1 ft. 6 in.

(c)   20 and 3.5.

(frame 3)

8. Find the area of a triangle if two adjacent sides are, respectively:

(a)   8 and 5, and include an angle of $30°$

(b)   8 and 12, and include an angle of $60°$

(frame 4)

9. Find the area of trapezoid $ABCD$ if $b = 25$, $b' = 15$, and $h = 7$.
(frame 5)

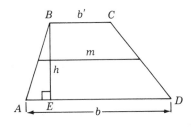

10. Find the area of isosceles trapezoid $ABCD$ if:

(a) $b = 22$, $b' = 12$, and $l = 13$

(b) $b = 20$, $l = 8$, and $\angle A = 60°$.
(frame 5)

11.   Find the area of a rhombus if:

(a)   the diagonals are 8 and 9

_____

(b)   the diagonals are $3x$ and $8x$

_____

(c)   the perimeter is 28 and an angle is $45°$.

_____

(frame 6)

12.   In a rhombus find:

(a)   a diagonal if the other diagonal is 7 and the area is 35;

_____

(b)   the diagonals, if their ratio is 4:3 and the area is 54.

_____

(frame 6)

Answers to Self-Test

1.   (a) 99 sq. in.;   (b) 500;   (c) $36\sqrt{3}$
2.   (a) 48;   (b) $25\sqrt{3}$
3.   (a) 7 and 4;   (b) 12 and 6;   (c) $b = 6$ or 2; $h = 2$ or 6
4.   (a) 1296 sq. in.;   (b) 100 square decimeters
5.   (a) 225;   (b) 121;   (c) 32
6.   (a) 128;   (b) 100
7.   (a) 16 sq. ft.;   (b) 6 sq. ft. or 864 sq. in.;   (c) 70
8.   (a) 10;   (b) $24\sqrt{3}$
9.   140
10. (a) 204;   (b) $64\sqrt{3}$
11. (a) 36;   (b) $12x^2$;   (c) $\frac{49}{2}\sqrt{2}$

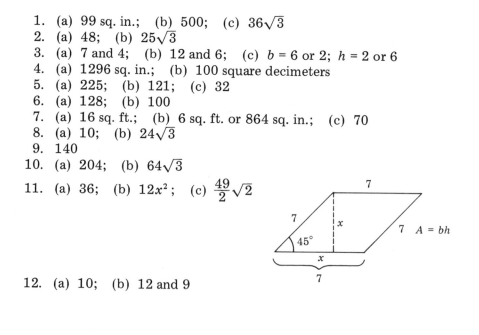

12. (a) 10;   (b) 12 and 9

REGULAR POLYGONS AND THE CIRCLE

8. A *regular polygon* (as we learned in frame 20, Chapter 2) is an equilateral and equiangular polygon.

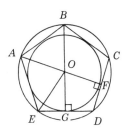

The *center of a regular polygon* is the common center of its inscribed and circumscribed circles. *O* is the center in the figure shown.

A *radius of a regular polygon* is a line joining its center to any vertex. A radius of a regular polygon is also a radius of the circumscribed circle. Thus, in the figure at the right, *OA* and *OB* are its radii.

A *central angle of a regular polygon* is an angle included between two radii drawn to successive vertices. Thus, $\angle AOB$ is a central angle.

An *apothem of a regular polygon* is a line from its center perpendicular to one of its sides. Thus, *OG* and *OF* are apothems. An apothem also is a radius of the inscribed circle.

Following are some useful principles relating to regular polygons.

Pr. 1:   If a regular polygon of *n* sides has a side *s*, the perimeter is
$p = ns.$

Pr. 2:   A circle may be circumscribed about any regular polygon.

Pr. 3:   A circle may be inscribed in any regular polygon.

Pr. 4:   The center of the circumscribed circle of a regular polygon is also the center of its inscribed circle.

Pr. 5:   An equilateral polygon inscribed in a circle is a regular polygon.

Pr. 6:   Radii of a regular polygon are congruent.

Pr. 7:   A radius of a regular polygon bisects the angle to which it is drawn. (Thus, in the above figure *OB* bisects $\angle ABC$.)

Pr. 8:   Apothems of a regular polygon are congruent.

Pr. 9:   An apothem of a regular polygon bisects the side to which it is drawn. (Thus, in the figure above *OF* bisects *CD* and *OG* bisects *ED*.)

Pr. 10: For a regular polygon of n sides:

(1) each central angle c equals $\dfrac{360°}{n}$.

(2) each interior angle $i = \dfrac{(n-2)180°}{n}$.

(3) each exterior angle e equals $\dfrac{360°}{n}$.

Thus, for the regular pentagon $ABCDE$,

$$\angle AOB = \angle ABS = \frac{360°}{n} = \frac{360°}{5} = 72°;$$

$$\angle ABC = \frac{(n-2)180°}{n} = \frac{(5-2)180°}{5} = 108°, \text{ and } \angle ABC + \angle ABS = 180°.$$

(You were introduced to this principle in frame 21, Chapter 2.)

Now let's apply Principles 1 and 10 (the only ones containing formulas) in solving a few problems.

(a) Find a side s of a regular pentagon if the perimeter p is 35.

(b) Find the apothem r of a regular pentagon if the radius of the inscribed circle is 21. (Check your definition of an apothem again before trying this one.)

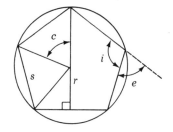

(c) In a regular polygon of 5 sides, find the central angle c, the exterior angle e, and the interior angle i.

(d) If an interior angle of a regular polygon is 165°, find the exterior angle, the central angle, and the number of sides.

— — — — — — — — — — — — — — —

(a) $p = 35$. Since $p = ns$, $35 = 5s$ and $s = 7$. (Pr. 1.)

(b) Since an apothem r is a radius of the inscribed circle, it equals 21. (Definition.)

(c) $n = 5$. Then $c = \dfrac{360°}{n} = \dfrac{360°}{5} = 72°$, $e = \dfrac{360°}{n} = 72°$,

$i = 180° - e = 108°$. (Pr. 10.)

(d) $i = 165°$. Then $c = e = 180° - i = 15°$. Since $c = \dfrac{360°}{n}$, $n = 24$.

9. Another handy formula for the regular polygon is this one.

The area of a regular polygon equals one-half the product of its perimeter and apothem.

As shown, by drawing radii a regular polygon of n sides and perimeter $p = ns$ can be divided into n triangles, each of area $\frac{1}{2}rs$. Hence the area of the regular polygon =

$$n(\tfrac{1}{2}rs) = \tfrac{1}{2}(ns)r = \tfrac{1}{2}pr.$$

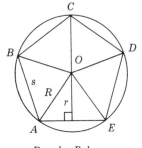

Regular Polygon
$A = \frac{1}{2}nsr = \frac{1}{2}pr$

Use this formula to help solve the following problems.

(a) Find the area of a regular hexagon if the apothem is $5\sqrt{3}$. (Hint: In a regular hexagon the central angles are all $60°$, hence the radius, R, equals the length of a side, s; also, you will need the formula from frame 34, Chapter 3, relating the apothem, r, to the length of a side.)

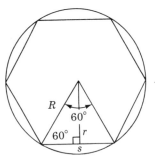

(b) Find the area of a regular hexagon, in radical form, if the side is 6.

― ― ― ― ― ― ― ― ― ― ― ― ― ― ―

(a) From frame 34, Chapter 3, Pr. 2, $r = \frac{1}{2}R\sqrt{3}$, but since $R = s$, we can write this $r = \frac{1}{2}s\sqrt{3}$. Substituting $5\sqrt{3}$ for r gives us

$5\sqrt{3} = \frac{1}{2}s\sqrt{3}$ or $s = 10$. And since $p = sn$, then $p = 10 \cdot 6 = 60$.

Therefore $A = \frac{1}{2}pr = \frac{1}{2}60(5\sqrt{3}) = 150\sqrt{3}$.

(b) $s = 6$. Therefore $r = \frac{1}{2}s\sqrt{3} = \frac{1}{2}(6\sqrt{3}) = 3\sqrt{3}$. Also

$p = sn = 6 \cdot 6 = 36$. Hence $A = \frac{1}{2}pr = \frac{1}{2}(36)(3\sqrt{3}) = 54\sqrt{3}$.

10. Our study of regular polygons leads us very logically to a consideration of the area of a circle, since a circle may be regarded as a regular polygon having an infinite number of sides.

The Greek letter π (pi) no doubt is familiar to you as the symbol for the ratio of the circumference (perimeter) of a circle to its diameter. That is, $\pi = \dfrac{C}{d}$. Hence $C = \pi d$ or $C = 2\pi r$. Approximate values for π are 3.1416, 3.14, or $\dfrac{22}{7}$. Unless otherwise indicated

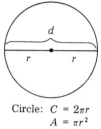

Circle: $C = 2\pi r$
$A = \pi r^2$

use 3.14 for π in solving problems in this book.
(You may recall from your study of algebra that π is an irrational number; that is, its value cannot be exactly represented as the ratio of two integers.)

If a square is inscribed in a circle and the number of sides continually doubled to form, successively, an octagon, a 16-gon, etc., the perimeters of the resulting polygons will very closely approximate the circumference of the circle. Thus, to find the area of a circle, the formula $A = \frac{1}{2}pr$ can be used with C (circumference) substituted for p (perimeter). Hence,

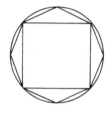

$A = \frac{1}{2}Cr = \frac{1}{2}(2\pi r)(r) = \pi r^2$. This gives us the familiar formula for the area of a circle, namely, $A = \pi r^2$.

Circles are similar figures since they have the same shape. As similar figures, (1) corresponding lines of circles are in proportion, and (2) the areas of two circles are to each other as the squares of their radii or circumferences.

Now let's apply what we have learned so far. Answer these in terms of π and also rounded to the nearest integer.

(a) Find the circumference and area of a circle if its radius is 6.

_____

(b) Find the radius and area of a circle if its circumference is 18π.

_____

(c) Find the radius and circumference of a circle if the area is 144π.

_____

– – – – – – – – – – – – – – – – –

(a) Given: $r = 6$. Therefore $C = 2\pi r = 12\pi$ and $A = \pi r^2 = 36\pi = 36(3.14) \rightarrow 113$.

(b) Given: $C = 18\pi$. Since $C = 2\pi r$, $18\pi = 2\pi r$ and $r = 9$, hence
$A = \pi r^2 = 81\pi \rightarrow 254$.

(c) Given: $A = 144\pi$. Since $A = \pi r^2$ and $r = 12$, then
$C = 2\pi r = 24\pi \rightarrow 75$.

11. Now let's combine some of the things we have discovered about regular polygons and circles.

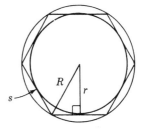

Example: Find the circumference and area
of the circumscribed and inscribed circles
of a regular hexagon if the side is 8.
Solution: Since in a hexagon $R = s$, then
$R = s = 8$. Hence for the circumscribed circle,
$C = 2\pi R = 16\pi$, and $A = \pi R^2 = 64\pi$. For the
inscribed circle, $r = \frac{1}{2}R\sqrt{3} = 4\sqrt{3}$. Then
$C = 2\pi r = 8\pi\sqrt{3}$ and $A = \pi r^2 = 48\pi$.

Here is a similar problem for you to work: Find the circumferences
and areas of the circumscribed and inscribed circles of a regular hexagon
if the side is 4. (Answers may be given in terms of π.)

––––––––––––––––––

Circumscribed: $C = 8\pi$, $A = 16\pi$; inscribed: $C = 4\sqrt{3}\pi$, $A = 12\pi$

12. A *sector of a circle* is the part of a circle
bounded by two radii and their intercepted
arc. Thus, the shaded portion of circle O
is sector OAB.

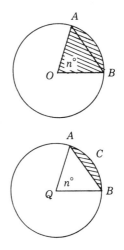

A *segment of a circle* is the part of a circle
bounded by a chord and its arc. Thus, the
shaded portion of circle Q is segment ACB.

The following principles relate to lengths of arcs and the areas of
sectors and segments of circles.

Pr. 1:   In a circle of radius r, the length l of an arc of $n°$ equals $\frac{n}{360}$

of the circumference of the circle, or $l = \frac{n}{360}(2\pi r) = \frac{\pi n r}{180}$.

Example: Find the length of an arc of $36°$ in a circle whose circumference is 45π.

Solution: $n° = 36$, $C = 2\pi r = 45\pi$. Then $l = \frac{n}{360}(2\pi r) =$

$\frac{36}{360}(45\pi) = \frac{9}{2}\pi$.

Find the radius of a circle if a $40°$ arc has a length of 4π.

– – – – – – – – – – – – – –

$l = 4\pi$, $n° = 40$. Then $l = \frac{n}{360}(2\pi r)$, or $4\pi = \frac{40}{360}(2\pi r)$ and $r = 18$.

13.   *Pr. 2:*   In a circle of radius r, the area K of a sector of $n°$ equals

$\frac{n}{360}$ of the area of the circle, or $K = \frac{n}{360}(\pi r^2)$.

Pr. 3: $\dfrac{\text{Area of a sector of } n°}{\text{Area of the circle}} = \dfrac{\text{Length of an arc of } n°}{\text{Circumference of circle}} = \dfrac{n}{360}$

Example 1:   Find the area K of a $300°$ sector of a circle whose radius is 12.
Solution: $n° = 300°$, $r = 12$. Then $K = \frac{n}{360}(\pi r^2) = \frac{300}{360}(144\pi) = 120\pi$.

Example 2:   Find the central angle of a sector whose area is 6π if the area of the circle is 9π.
Solution: $\dfrac{\text{Area of sector}}{\text{Area of circle}} = \dfrac{n}{360}, \dfrac{6\pi}{9\pi} = \dfrac{n}{360}$, and $n = 240$, hence the central angle is $240°$.

Find the radius of a circle if an arc of length 2π has a sector of area 10π.

– – – – – – – – – – – – – –

$\dfrac{\text{Length of arc}}{\text{Circumference}} = \dfrac{\text{Area of sector}}{\text{Area of circle}}, \dfrac{2\pi}{2\pi r} = \dfrac{10\pi}{\pi r^2}$ or $r = 10$.

14. *Pr. 4:*   The area of a minor segment of a circle
equals the area of its sector less the
area of the triangle formed by its radii
and chord.

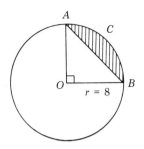

Example: Find the area of a segment if its
central angle is 90° and the radius of the
circle is 8.
Solution: $n° = 90°. r = 8.$ Area of sector

$OAB = \frac{n}{360}(\pi r^2) = \frac{90}{360}(64\pi) = 16\pi.$

Area of right $\triangle OAB = \frac{1}{2}bh = \frac{1}{2}(8)(8) = 32.$ Hence the area of segment
$ACB = 16\pi - 32.$

Find the area of a segment of a circle if the radius of the circle is 4 and
the central angle is 90°.

- - - - - - - - - - - - - - -

Area of segment $= 4\pi - 8$

To help you check up on your ability to apply these concepts and formulas relating to regular polygons and circles, below is another short self-test.
As always, be sure to review any portions of the material you find difficult.

SELF-TEST

1. In a regular polygon, find:

(a)   the perimeter if a side is 8 and the number of sides is 25.

_____

(b)   the perimeter if a side is 2.45 and the number of sides is 10.

_____

(c)   the number of sides if the perimeter is 325 and a side is 25.

_____

(d)   the side if the number of sides is 30 and the perimeter is 100.

_____

(frame 8)

2. Find the area of a regular hexagon, in radical form, if:

    (a)   its radius is 8. _____

    (b)   the apothem is $10\sqrt{3}$. _____

                                                       (frame 9)

3. In a circle find: (Note: You may leave π in your answers.)

    (a)   the circumference and area if the radius is 5.

       _____

    (b)   the radius and area if the circumference is 16π.

       _____

    (c)   the radius and circumference if the area is 16π.

       _____

                                                       (frame 10)

4. Find the circumference and area of the circumscribed and inscribed circles of a regular hexagon if the apothem is $4\sqrt{3}$.   (frame 11)

5. (Here is a problem to test your ingenuity. Hint: Find the areas of the circular cross-sections of the two pipes first.) Find the radius of a pipe having the same capacity (that is, cross-section area) as two pipes whose radii are 6 ft. and 8 ft.

6. In a circle, find the length of a 90° arc if:

    (a)   the radius is 4. _____

    (b)   the diameter is 40. _____

    (c)   the circumference is 32. _____

                                                       (frame 12)

7. In a circle, find the area of a 60° sector if:

    (a)   the radius is 6. _____

    (b)   the diameter is 2. _____

    (c)   the circumference is 10π. _____

                                                        (frame 13)

8. Find the area of a segment of a circle if the central angle is 90° and the length of the arc is 4π.   (frame 14)

Answers to Self-Test

1. (a) 200;   (b) 24.5;   (c) 13;   (d) $3\frac{1}{3}$
2. (a) $96\sqrt{3}$;   (b) $600\sqrt{3}$
3. (a) $C = 10\pi$, $A = 25\pi$
    (b) $r = 8$, $A = 64\pi$
    (c) $r = 4$, $C = 8\pi$
4. $C = 16\pi$, $A = 64\pi$; $C = 8\sqrt{3}\pi$, $A = 48\pi$.
5. 10 ft.
6. (a) 2π;   (b) 10π;   (c) 8
7. (a) 6π;   (b) $\frac{\pi}{6}$;   (c) $\frac{25\pi}{6}$
8. $16\pi - 32$

LOCUS

15. The word locus, in Latin, means location. The plural is loci. The *locus of a point* is the set of points, and only those points, that satisfy given conditions. Thus, the locus of a point that is 1 inch from a given point P is the set of points 1 inch from P. These points lie on a circle with its center at P and a radius of 1 inch. (Remember, we are dealing only with *plane* surfaces.)

To determine a locus, (1) state the given condition to be satisfied, (2) find several points satisfying the condition which indicate the shape of the locus, and (3) connect the points and describe the locus fully.

All geometric constructions require the use of a straightedge and compass, so make sure yours are available. Shown at the right is a compass.

Following are the fundamental locus theorems.

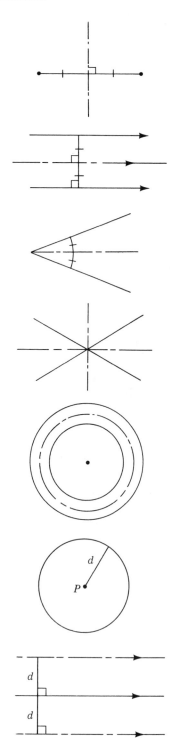

Pr. 1:   The locus of a point equidistant
from two given points is the
perpendicular bisector of the line
joining the two points.

Pr. 2:   The locus of a point equidistant
from two given parallel lines is a
line parallel to the two lines and
midway between them.

Pr. 3:   The locus of a point equidistant
from the sides of a given angle
is the bisector of the angle.

Pr. 4:   The locus of a point equidistant
from two given intersecting lines
is the bisectors of the angles
formed by the lines.

Pr. 5:   The locus of a point equidistant
from two concentric circles is the
circle concentric with the given
circles and midway between them.

Pr. 6:   The locus of a point at a given
distance from a given point is a
circle whose center is the given
point and whose radius is the
given distance.

Pr. 7:   The locus of a point at a given
distance from a given line is a
pair of lines, parallel to the given
line and at the given distance
from the given line.

Pr. 8:   The locus of a point at a given
distance from a given circle whose
radius is greater than that distance
is a pair of concentric circles, one
on either side of the given circle
and at the given distance from it.

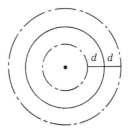

Pr. 9:   The locus of a point at a given
distance from a given circle whose
radius is less than the distance, is
a circle outside the given circle
and concentric to it. Note: If
$r = d$, the locus also includes the
center of the given circle.

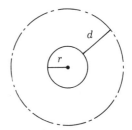

Now let's see how we can apply these principles.

Example: Determine the locus of a runner
moving equidistant from the sides of a
straight track.
Solution: By Pr. 2, the locus is a line
parallel to the two given lines (sides of the
track) and midway between them.

Here are a few similar problems. Compare the conditions given in each
with the principles above and decide which applies. In the problems
which follow in this section, we will ask you to draw a locus. We have
left you some space to do this but you might prefer to use a separate
sheet of paper. Determine (i.e., draw a figure showing) the locus of:

(a)   a plane flying equidistant from two
separated ground aircraft batteries.

(b)   a satellite 100 miles above earth.

(c)   the furthermost point reached by a
gun with a range of 10 miles.

_ _ _ _ _ _ _ _ _ _ _ _ _ _ _ _

(a)   By Pr. 1, the locus is the perpendicular
bisector of the line joining the two
points.

(b)   By Pr. 8, the locus is a circle concentric
with the earth and of radius 100 miles
greater than the earth.

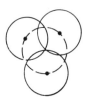

(c)   By Pr. 6, the locus is a circle of radius
10 miles with its center at the gun.

16.   Consider next the problem of determining the locus of the center of a
circle.

Example: Determine the locus of the center of a circular disk moving
so that it touches each of two parallel lines.
Solution: From Pr. 2, the locus is a
line parallel to the two given lines
and midway between them.

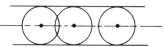

Determine the locus of the center of a circle:

(a)   moving tangentially to two
concentric circles.

(b)   moving so that its rim passes
through a fixed point.

(c)   rolling along the outside of a
large fixed circular hoop.

— — — — — — — — — — — — —

(a)   From Pr. 5, the locus is a circle
concentric to the given circles and
midway between them.

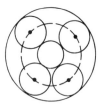

(b)   From Pr. 6, the locus is a circle
whose center is the given point
and whose radius is the radius of
the moving circle.

(c)   From Pr. 9, the locus is a circle
       outside the given circle and
       concentric to it.

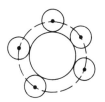

17.   A point or points that satisfy *two* conditions can be found by drawing
       the locus for each condition. The required points are the points of inter-
       section of the two loci. We won't go into this in detail, but here is an
       interesting example.

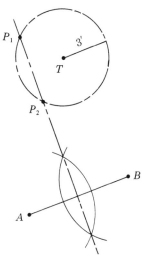

Example: On a map, locate buried treasure
that is 3 feet from a tree (T) and equidistant
from two points (A and B).
Solution: The required loci are (1) the
perpendicular bisector of AB (representing
the locus of points equidistant from A and
B), and (2) a circle with its center at T and
a radius of 3 feet (representing the locus of
points 3 feet from the tree). As you can see,
these loci meet at points P_1 and P_2, which
represent two possible locations of the
treasure.

Although this is valid as far as it goes, it is
important to recognize that there are three
possible answers depending on the location
of T with respect to A and B. Thus,
(a) there are two points if the loci intersect; (2) there is one point if the
perpendicular bisector is tangent to the circle; (3) there is *no* point if
the perpendicular bisector does not meet the circle.

SELF-TEST

1.   Determine the locus of:

      (a)   the midpoint of a radius of a
             given circle.

      (b)   the midpoint of a chord of fixed
             length in a given circle.

      (c)   the vertex of an isosceles triangle
             having a given base.
                           (frame 15)

2. Determine the locus of:

    (a)   a boat moving so that it is
          equidistant from the parallel
          banks of a stream.

    (b)   a swimmer maintaining the
          same distance from two
          floats.

    (c)   a police helicopter in pursuit
          of a car that has just passed
          the junction of two straight
          roads and which may be on
          either one of them.
                        (frame 15)

3. Determine the locus of:

    (a)   a planet moving at a fixed
          distance from its sun.

    (b)   a boat moving at a fixed
          distance from the coast of
          a circular island.

    (c)   plants being laid at a distance
          of 20 ft. from (on either side of)
          a row of other plants.
                        (frame 15)

4. Describe the locus of a point in rhombus *ABCD* that is equidistant from:

    (a)   *AB* and *AD* _____

    (b)   *AB* and *BC* _____

    (c)   *A* and *C* _____

    (d)   *B* and *D* _____

    (e)   Each of the four sides.

          _____
                        (frame 15)

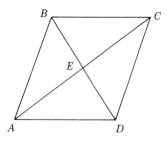

5. Determine the locus of the center of:

    (a)   a coin rolling around and
          touching a smaller coin.

(b)   a wheel moving between two
parallel bars and touching
both of them.

(c)   a wheel moving along a
straight metal bar and
touching it.

(frame 16)

6.   Locate each of the following.

(a)   Treasure that is buried 5 ft.
from a straight fence and
equidistant from two given
points where the fence meets
the ground.

(b)   Points that are 3 ft. from a
circle whose radius is 2 ft.
and also equidistant from
two lines that are parallel to
each other and tangent to the
circle.

(c)   A point equidistant from the
three vertices of a given
triangle.

(frame 17)

Answers to Self-Test

1.   (a)   A circle equidistant from the given
circle and its center and concentric
to the given circle.

(b)   A circle at a given distance from a
given circle and lying between the
given circle and its center.

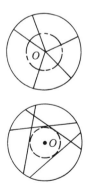

(c)   The perpendicular bisector of the
given base.

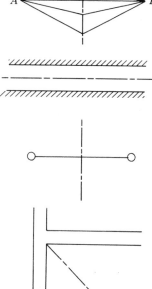

2. (a)   The line parallel to the banks and
midway between them.

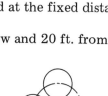

(b)   The perpendicular bisector of the
line joining the two floats.

(c)   The bisector of the angle between
the roads.

3. (a)   A circle having the sun as its center and the fixed distance as its
radius.
   (b)   A circle concentric to the coast, outside it and at the fixed distance
from it.
   (c)   A pair of parallel lines on either side of the row and 20 ft. from it.

4. (a) AC;   (b) BD;   (c) BD;   (d) AC;   (e) E

5. (a)   A circle concentric to the circumference
of the smaller coin and at a fixed
distance from it.

(b)   A line parallel to the two given bars
and midway between them.

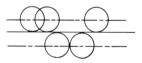

(c)   Two lines parallel to the given bar
and equidistant from it (the distance
being equal to the radius of the
wheel).

6. (Supply your own explanation for each of these.)

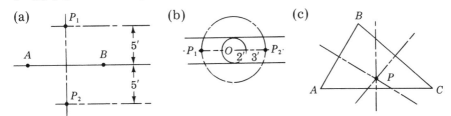

CONSTRUCTIONS

18. Closely allied to our work with loci is the topic of construction, for it
enables us to *draw* accurately the locus by using certain instruments.
    It may have occurred to you that the number of curves considered
in these first four chapters has been rather limited. In fact, they have
been limited to just two: the line (considered as a special type of curve)
and the circle. No other curve was examined — not because of lack of
space but because of the definition of plane geometry as agreed upon
by Greek mathematicians:

> Plane geometry is that branch of mathematics studying
> figures constructed only by the straightedge and the
> compass.

And this definition is still our guide today. By sticking to it we not
only become much more resourceful in our ability to construct geometric figures with a minimum of tools, but we gain a great deal of
insight into the relationships between lines, arcs, and angles.
    As you well know, the compass is the instrument used for drawing
circles or arcs of circles. The straightedge, on the other hand, is the
instrument for drawing lines; it looks like a ruler except that it has no
markings on it. (You can *use* a ruler as a straightedge, of course, as long
as you ignore the markings.)
    A question such as "Can the bisector of an angle be constructed?"
is largely meaningless until we are told what instruments we are permitted to use. If we are not permitted any, then the construction is not
possible. However, in our work we will accept the Greek definition of
geometry and consider that the only instruments available and permissible are the straightedge and the compass. Thus, the above question
should be interpreted as "Can the bisector of an angle be constructed
by using straightedge and compass only?" You will find the answer to
this question in the pages that follow.
    Although a great many constructions are possible, we will limit
ourselves only to the basic ones, and just enough of those to give you
practice in the techniques and methodology of geometric construction.

See if you can complete the following statements about your basic instruments.

The straightedge is used for _____

The compass is used for _____

_ _ _ _ _ _ _ _ _ _ _ _ _ _

constructing straight lines; drawing circles or arcs of circles

19. Although one or two steps sometimes are combined, every construction problem can be solved by these six steps:

   (1) *A general statement of the problem* that tells what is to be constructed.
   (2) *A figure* representing the given parts.
   (3) *A statement of what is given* in the representation of step 2.
   (4) *A specific statement of what is to be constructed* (result to be obtained).
   (5) *The construction*, with a description of each step, including the authority (reason) for each step.
   (6). *Statement of the conclusion* or proof that the desired result was obtained.

   You will find it helpful also if, in making your constructions, you use the following distinguishing lines:

   Given lines, drawn as heavy, full lines.   ———————

   Construction lines, drawn as light lines.   ———————

   Required lines, drawn as heavy dashed lines.   _ _ _ _ _ _

   Now to our first construction.

Construction 1: Construct a line segment congruent to a given line segment.
   Given: Line segment *AB*.
   To construct: A line segment congruent to *AB*.
   Construction: On working line *w*, with any point *C* as a center and a radius equal to *AB*, construct an arc intersecting *w* at *D*. Then *CD* is the required line segment.

Apply this procedure in the following constructions. Given line segments *a* and *b*, construct line segments with lengths equal to:

(a) *a* + 2*b*

(b) $2(a + b)$

_ _ _ _ _ _ _ _ _ _ _ _ _ _

(a) On a working line, w, construct a
line segment AB to line segment a.
From B, construct a line segment
equal to b, to point C; and from C
construct a line segment equal to b, to point D. Then AD is the
required line segment.

(b) Construct similarly to (a); $AD = 2(a + b)$.

20. *Construction 2:* Construct an angle
equal to a given angle.

    Given: $\angle A$

    To construct: An angle congruent to $\angle A$.

    Construction: With A as center and a
convenient radius, construct (swing,
draw) an arc (1) intersecting the sides
of $\angle A$ at B and C. With A', a point on
working line w, as center *and the
same radius*, construct arc (2) intersecting w at B'. With B' as center and
a radius equal to BC, construct arc (3)
intersecting arc (2) at C'. Draw $A'C'$. Then $\angle A'$ is the required
angle. ($\triangle ABC \cong \triangle A'B'C'$ by SSS, hence $\angle A \cong \angle A'$.)

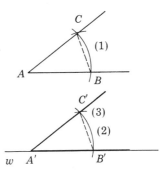

Remember not to change your compass setting when making the arcs in
a construction and try these problems. Given $\triangle ABC$, construct angles
equal to:

(a) $A + B + C$

(b) $2A$

_ _ _ _ _ _ _ _ _ _ _ _ _ _

(a) Using working line w as one side, duplicate
$\angle A$ (as you learned to do above in
Construction 2). Construct $\angle B$ adjacent
to $\angle A$ similarly, as shown. Then construct
$\angle C$ adjacent to $\angle B$. The exterior sides of
the copied angles A and C form the required
angle. Note that the angle is a straight angle.

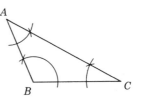

(b)   Constructed similarly.

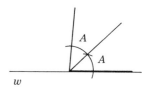

21. *Construction 3:* Bisect a given angle.
     Given: $\angle A$
     To construct: The bisector of $\angle A$.
     Construction: With A as center and
     a convenient radius, construct an arc
     intersecting the sides of $\angle A$ at B and
     C. With B and C as centers and using
     equal radii, construct arcs intersecting
     at a point, which we will call D.
     Draw AD. AD is then the required
     bisector. ($\triangle ABD \cong \triangle ADC$ by SSS,
     hence $\angle 1 \cong \angle 2$.)

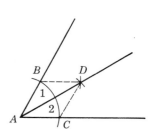

In $\triangle DEF$, D is an obtuse angle. Construct
the bisector of $\angle E$.

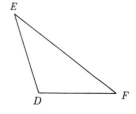

Use Construction 3 to bisect $\angle E$. *EH* is
the required line.

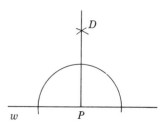

22. *Construction 4:* Construct a line perpen-
     dicular to a given line through a given
     point on the line.
         Given: Line w and point P on w.
         To construct: A perpendicular to w and P.
         Construction: Using Construction 3,
         bisect the straight angle at P. DP is the
         required perpendicular.

Construct angles of 90° and 45°.

- - - - - - - - - - - - - - -

Using Construction 4, construct the
perpendicular AD, from which
$\angle DAB = 90°$. Then using Construction 3, bisect $\angle CAD$ to obtain
$\angle CAE = 45°$.

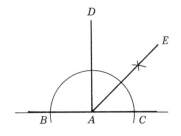

23. *Construction 5:* Bisect a given line segment. (Construct the perpendicular bisector of a given line segment.)
    Given: Line segment AB.
    To construct: Perpendicular bisector of AB.
    Construction: With A as center and a radius
of more than half AB, construct arc (1).
Then, with B as center and the same radius,
construct arc (2) intersecting arc (1) at C
and D. Draw CD. CD is the required
perpendicular bisector of AB. (Two points
each equidistant from the ends of a line
segment determine the perpendicular
bisector of the segment.)

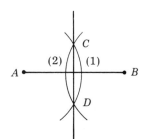

In scalene triangle ABC construct a
perpendicular bisector of AB.

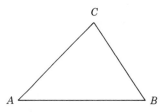

- - - - - - - - - - - - - - -

Use Construction 5 to obtain PQ, the
perpendicular bisector of AB.

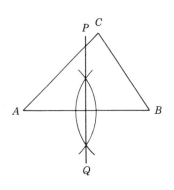

24. *Construction 6:* Construct a line perpendicular
    to a given line from a point not on the line.
    Given: Line w and point P outside
                    of w.
    To construct: A perpendicular to w from P.
    Construction: With P as center and a
    sufficiently long radius, construct an arc
    intersecting w at B and C. With B and C
    as centers and equal radii more than half of
    BC, construct arcs intersecting at A. Draw
    PA. PA is the required perpendicular.
    (Points P and A are each equidistant from
    B and C.)

In $\triangle DEF$, D is an obtuse angle. Construct
the altitude to DF.

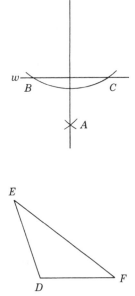

- - - - - - - - - - - - - - - - -

Use Construction 6 to obtain EG, the
altitude of DF (extended). (Note:
Bear in mind, from our definition of
a line in frame 3, Chapter 1, that a
line can be extended in either
direction indefinitely. We will have
occasion to use this property on a
number of future occasions.)

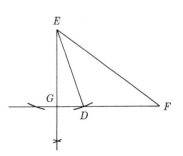

25. *Construction 7:* Construct a line parallel
    to a given line through a given point.
    Given: AB and point P.
    To construct: A line through P
                    parallel to AB.
    Construction: Draw a line, RS, through
    point P intersecting AB at Q. Con-
    struct $\angle SPD \cong \angle PQB$. Then CD is the
    required parallel. (If two corresponding
    angles are congruent, the lines cut by
    the transversal are parallel.)

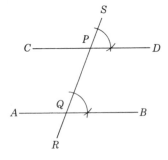

Construct a parallelogram given two adjacent sides and a diagonal. (*a* and *b* are the adjacent sides, *d* is the diagonal.)

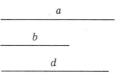

- - - - - - - - - - - - - -

Using *A* and *D* as centers and *b* and *d* as radii, construct arcs intersecting at *B*. Then, using Construction 7, construct *BC* parallel to *AD*. Using *B* and *D* as centers and *a* and *b* as radii, construct arcs intersecting at *C*. Draw *DC* to complete the parallelogram. (Vertex *C* also can be obtained by constructing *BC* ∥ *AD* and *DC* ∥ *AB*.)

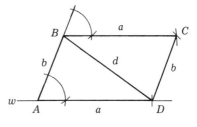

26. *Construction 8:* Divide a line segment into any number of congruent parts.

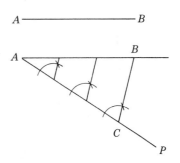

   Given: Line segment *AB*.
   To construct: Divide *AB* into any number of congruent parts.

Construction: On a line *AP*, cut off the required number of congruent segments. Then connect *B* to the endpoint of the last segment and construct parallels to *BC*. The points of intersection of these parallels and *AB* divide *AB* into the required number of segments. (If three or more parallel lines cut off congruent segments on one transversal, they cut off congruent segments on any other transversal.)

Find two-fifths of line segment *AB*.

- - - - - - - - - - - - - -

On another line, *AP*, construct five
congruent segments. Draw *BH*.
Through the endpoint *E* of the
second segment of *AH*, construct
a line parallel to *BH*, meeting *AB*
at *C*. Then *AC* is two-fifths of *AB*.

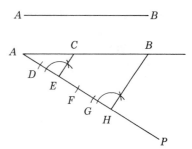

27. *Construction 9:* Circumscribe a circle about a triangle.

      Given: $\triangle ABC$

      To construct: The circumscribed circle
                of $\triangle ABC$.

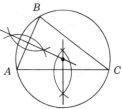

Construction: Construct the perpendicular
bisectors of two sides of the triangle.
Their intersection is the center of the
required circle, and the distance to any
vertex is the radius. (Any point on the
perpendicular bisector of a line is equidistant from the ends of the line.)

Construction 10: Inscribe a circle in a given triangle.

      Given: $\triangle ABC$

      To construct: The circle inscribed in
                $\triangle ABC$.

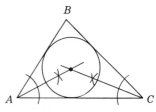

Construction: Construct the bisectors
of two of the angles of $\triangle ABC$. Their
intersection is the center of the
required circle and the distance (perpendicular) to any side is the radius.
(Any point on the bisector of an angle
is equidistant from the sides of the angle.)

Use Constructions 9 and 10 to construct the circumscribed and inscribed
circles of an isosceles triangle.

Since $\triangle DEF$ is isosceles, the bisector of $\angle E$ also is the perpendicular bisector of DF. Therefore, the center of each circle is on EG. I, the center of the inscribed circle, is found by constructing the bisector of $\angle D$ or $\angle F$ and extending it until it intersects EG. C, the center of the circumscribed circle, is found by constructing the perpendicular bisector of DE or EF and extending it until it intersects EG.

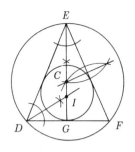

SELF-TEST

1. Given line segments a and b, construct a line segment whose length equals:

    (a) $a + b$

    (b) $a - b$

    (c) $2a + b$

    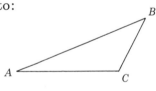

    (frame 19)

2. Given triangle ABC, construct an angle equal to:

    (a) $A + B$

    (b) $C - A$

    (c) $2B$

    (frame 20)

3. For each kind of triangle (acute, right, and obtuse), show that the following sets of lines are concurrent (that is, intersect in one point).

    (a)   the angle bisectors

    (b)   the medians

    (c)   the perpendicular bisectors   (frames 21–24)

4. Given △ABC, construct: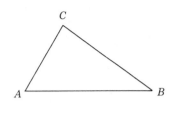

   (a) the supplement of ∠A

   (b) the complement of ∠B

   (c) the complement of one-half ∠C
                    (frames 21–24)

5. Given an acute angle, construct its:

   (a) supplement

   (b) complement

   (c) half of its supplement

   (d) half of its complement
                    (frames 21–24)

6. Construct a parallelogram, given:

   (a) two adjacent sides and an angle

   (b) the diagonals and a side

   (c) a side, an angle, and the
        altitude to the given side.
                    (frame 25)

7. Divide a line into two parts such that:

   (a) one part is three times the other

   (b) one part is three-fifths of the
        given line.
                    (frame 26)

8. Circumscribe a circle about:

   (a) a right triangle

   (b) a rectangle

   (c) a square.
                    (frame 27)

Answers to Self-Test

This time it will be left to you to check your own work.

This brings us to the conclusion of our work with the elements of Euclidean plane geometry. Since there is not sufficient space in a condensed presentation such as this to include as many problems as either the reader or the author might wish, you are urged to refer to any good geometry text-book (see the selected references in the front of this book) for more practice with proofs and applications of the fundamentals we have covered. Now on to the next topic: trigonometry.

CHAPTER FIVE

Numerical Trigonometry

Numerical trigonometry is principally concerned with finding the lengths of the sides and the sizes of the angles of triangles. It is the next logical subject for us to study in our "geometric" approach to preparation for the study of calculus since it provides some extremely useful techniques for solving a large category of problems. Also, it follows naturally from the study of geometry and our work with triangles.

It would not be correct, however, to leave you with the impression that the study of trigonometry is limited to its applications to triangles. Its modern uses are many, in both theoretical and applied fields of knowledge. Inevitably you will meet, and find it necessary to use, the trigonometric functions when you study the calculus of certain algebraic functions. You also will meet the trig functions when you study wave motion, vibrations, alternating current, and sound.

In this chapter we will be dealing only with the solution of right triangles, that is, finding the numerical values of the sides and angles when some of these elements are known. You already have learned how to find the lengths of the sides by using the Law of Pythagoras (or Pythagorean Theorem, as it also is known). Now you will learn some additional methods that will enable you not only to find the lengths of the sides but the sizes of the angles as well.

Specifically, when you have completed this chapter you will be familiar with and be able to use:

- the six trigonometric functions: the sine, cosine, tangent, secant, cosecant, and cotangent;

- cofunctions;

- functions of 30°, 45°, and 60° angles;

- vectors;

- both angular and circular measure, that is, both degrees and radians as measures of angle size.

TRIGONOMETRIC FUNCTIONS OF ACUTE ANGLES

1. Trigonometry deals with triangles, that is, geometric figures bounded
    by three lines. *Plane* trigonometry (the kind we will be concerned with
    in this book) deals with *plane* triangles formed by the intersection of
    three straight lines (as distinguished from spherical triangles, which lie
    on the surface of a sphere and therefore are bounded by curved lines).
    A *plane* is, of course, simply a flat (two-dimensional) surface.

    As you also learned from geometry, a *right triangle* is a triangle con-
    taining a right (90°) angle.

    Just to make sure you are clear as to what a right triangle looks like,
    indicate in the spaces provided which of the following are right triangles
    (mark them with an X):

    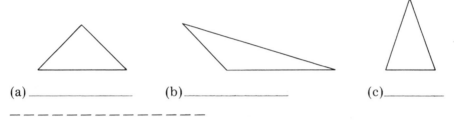

    (a) _____   (b) _____   (c) _____

    Triangle (a) is the only possible right triangle. The angles in triangle (c)
    obviously are all less than 90°. And of those in triangle (b), one obvi-
    ously is too large and the others too small to be 90°. This leaves only
    triangle (a), and although the two angles at the base are evidently less
    than 90°, the angle at the top *appears* to be a right angle. You can
    check this by placing any corner of a sheet of note paper in this angle.
    Also, if you turn the book so that one of the sides of this triangle forms
    the base it becomes readily apparent that it *is* a right angle.

2. That certain relationships exist between the sides and angles of a right
    triangle can readily be shown by the following illustration.

    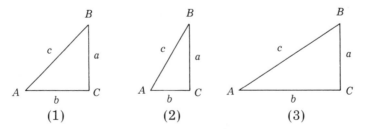

    (1)   (2)   (3)

    Note that the length of side *a* is the same in all three triangles.
    However, in example 2 side *b* is shorter than in example 1 and angle *A*
    is correspondingly larger. In example 3 side *b* is larger than in example

1, and $\angle A$ is correspondingly smaller. So it is apparent that as the length of side b increases, and side a remains the same, the size of $\angle A$ decreases. By using other diagrams we could see that if side a increases and side b remains constant, $\angle A$ increases. The reverse also is true: if side a decreases, $\angle A$ decreases.

These are but a few of all the relationships existing between the sides and angles of a right triangle. They are enough to suggest, however, that *the size of an angle in a right triangle depends upon the ratio existing between any two sides of the triangle.*

Look again at the triangles above and answer this question: What happens to the size of angle B as side b increases, if side a remains constant?

_____

– – – – – – – – – – – – – – –

Angle B increases in size.

3. Both from your study of algebra and also from the earlier discussions in this book, you should be familiar with ratios. However, since "ratio" is an important concept and a short review won't hurt, let's run over it again.

In the preceding frame we stated that the size of an angle in a right triangle depends upon the ratio existing between any two sides of the triangle. And in Chapter 3, frame 12, we indicated that the ratio of two quantities is the first divided by the second. We can either simply *indicate* the intended division by use of a fraction bar ($\frac{1}{9}$, for example) or we can *perform* the division, in which case the resulting number is said to be a decimal fraction, or simply a decimal. Thus the decimal equivalent of $\frac{1}{9}$ would be 0.1111 . . ., carried to as many decimal places as the problem required.

To make sure you have the right idea of "ratio," mark the phrase below that completes correctly this statement: The ratio of one number (length of the side of a triangle, for instance) to another number (length of another side) is the result of dividing the first number by the second. This division:

_____(a)   must be indicated by a fraction bar.

_____(b)   must be performed and shown as a decimal fraction.

_____(c)   may either be shown by use of a fraction bar or performed and shown as a decimal.

– – – – – – – – – – – – – – – –

Choice (c). (*Note:* The division could, of course, be indicated by the division symbol, but use of the fraction bar is much more common.)

4. You should also be aware that our "decimal fraction" may occasionally be a whole number, although it seldom is in trigonometry. It may also be composed of a whole number *and* a decimal fraction. We'll see some examples of this as we go along.

    Now back to our triangles again for a little practice in working with ratios.

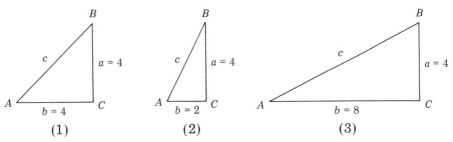

In triangle 1 above the ratio of side *a* to side *b* (that is, $\frac{a}{b}$) is $\frac{4}{4}$ or, if we divide, 1.

What will be the ratio of *a* to *b* in triangle 3? And what will happen to the size of angle *A* as compared with what it was in triangle 1?

– – – – – – – – – – – – – – – –

The ratio will be $\frac{4}{8}$ or 0.50, and angle *A* will have decreased in size. In this case the size of angle *A* varies directly with the value of the ratio between sides *a* and *b*. That is, as the ratio becomes smaller, the size of angle *A* decreases; as the ratio becomes larger, the size of angle *A* increases.

5. Be careful not to fall into the error of thinking that the size of an angle always is *directly* proportional to the ratio between two particular sides. Consider, for example, what happens to the size of angle *B* as the ratio between sides *a* and *b* changes.

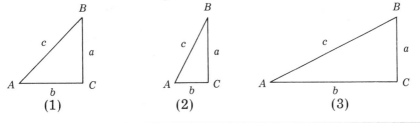

In example 1, angle B appears to be approximately equal to angle A. But in example 3, where the ratio of a to b has *decreased* by 50%, angle B has *increased* in size. We say, therefore, that the size of angle B is *inversely* proportional to the ratio of side a to side b.

Since the size of an angle depends upon the ratio of the sides of a triangle, we can correctly conclude that, conversely, the length of the sides will depend upon the size of the angle. Thus, in the triangles above, if angle A increases in size, side a will increase; if angle B decreases in size, side b will decrease; and so on.

This matter of the relationships between the sides and angles of a right triangle is pretty much the essence of plane trigonometry — although this statement in no way minimizes these relationships, for they are all-important. They make possible the solution of a great many problems that otherwise would be very difficult to solve.

Because these relationships are so important they have been carefully defined, and each has been given a name. Thus, in the triangle at the right, the ratio of the side opposite angle A to the side opposite the right angle is called the *sine of angle A*. (The side opposite the right angle is, as you learned in geometry, called the hypotenuse.) So the sine of angle A may be expressed as a ratio:

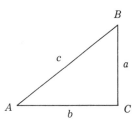

$$\text{sine } A = \frac{\text{opposite side}}{\text{hypotenuse}} \quad \text{or} \quad \sin A = \frac{a}{c}.$$

The term "sine" usually is abbreviated, as shown, to "sin" but pronounced as though it still had the "e" on the end.

If the sine of an angle is the ratio of the side *opposite* an angle to the *hypotenuse* of a right triangle, what would be the sine of angle B?

_____

— — — — — — — — — — — — — —

The sine of angle B would be $\frac{b}{c}$.

6. In trigonometry when we talk about variations in the size of an angle with changes in the lengths of the sides of a triangle, we have to talk about a *specific angle* and *two specific sides*, otherwise our discussion would be meaningless. Not only must we be sure which two sides we are referring to, but we also must be sure (since we're forming their ratio) that we have them in the correct order ($\frac{3}{5}$ obviously is quite different from $\frac{5}{3}$). Thus, when we say "hypotenuse" we always mean the

side opposite the right angle. When we say "adjacent side" we are referring to the side next to the angle we're interested in. Similarly, when we use the term "opposite side" we mean the side opposite the angle of interest.

In addition to the sine, which we have just discussed, there are five additional relationships, or ratios, which we use in trigonometry. These relationships and their abbreviations are as follows:

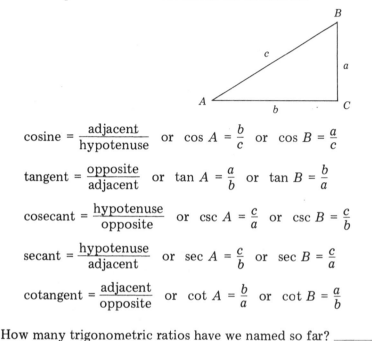

$$\text{cosine} = \frac{\text{adjacent}}{\text{hypotenuse}} \quad \text{or} \quad \cos A = \frac{b}{c} \quad \text{or} \quad \cos B = \frac{a}{c}$$

$$\text{tangent} = \frac{\text{opposite}}{\text{adjacent}} \quad \text{or} \quad \tan A = \frac{a}{b} \quad \text{or} \quad \tan B = \frac{b}{a}$$

$$\text{cosecant} = \frac{\text{hypotenuse}}{\text{opposite}} \quad \text{or} \quad \csc A = \frac{c}{a} \quad \text{or} \quad \csc B = \frac{c}{b}$$

$$\text{secant} = \frac{\text{hypotenuse}}{\text{adjacent}} \quad \text{or} \quad \sec A = \frac{c}{b} \quad \text{or} \quad \sec B = \frac{c}{a}$$

$$\text{cotangent} = \frac{\text{adjacent}}{\text{opposite}} \quad \text{or} \quad \cot A = \frac{b}{a} \quad \text{or} \quad \cot B = \frac{a}{b}$$

How many trigonometric ratios have we named so far? _____

— — — — — — — — — — — — — — — —

Six (although we have done each for two angles).

7. Since there are only six possible combinations of the three sides of a right triangle, taken two at a time, there are just six trigonometric ratios. Hence these are all the relationships — together with their odd names — you need to learn about in order to work problems in trigonometry. Once you have *memorized* these ratios (and it is very important that you do so), the hardest part of the job will be over. Of course you will need practice using them to solve problems, but it will be necessary to introduce very few additional concepts in this chapter. Memorize the six ratios now, and then return here.

Now notice this about the six relationships we have been discussing: the last three given you are merely reciprocals (inversions) of the first three! Keep this in mind and you will find the memorizing much easier. Thus, to summarize,

$$\sin = \frac{\text{opposite}}{\text{hypotenuse}} \qquad\qquad \csc = \frac{\text{hypotenuse}}{\text{opposite}}$$

$$\cos = \frac{\text{adjacent}}{\text{hypotenuse}} \qquad\qquad \sec = \frac{\text{hypotenuse}}{\text{adjacent}}$$

$$\tan = \frac{\text{opposite}}{\text{adjacent}} \qquad\qquad \cot = \frac{\text{adjacent}}{\text{opposite}}$$

To make sure you remember what the abbreviations above stand for, write in the full name of each of the trigonometric relationships in the spaces below.

sin _____   csc _____

cos _____   sec _____

tan _____   cot _____

– – – – – – – – – – – –

sine   cosecant
cosine   secant
tangent   cotangent

8. If you had any difficulty with any of the above terms, be sure to review them before going ahead. You should become thoroughly familiar with them as soon as possible.

Did you memorize the six trigonometric functions (ratios)? Let's see. Fill in the missing information below in terms of the sides of a right triangle (i.e., opposite, adjacent, and hypotenuse).

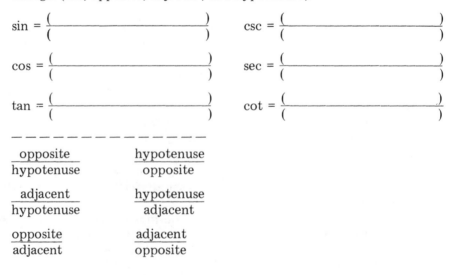

$$\sin = \frac{(\qquad\qquad)}{(\qquad\qquad)} \qquad\qquad \csc = \frac{(\qquad\qquad)}{(\qquad\qquad)}$$

$$\cos = \frac{(\qquad\qquad)}{(\qquad\qquad)} \qquad\qquad \sec = \frac{(\qquad\qquad)}{(\qquad\qquad)}$$

$$\tan = \frac{(\qquad\qquad)}{(\qquad\qquad)} \qquad\qquad \cot = \frac{(\qquad\qquad)}{(\qquad\qquad)}$$

– – – – – – – – – – – –

$$\frac{\text{opposite}}{\text{hypotenuse}} \qquad\qquad \frac{\text{hypotenuse}}{\text{opposite}}$$

$$\frac{\text{adjacent}}{\text{hypotenuse}} \qquad\qquad \frac{\text{hypotenuse}}{\text{adjacent}}$$

$$\frac{\text{opposite}}{\text{adjacent}} \qquad\qquad \frac{\text{adjacent}}{\text{opposite}}$$

Again, if you missed any of these be sure to study them once more before going on.

SOLUTION OF RIGHT TRIANGLES

9. Now, restating the functions in terms of angle A we get:

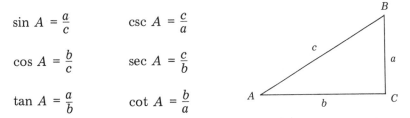

$$\sin A = \frac{a}{c} \qquad \csc A = \frac{c}{a}$$

$$\cos A = \frac{b}{c} \qquad \sec A = \frac{c}{b}$$

$$\tan A = \frac{a}{b} \qquad \cot A = \frac{b}{a}$$

Lest these terms and equations become too terrifying, it is well to keep in mind exactly what they mean.

Look at the first equation. Putting it into words it reads as follows: The sine of angle A is equal to the ratio of side a to side c; or, the sine of angle A is equal to the ratio of the side opposite angle A to the hypotenuse of the right triangle. The terms sine, cosine, tangent, cosecant, secant, and cotangent are called *trigonometric functions*, a term we will use quite frequently.

Actually the size of angle A (or angle B) is dependent upon the ratios existing between three sets of sides which, with their reciprocals, constitute six separate trigonometric functions, as shown above. The use of the term "sine," for example, merely indicates which function we're talking about. A few examples should help clarify this for you. So let's see how the different trigonometric functions can be used in a practical way to solve mathematical problems.

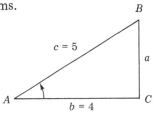

Referring to the familiar right-triangle figure at the right, let us assume that the lengths of sides b and c are known and that we need to find the size of angle A; $b = 4$, $c = 5$, and $A = ?$

Our first step is to choose a trigonometric function involving the two known values and the unknown value under consideration. Here again are the choices open to us.

$$\sin A = \frac{a}{c} \qquad \csc A = \frac{c}{a}$$

$$\cos A = \frac{b}{c} \qquad \sec A = \frac{c}{b}$$

$$\tan A = \frac{a}{b} \qquad \cot A = \frac{b}{a}$$

Which function would you choose to help you solve this problem?

_____

— — — — — — — — — — — —

The cos function. Why? Because the cosine function involves the use of two *known* values, namely, the lengths of the sides *b* and *c*. You also could have selected the secant instead of the cosine and still been correct since one is merely the reciprocal of the other and both involve the same two sides of the triangle, sides *b* and *c*. However, since our tables of the trigonometric functions (which we will discuss in the next frame) do not include values for the secant and the cosecant, we will stick to using the sin and cos in problem solving.

10.   Here is our triangle again so that you will have it conveniently at hand while we discuss it.

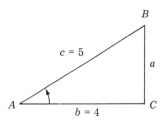

In choosing an appropriate function for solving a right triangle problem, you always have a choice of two, just as we did above. That is, for any given combination of two sides and one angle of a right triangle you always have a choice between either a primary function (sine, cosine, or tangent) or its reciprocal (cosecant, secant, or cotangent). Sometimes the reciprocal function works out a little more conveniently than the primary function, or vice versa. The general rule is that where there is no apparent advantage of one over the other, *choose the primary function.*

Let's see if you have caught the point. Place an X by the statement below that best summarizes what we have just said.

_____(a)   The reciprocal can always be used in place of the primary function.

_____(b)   The reciprocal can sometimes be used in place of the primary function.

_____(c)   Either the primary or secondary function can be used, but where the choice appears even, use the primary function.

— — — — — — — — — — — —

Choice (c). The first answer is a correct statement also, but it is not the best (most complete) answer. In (b), "sometimes" is incorrect — it can *always* be used.

11. Getting back to our triangle again, then, we are agreed that, although either the cosine or the secant could be used, we will follow the general rule of using the primary function where the choice is even. This gives us:

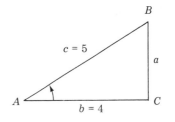

$$\cos A = \frac{b}{c}$$

and, by substituting the given values of the sides,

$$\cos A = \frac{4}{5} = 0.8000.$$

All that remains, therefore, is to discover what angle the value 0.8000 represents. To find this out we refer to the Table of Trigonometric Functions, found at the back of this book.

Before delving into the table of natural trig functions, let's make sure you're clear as to what the number 0.8000 represents. Select the correct statement below.

_____(a) It simply represents the ratio of 4 to 5.

_____(b) It is the size of angle A in inches.

_____(c) It is the size of angle A in degrees.

- - - - - - - - - - - - - - - - -

Choice (a)

12. Perhaps this was obvious to you. It isn't to everyone. Probably because we tend to want to assign some dimensional characteristic to most numbers, especially if they relate to other numbers that *do* have dimensional properties. However, you should recall from your study of algebra that ratios do *not* have to have dimensional units attached to them. In the present case 4 over 5 is a ratio expressed as a simple fraction. If we divide 5 into 4, we get the decimal fraction 0.8000, which represents the same ratio.

Now we're ready to look at the table of trigonometric functions (see Appendix) and find out what angle it is that the ratio 0.8000 represents. Notice (on page 395) the column labeled "cos" and observe that the cosine begins with a value of 1.00000 at 0° and *decreases* as the angle *increases*. If we read down this column, turning the pages as we go, we arrive eventually at the value 0.80003, which is as close as we can come to 0.80000 without interpolation (the process of finding intermediate values — that is, between those given in the table — through the use of proportional parts). This value is shown in the sample table page on page 203.

What is the corresponding angular value? _____

— — — — — — — — — — — — — —

$A = 36°52'$

If you didn't get the above answer, look carefully at the circled parts of the figure on the next page, then go back and find this page in the table, checking carefully again until you see how to find the correct answer. Be careful to stay in the "cos" column and to read from the top down, not up from the bottom. There will be occasions when it is appropriate to read from the bottom up, and we will be talking about them soon. But for now remember that we started at the top of the first page of the table, noting that the cosine begins with a value of 1.00000 at 0° and decreases gradually as the angular values increase. And these increasing angular values appear along the left side of the table. The angular values shown at the bottom and along the right side of the table should be ignored for the moment.

13. There is another feature of these tables you should be aware of if you have not used trig tables before. In order to save space publishers take advantage of the fact that the sine and *co*sine (also the tangent and *co*tangent, secant and *co*secant) are what we call "co-functions." We will go into the matter of just what a co-function is a little later. For the present it is sufficient to point out that the sine of an angle is numerically equal to the cosine of 90° minus the angle. Similarly, the tangent of an angle is equal to the cotangent of 90° minus the angle, and the secant is equal to the cosecant of 90° minus the angle. What this means, in terms of the trigonometric tables, is that the entire range of natural trigonometric functions up to 90° can be covered simply by printing tables up to 45°. Beyond 45° it is necessary only to change the function names at the bottom of the columns, print the degree values at the bottom of the page, reverse the minute values to read up instead of down (along the right-hand edge), and the job is done.

Use the tables to find the sine of 53°08'. _____

— — — — — — — — — — — — — —

$\sin 53°08' = 0.80003$.

If you did not get the correct answer, check the circled parts of the figure on page 204 and then refer back to the table until you are sure you see how to find the correct answer.

36°

'	sin	tan	cot	cos	
0	.58779	.72654	1.3764	.80902	60
1	802	699	.3755	885	59
2	826	743	.3747	867	58
3	849	788	.3739	850	57
4	873	832	.3730	833	56
5	.58896	.72877	1.3722	.80816	55
6	920	921	.3713	799	54
7	943	.72966	.3705	782	53
8	967	.73010	.3697	765	52
9	.58990	055	.3688	748	51
10	.59014	.73100	1.3680	.80730	50
11	037	144	.3672	713	49
12	061	189	.3663	696	48
13	084	234	.3655	679	47
14	108	278	.3647	662	46
15	.59131	.73323	1.3638	.80644	45
16	154	368	.3630	627	44
17	178	413	.3622	610	43
18	201	457	.3613	593	42
19	225	502	.3605	576	41
20	.59248	.73547	1.3597	.80558	40
21	272	592	.3588	541	39
22	295	637	.3580	524	38
23	318	681	.3572	507	37
24	342	726	.3564	489	36
25	.59365	.73771	1.3555	.80472	35
26	389	816	.3547	455	34
27	412	861	.3539	438	33
28	436	906	.3531	420	32
29	459	951	.3522	403	31
30	.59482	.73996	1.3514	.80386	30
31	506	74041	.3506	368	29
32	529	086	.3498	351	28
33	552	131	.3490	334	27
34	576	176	.3481	316	26
35	.59599	.74221	1.3473	.80299	25
36	622	267	.3465	282	24
37	646	312	.3457	264	23
38	669	357	.3449	247	22
39	693	402	.3440	230	21
40	.59716	.74447	1.3432	.80212	20
41	739	492	.3424	195	19
42	763	538	.3416	178	18
43	786	583	.3408	160	17
44	809	628	.3400	143	16
45	.59832	.74674	1.3392	.80125	15
46	856	719	.3384	108	14
47	879	764	.3375	091	13
48	902	810	.3367	073	12
49	926	855	.3359	056	11
50	59949	.74900	1.3351	.80038	10
51	972	946	.3343	021	9
52	.59995	.74991	.3335	80003	8
53	.60019	.75037	.3327	.79986	7
54	042	082	.3319	968	6
55	.60065	.75128	1.3311	.79951	5
56	089	173	.3303	934	4
57	112	219	.3295	916	3
58	135	264	.3287	899	2
59	158	310	.3278	881	1
60	.60182	.75355	1.3270	.79864	0
	cos	cot	tan	sin	'

53°

37°

'	sin	tan	cot	cos	
0	.60182	.75355	1.3270	.79864	60
1	205	401	.3262	846	59
2	228	447	.3254	829	58
3	251	492	.3246	811	57
4	274	538	.3238	793	56
5	.60298	.75584	1.3230	.79776	55
6	321	629	.3222	758	54
7	344	675	.3214	741	53
8	367	721	.3206	723	52
9	390	767	.3198	706	51
10	.60414	.75812	1.3190	.79688	50
11	437	858	.3182	671	49
12	460	904	.3175	653	48
13	483	950	.3167	635	47
14	506	.75996	.3159	618	46
15	.60529	.76042	1.3151	.79600	45
16	553	088	.3143	583	44
17	576	134	.3135	565	43
18	599	180	.3127	547	42
19	622	226	.3119	530	41
20	.60645	.76272	1.3111	.79512	40
21	668	318	.3103	494	39
22	691	364	.3095	477	38
23	714	410	.3087	459	37
24	738	456	.3079	441	36
25	.60761	.76502	1.3072	.79424	35
26	784	548	.3064	406	34
27	807	594	.3056	388	33
28	830	640	.3048	371	32
29	853	686	.3040	353	31
30	.60876	.76733	1.3032	.79335	30
31	899	779	.3024	318	29
32	922	825	.3017	300	28
33	945	871	.3009	282	27
34	968	918	.3001	264	26
35	.60991	.76964	1.2993	.79247	25
36	.61015	.77010	.2985	229	24
37	038	057	.2977	211	23
38	061	103	.2970	193	22
39	084	149	.2962	176	21
40	.61107	.77196	1.2954	.79158	20
41	130	242	.2946	140	19
42	153	289	.2938	122	18
43	176	335	.2931	105	17
44	199	382	.2923	087	16
45	.61222	.77428	1.2915	.79069	15
46	245	475	.2907	051	14
47	268	521	.2900	033	13
48	291	568	.2892	.79016	12
49	314	615	.2884	.78998	11
50	.61337	.77661	1.2876	.78980	10
51	360	708	.2869	962	9
52	383	754	.2861	944	8
53	406	801	.2853	926	7
54	429	848	.2846	908	6
55	.61451	.77895	1.2838	.78891	5
56	474	941	.2830	873	4
57	497	.77988	.2822	855	3
58	520	.78035	.2815	837	2
59	543	082	.2807	819	1
60	.61566	.78129	1.2799	.78801	0
	cos	cot	tan	sin	'

52°

36°

′	sin	tan	cot	cos	
0	.58779	.72654	1.3764	.80902	60
1	802	699	.3755	885	59
2	826	743	.3747	867	58
3	849	788	.3739	850	57
4	873	832	.3730	833	56
5	.58896	.72877	1.3722	.80816	55
6	920	921	.3713	799	54
7	943	.72966	.3705	782	53
8	967	.73010	.3697	765	52
9	.58990	055	.3688	748	51
10	.59014	.73100	1.3680	.80730	50
11	037	144	.3672	713	49
12	061	189	.3663	696	48
13	084	234	.3655	679	47
14	108	278	.3647	662	46
15	.59131	.73323	1.3638	.80644	45
16	154	368	.3630	627	44
17	178	413	.3622	610	43
18	201	457	.3613	593	42
19	225	502	.3605	576	41
20	.59248	.73547	1.3597	.80558	40
21	272	592	.3588	541	39
22	295	637	.3580	524	38
23	318	681	.3572	507	37
24	342	726	.3564	489	36
25	.59365	.73771	1.3555	.80472	35
26	389	816	.3547	455	34
27	412	861	.3539	438	33
28	436	906	.3531	420	32
29	459	951	.3522	403	31
30	.59482	.73996	1.3514	.80386	30
31	506	74041	.3506	368	29
32	529	086	.3498	351	28
33	552	131	.3490	334	27
34	576	176	.3481	316	26
35	.59599	.74221	1.3473	.80299	25
36	622	267	.3465	282	24
37	646	312	.3457	264	23
38	669	357	.3449	247	22
39	693	402	.3440	230	21
40	.59716	.74447	1.3432	.80212	20
41	739	492	.3424	195	19
42	763	538	.3416	178	18
43	786	583	.3408	160	17
44	809	628	.3400	143	16
45	.59832	.74674	1.3392	.80125	15
46	856	719	.3384	108	14
47	879	764	.3375	091	13
48	902	810	.3367	073	12
49	926	855	.3359	056	11
50	59949	.74900	1.3351	.80038	10
51	972	946	.3343	021	9
52	.59995	.74991	.3335	80003	8
53	.60019	.75037	.3327	.79986	7
54	042	082	.3319	968	6
55	.60065	.75128	1.3311	.79951	5
56	089	173	.3303	934	4
57	112	219	.3295	916	3
58	135	264	.3287	899	2
59	158	310	.3278	881	1
60	.60182	75355	1.3270	.79864	0
	cos	cot	tan	sin	′

53°

37°

′	sin	tan	cot	cos	
0	.60182	.75355	1.3270	.79864	60
1	205	401	.3262	846	59
2	228	447	.3254	829	58
3	251	492	.3246	811	57
4	274	538	.3238	793	56
5	.60298	.75584	1.3230	.79776	55
6	321	629	.3222	758	54
7	344	675	.3214	741	53
8	367	721	.3206	723	52
9	390	767	.3198	706	51
10	.60414	.75812	1.3190	.79688	50
11	437	858	.3182	671	49
12	460	904	.3175	653	48
13	483	950	.3167	635	47
14	506	.75996	.3159	618	46
15	.60529	.76042	1.3151	.79600	45
16	553	088	.3143	583	44
17	576	134	.3135	565	43
18	599	180	.3127	547	42
19	622	226	.3119	530	41
20	.60645	.76272	1.3111	.79512	40
21	668	318	.3103	494	39
22	691	364	.3095	477	38
23	714	410	.3087	459	37
24	738	456	.3079	441	36
25	.60761	.76502	1.3072	.79424	35
26	784	548	.3064	406	34
27	807	594	.3056	388	33
28	830	640	.3048	371	32
29	853	686	.3040	353	31
30	.60876	.76733	1.3032	.79335	30
31	899	779	.3024	318	29
32	922	825	.3017	300	28
33	945	871	.3009	282	27
34	968	918	.3001	264	26
35	.60991	.76964	1.2993	.79247	25
36	.61015	.77010	.2985	229	24
37	038	057	.2977	211	23
38	061	103	.2970	193	22
39	084	149	.2962	176	21
40	.61107	.77196	1.2954	.79158	20
41	130	242	.2946	140	19
42	153	289	.2938	122	18
43	176	335	.2931	105	17
44	199	382	.2923	087	16
45	.61222	.77428	1.2915	.79069	15
46	245	475	.2907	051	14
47	268	521	.2900	033	13
48	291	568	.2892	.79016	12
49	314	615	.2884	.78998	11
50	.61337	.77661	1.2876	.78980	10
51	360	708	.2869	962	9
52	383	754	.2861	944	8
53	406	801	.2853	926	7
54	429	848	.2846	908	6
55	.61451	.77895	1.2838	.78891	5
56	474	941	.2830	873	4
57	497	.77988	.2822	855	3
58	520	.78035	.2815	837	2
59	543	082	.2807	819	1
60	.61566	.78129	1.2799	.78801	0
	cos	cot	tan	sin	′

52°

14. Do you recognize the answer in frame 13 as being the same as the cosine of $36°52'$? If you got this answer you must have done everything right. Remember, when we are looking up an angle greater than $45°$ we look for the angle values at the bottom of the page rather than at the top. Also we read the minute values in the right-hand minutes column, starting at the bottom. So if you got it right you now know how to find function values of angles greater than $45°$. Try these for more practice.

sin $16°11'$ = _____   cos $1°55'$ = _____

cos $33°30'$ = _____   sin $81°48'$ = _____

sin $45°15'$ = _____   cos $73°49'$ = _____

Which pairs have the same values? _____
(In rotation the answers are: 0.27871, 0.55194, 0.71019, 0.99944, 0.98978, 0.27871; the first and last.)

Let's try another problem. Suppose we know angle B and side b in our right triangle but wish to find the length of side a. Again we must choose a trigono-metric function involving the two known values as well as the unknown value desired. For your convenience the six trigonometric functions are restated below, but this time in terms of angle B (since that's our known angle in this case) rather than in terms of angle A.

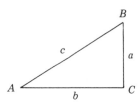

$$\sin B = \frac{b}{c} \qquad \csc B = \frac{c}{b}$$

$$\cos B = \frac{a}{c} \qquad \sec B = \frac{c}{a}$$

$$\tan B = \frac{b}{a} \qquad \cot B = \frac{a}{b}$$

With these functions before you, see if you can select the one most appropriate to the solution of the problem.

Your choice: _____

— — — — — — — — — — — — — —

The best choice would be: $\cot B = \frac{a}{b}$. Tan $B = \frac{b}{a}$ is also an acceptable choice since it contains the two known values as well as the unknown value (side a) we are seeking. Either the tangent or the cotangent would serve since both contain the necessary terms. Although it may not be apparent to you at the moment, the cotangent actually will be a little easier to use since we will need to transpose fewer terms in solving the equation.

15. Using the cotangent function, then, we can set up the problem as follows:

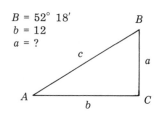

$$\cot B = \frac{\text{adjacent side}}{\text{opposite side}} = \frac{a}{b}$$

By substitution of the known values shown in the figure we get

$$\cot 52°18' = \frac{a}{12}$$

or, multiplying,

$a = 12(\cot 52°18')$ or simply $12 \cot 52°18'$.

In other words, to solve the equation for a, we must multiply 12 times the cotangent of $52°18'$. But what *is* the cotangent of $52°18'$? Look it up in the table of trigonometric functions. What you find should enable you to select the correct answer below.

____(a) 0.77289   ____(b) 1.2938   ____(c) 1.3127

— — — — — — — — — — — — — — — —

Choice (a). (See the sample table portion on the next page.)

16. Now that we know the value of the cotangent, we can substitute as follows:

$$a = 12 \cot 52°18'$$
$$\text{or} \quad a = 12(0.77289)$$
$$a = 9.27.$$

"Yes, but 9.27 *what*?" you may wonder. The answer is, 9.27 *anything*, that is, any kind of *linear* measure: feet, inches, miles, furlongs, centi-meters — the choice is yours. If no unit was specified in the problem (and it wasn't in this case) for the length of side b, then you can think of both side a and side b as being in any convenient unit of measurement. Usually, in a practical problem, the type of unit would be specified, in which case all sides would be in the same unit, whatever it was.

Now let's try another problem, one that involves finding an angular value. Consider the triangle shown at the right. The lengths of the two sides are given and, if necessary, we could find the length of the third side by use of the Pythagorean Theorem (chapter 3, frame 30). But this wouldn't help us find out anything about the size of the angles.

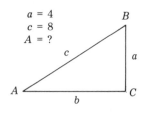

So, turning to trigonometry for help, we first have to select a function that includes the two known quantities (lengths of the two sides) plus one of the unknown angles — angle A, for instance.

36°

'	sin	tan	cot	cos	
0	.58779	.72654	1.3764	.80902	60
1	802	699	.3755	885	59
2	826	743	.3747	867	58
3	849	788	.3739	850	57
4	873	832	.3730	833	56
5	.58896	.72877	1.3722	.80816	55
6	920	921	.3713	799	54
7	943	.72966	.3705	782	53
8	967	.73010	.3697	765	52
9	.58990	055	.3688	748	51
10	.59014	.73100	1.3680	.80730	50
11	037	144	.3672	713	49
12	061	189	.3663	696	48
13	084	234	.3655	679	47
14	108	278	.3647	662	46
15	.59131	.73323	1.3638	.80644	45
16	154	368	.3630	627	44
17	178	413	.3622	610	43
18	201	457	.3613	593	42
19	225	502	.3605	576	41
20	.59248	.73547	1.3597	.80558	40
21	272	592	.3588	541	39
22	295	637	.3580	524	38
23	318	681	.3572	507	37
24	342	726	.3564	489	36
25	.59365	.73771	1.3555	.80472	35
26	389	816	.3547	455	34
27	412	861	.3539	438	33
28	436	906	.3531	420	32
29	459	951	.3522	403	31
30	.59482	.73996	1.3514	.80386	30
31	506	74041	.3506	368	29
32	529	086	.3498	351	28
33	552	131	.3490	334	27
34	576	176	.3481	316	26
35	.59599	.74221	1.3473	.80299	25
36	622	267	.3465	282	24
37	646	312	.3457	264	23
38	669	357	.3449	247	22
39	693	402	.3440	230	21
40	.59716	.74447	1.3432	.80212	20
41	739	492	.3424	195	19
42	763	538	.3416	178	18
43	786	583	.3408	160	17
44	809	628	.3400	143	16
45	.59832	.74674	1.3392	.80125	15
46	856	719	.3384	108	14
47	879	764	.3375	091	13
48	902	810	.3367	073	12
49	926	855	.3359	056	11
50	59949	.74900	1.3351	.80038	10
51	972	946	.3343	021	9
52	.59995	.74991	.3335	.80003	8
53	.60019	.75037	.3327	.79986	7
54	042	082	.3319	968	6
55	.60065	.75128	1.3311	.79951	5
56	089	173	.3303	934	4
57	112	219	.3295	916	3
58	135	264	.3287	899	2
59	158	310	.3278	881	1
60	.60182	.75355	1.3270	.79864	0
	cos	cot	tan	sin	'

53°

37°

'	sin	tan	cot	cos	
0	.60182	.75355	1.3270	.79864	60
1	205	401	.3262	846	59
2	228	447	.3254	829	58
3	251	492	.3246	811	57
4	274	538	.3238	793	56
5	.60298	.75584	1.3230	.79776	55
6	321	629	.3222	758	54
7	344	675	.3214	741	53
8	367	721	.3206	723	52
9	390	767	.3198	706	51
10	.60414	.75812	1.3190	.79688	50
11	437	858	.3182	671	49
12	460	904	.3175	653	48
13	483	950	.3167	635	47
14	506	.75996	.3159	618	46
15	.60529	.76042	1.3151	.79600	45
16	553	088	.3143	583	44
17	576	134	.3135	565	43
18	599	180	.3127	547	42
19	622	226	.3119	530	41
20	.60645	.76272	1.3111	.79512	40
21	668	318	.3103	494	39
22	691	364	.3095	477	38
23	714	410	.3087	459	37
24	738	456	.3079	441	36
25	.60761	.76502	1.3072	.79424	35
26	784	548	.3064	406	34
27	807	594	.3056	388	33
28	830	640	.3048	371	32
29	853	686	.3040	353	31
30	.60876	.76733	1.3032	.79335	30
31	899	779	.3024	318	29
32	922	825	.3017	300	28
33	945	871	.3009	282	27
34	968	918	.3001	264	26
35	.60991	.76964	1.2993	.79247	25
36	.61015	.77010	.2985	229	24
37	038	057	.2977	211	23
38	061	103	.2970	193	22
39	084	149	.2962	176	21
40	.61107	77196	1.2954	.79158	20
41	130	242	.2946	140	19
42	153	289	.2938	122	18
43	176	335	.2931	105	17
44	199	382	.2923	087	16
45	.61222	.77428	1.2915	.79069	15
46	245	475	.2907	051	14
47	268	521	.2900	033	13
48	291	568	.2892	.79016	12
49	314	615	.2884	.78998	11
50	.61337	.77661	1.2876	.78980	10
51	360	708	.2869	962	9
52	383	754	.2861	944	8
53	406	801	.2853	926	7
54	429	848	.2846	908	6
55	.61451	.77895	1.2838	.78891	5
56	474	941	.2830	873	4
57	497	.77988	.2822	855	3
58	520	.78035	.2815	837	2
59	543	082	.2807	819	1
60	.61566	.78129	1.2799	.78801	0
	cos	cot	tan	sin	'

52°

What equation, or function, would you choose? _____

– – – – – – – – – – – – – – – –

Either the sine or the cosecant of angle A. Since there is no real prefer-
ence in this case from the standpoint of making the work easier, following our basic rule we would select the sine of angle A. We would
also have to use the sine because, as pointed out earlier, the cosecant
function does not appear in our tables.

17. We now have

$$\sin A = \frac{a}{c}$$

or, substituting the known values,

$$\sin A = \frac{4}{8} = 0.50000.$$

To look up the angular value of A corresponding to a sine value of
0.50000, we begin at the beginning, noting that the sine of $0°$ is zero
but that it gradually increases as we proceed into higher angular values.
Continuing, then, we come eventually to the sine value of 0.50000 and
find it opposite $30°$. Hence the solution of our problem is $A = 30°$.
(See the table portion on the next page.)
   You should be ready now for some applied problems. Although
some of these problems may involve heights or horizontal distances,
each one is based upon the requirement to solve a right triangle by
means of the six trigonometric functions. Remember, however, to
avoid using the secant and cosecant functions since these are not shown in
our tables. The solution requirements will not differ essentially from the
"abstract" problems you have been solving, so don't let the details confuse you.

In the first three problems below use the
figure at the right to assist you.

(a) $a = 6.4$ ft., $b = 6.4$ ft., $A = $ _____

(b) $b = 12$ in., $A = 15° 39'$, $c = $ _____

(c) $b = 27.1$ mi., $c = 29.1$ mi., $B = $ _____

(d)   A mountain climber stretches a rope
      from the top of a sheer cliff to a
      point on the level ground below,
      making an angle of $48°$ with the
      ground. If the rope is 86 feet long,
      how high is the cliff?

_____

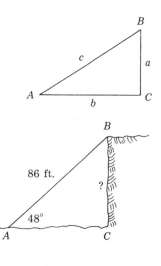

30°

	sin	tan	cot	cos	
0	.50000	.57735	1.7321	.86603	60
1	025	774	.7309	588	59
2	050	813	.7297	573	58
3	076	851	.7286	559	57
4	101	890	.7274	544	56
5	.50126	.57929	1.7262	.86530	55
6	151	.57968	.7251	515	54
7	176	.58007	.7239	501	53
8	201	046	.7228	486	52
9	227	085	.7216	471	51
10	.50252	.58124	1.7205	.86457	50
11	277	162	.7193	442	49
12	302	201	.7182	427	48
13	327	240	.7170	413	47
14	352	279	.7159	398	46
15	.50377	.58318	1.7147	.86384	45
16	403	357	.7136	369	44
17	428	396	.7124	354	43
18	453	435	.7113	340	42
19	478	474	.7102	325	41
20	.50503	.58513	1.7090	.86310	40
21	528	552	.7079	295	39
22	553	591	.7067	281	38
23	578	631	.7056	266	37
24	603	670	.7045	251	36
25	.50628	.58709	1.7033	.86237	35
26	654	748	.7022	222	34
27	679	787	.7011	207	33
28	704	826	.6999	192	32
29	729	865	.6988	178	31
30	.50754	.58905	1.6977	.86163	30
31	779	944	.6965	148	29
32	804	.58983	.6954	133	28
33	829	.59022	.6943	119	27
34	854	061	.6932	104	26
35	.50879	.59101	1.6920	.86089	25
36	904	140	.6909	074	24
37	929	179	.6898	059	23
38	954	218	.6887	045	22
39	.50979	258	.6875	030	21
40	.51004	.59297	1.6864	.86015	20
41	029	336	.6853	.86000	19
42	054	376	.6842	.85985	18
43	079	415	.6831	970	17
44	104	454	.6820	956	16
45	.51129	.59494	1.6808	.85941	15
46	154	533	.6797	926	14
47	179	573	.6786	911	13
48	204	612	.6775	896	12
49	229	651	.6764	881	11
50	.51254	.59691	1.6753	.85866	10
51	279	730	.6742	851	9
52	304	770	.6731	836	8
53	329	809	.6720	821	7
54	354	849	.6709	806	6
55	.51379	.59888	1.6698	.85792	5
56	404	928	.6687	777	4
57	429	.59967	.6676	762	3
58	454	.60007	.6665	747	2
59	479	046	.6654	732	1
60	.51504	.60086	1.6643	.85717	0
'	cos	cot	tan	sin	

59°

31°

'	sin	tan	cot	cos	'
0	.51504	.60086	1.6643	.85717	60
1	529	126	.6632	702	59
2	554	165	.6621	687	58
3	579	205	.6610	672	57
4	604	245	.5599	657	56
5	.51628	.60284	1.6588	.85642	55
6	653	324	.6577	627	54
7	678	364	.6566	612	53
8	703	403	.6555	597	52
9	728	443	.6545	582	51
10	.51753	.60483	1.6534	.85567	50
11	778	522	.6523	551	49
12	803	562	.6512	536	48
13	828	602	.6501	521	47
14	852	642	.6490	506	46
15	51877	.60681	1.6479	.85491	45
16	902	721	.6469	476	44
17	927	761	.6458	461	43
18	952	801	.6447	446	42
19	51977	841	.6436	431	41
20	52002	.60881	1.6426	.85416	40
21	026	921	.6415	401	39
22	051	60960	.6404	385	38
23	076	.61000	.6393	370	37
24	101	040	.6383	355	36
25	52126	.61080	1.6372	.85340	35
26	151	120	.6361	325	34
27	175	160	.6351	310	33
28	200	200	.6340	294	32
29	225	240	.6329	279	31
30	.52250	.61280	1.6319	.85264	30
31	275	320	.6308	249	29
32	299	360	.6297	234	28
33	324	400	.6287	218	27
34	349	440	.6276	203	26
35	.52374	.61480	1.6265	.85188	25
36	399	520	.6255	173	24
37	423	561	.6244	157	23
38	448	601	.6234	142	22
39	473	641	.6223	127	21
40	.52498	.61681	1.6212	.85112	20
41	522	721	.6202	096	19
42	547	761	.6191	081	18
43	572	801	.6181	066	17
44	597	842	.6170	051	16
45	.52621	.61882	1.6160	.85035	15
46	646	922	.6149	020	14
47	671	.61962	.6139	.85005	13
48	696	.62003	.6128	.84989	12
49	720	043	.6118	974	11
50	.52745	.62083	1.6107	.84959	10
51	770	124	.6097	943	9
52	794	164	.6087	928	8
53	819	204	.6076	913	7
54	844	245	.6066	897	6
55	.52869	.62285	1.6055	.84882	5
56	893	325	.6045	866	4
57	918	366	.6034	851	3
58	943	406	.6024	836	2
59	967	446	.6014	820	1
60	.52992	.62487	1.6003	.84805	0
'	cos	cot	tan	sin	'

58°

(e)   A 34-foot ladder is placed against the side of a house with the foot of the ladder 9 feet away from the building. What angle does the ladder make with the ground?

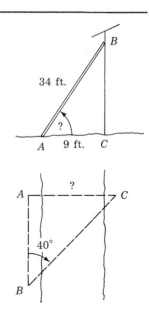

_____

(f)   In order to find the width of a river, a distance AB was measured along the bank, the point A being directly opposite a tree, C, on the other side. If the angle at B was observed to be $40°$ and the distance AB was 100 feet, how wide was the river?

_____

- - - - - - - - - - - - - - -

(a) $\tan A = \dfrac{a}{b} = \dfrac{6.4}{6.4} = 1.0000$; $A = 45°$. (Of course you could have gotten this answer from geometry since angles opposite equal sides are equal.)

(b) $\cos A = \dfrac{b}{c}$, $c = \dfrac{b}{\cos A}$, $c = \dfrac{12}{0.96293} = 12.5$ in.

(c) $\sin B = \dfrac{b}{c} = \dfrac{27.1}{29.1} = 0.93127$; $B = 68°\,38'$.

(d) $\sin A = \dfrac{a}{c}$, $a = c \sin A = 86 \sin 48° = 86(0.74314)$; $a = 63.9$ ft.

(e) $\cos A = \dfrac{b}{c} = \dfrac{9}{34} = 0.26471$; $A = 74°\,39'$.

(f) $\tan B = \dfrac{AC}{AB}$, $AC = AB \tan B = 100 \tan 40° = 100(0.83910)$; $AC = 83.9$ ft.

18.   Now you have a choice. If you would like to work some more problems, you will find them here. On the other hand if you feel you would like to go on, go directly to frame 19. Solve these problems using the table of trigonometric functions in the Appendix.

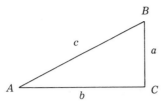

Given	*Find*
	(to the nearest minute of arc and three significant figures, i.e., three digits)

(a) $A = 30°18'$, $a = 3$ $B = $ _____ , $c = $ _____ , $b = $ _____

(b) $a = 6$, $c = 11.8$ $A = $ _____ , $B = $ _____ , $b = $ _____

(c) $a = 4$, $b = 3.9$ $A = $ _____ , $B = $ _____ , $c = $ _____

(d) $A = 36°$, $c = 1$ $B = $ _____ , $a = $ _____ , $b = $ _____

(e) $A = 75°32'$, $a = 80$ $B = $ _____ , $b = $ _____ , $c = $ _____

(f) $A = 25°48'$, $a = 30$ $B = $ _____ , $b = $ _____ , $c = $ _____

(g) $B = 15°19'$, $b = 20$ $A = $ _____ , $a = $ _____ , $c = $ _____

(h) $a = 36.4$, $b = 100$ $A = $ _____ , $B = $ _____ , $c = $ _____

(i) $B = 88°02'$, $b = .08$ $A = $ _____ , $a = $ _____ , $c = $ _____

(j) $a = 30.2$, $c = 33.3$ $A = $ _____ , $B = $ _____ , $b = $ _____

Draw a sketch for and solve each of the following problems.

(k) Two battleships are stationed 3 miles apart. From one of them an enemy submarine is observed due south, and from the other it is observed $40°15'$ east of south. How far is the submarine from the nearest battleship? _____

(l) The vertical central pole of a circular tent is 20 feet high and its top is fastened by ropes 38 feet long to stakes set in the ground. How far are the stakes from the foot of the pole, and what is the angle between the ropes and the ground? _____

(m) At a distance of 58.6 feet from the base of a tower, the angle of elevation of its top is observed to be $58°\,24'$. What is the height of the tower? _____

(n) If a tower casts a shadow that is three-fourths of its own length, what is the angle of elevation of the sun? _____

(o) From the top of a cliff 587 feet above sea level, the angles of depression (that is, below the horizontal) of two boats in line with the observer are $14°\,10'$ and $24°\,45'$ respectively. Find the distance between the boats. _____

- - - - - - - - - - - -

(a) $B = 59°\,42'$, $c = 5.95$, $b = 5.13$
(b) $A = 30°\,34'$, $B = 59°\,26'$, $b = 10.2$
(c) $A = 45°\,44'$, $B = 44°\,16'$, $c = 5.59$
(d) $B = 54°\,00'$, $a = 0.588$, $b = 0.809$
(e) $B = 14°\,28'$, $b = 20.6$, $c = 82.6$
(f) $B = 64°\,12'$, $b = 62.1$, $c = 68.9$
(g) $A = 74°\,41'$, $a = 73.0$, $c = 75.7$
(h) $A = 20°\,00'$, $B = 70°\,00'$, $c = 106$
(i) $A = 1°\,58'$, $a = 0.00275$, $c = 0.0802$
(j) $A = 65°\,05'$, $B = 24°\,55'$, $b = 14.0$
(k) 3.54 miles (Use the tangent function; see sketch.)

(l) $A = 31°\,45'$, $b = 32.3$ ft. (Use the sine function; see sketch.)

(m) 95.25 feet (Use the
tangent function; see
sketch.)

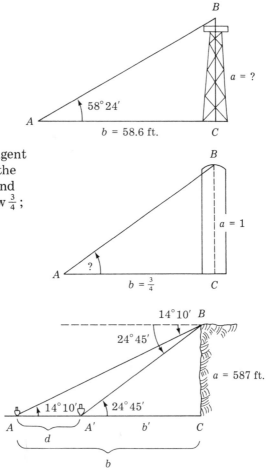

(n) 53° 08′ (Again, the tangent
is your function. Call the
height of the tower 1 and
the length of its shadow $\frac{3}{4}$;
see sketch.)

(o) 1,050 feet
(Recognize that
you have *two*
triangles to solve
here, having the
common side a,
and that what you
are seeking in each
case is the length of
the base—indicated
by b and b' in the
sketch. Once you have obtained these you then simply subtract
the distance of the boat nearest the cliff from that of the boat
farthest from shore to obtain the distance between the boats.)

CO-FUNCTIONS

19. In frame 13 we promised to go into a little more detail about co-
functions, and now is the time to do so.
    If you will recall for a moment the names of the six trigonometric
functions you will note that, basically, the six terms bear only three
names, the other three names being formed by the prefix "co." Thus
there is the sine and *co*sine, tangent and *co*tangent, secant and *co*secant.
    The cosine is said to be the co-function of the sine, the cotangent
the co-function of the tangent, and the cosecant the co-function of the
secant. What the term "co-function" means can best be shown by an

example. In the figure shown at the right, note that sin $A = \frac{a}{c}$, but that cos B also equals $\frac{a}{c}$.

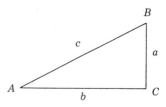

Another way of approaching this matter of co-functions is to remember that the sum of angles A and B (in a right triangle) is $90°$, hence they are complementary angles. The term "complementary," which you will recall from our study of geometry, means that the sum of two angles is $90°$. We know that the sum of the three angles in a plane triangle is $180°$, hence in a right triangle the sum of the two acute angles must equal $90°$. Therefore, if angle A has a certain value, then the value of angle B must be $90° - A$. Conversely, the value of angle A (if angle B is known) is $90° - B$. We come, then, to the interesting conclusion that

sin A = cos B

or, since

$B = 90° - A$

then

sin A = cos($90° - A$).

Similarly

tan A = cot B
        = cot($90° - A$)

and

sec A = csc B
sec A = csc($90° - A$).

And the same equations could be derived for angle B.

Example: A single example should be enough to show you just how this co-function relationship works and enable you to relate it to other values found in the trig tables. Let's consider the function sin A and assume that $A = 30°$. From what we have just learned we know that

sin A = cos B

or, since $A = 30°$,

sin $30°$ = cos B

but since $B = 90° - A$, then $B = 60°$. Therefore

sin $30°$ = cos $60°$.

But sin 30° is a definite numerical quantity. If we look up its value in the trig tables we find it to be 0.50000. And since sin 30° = cos 60°, we should expect to find that cos 60° also equals 0.50000. And so it does, as you can see for yourself in the tables. Obviously, therefore, there is no need to print this function value of 0.50000 more than once to suit the needs of both the sin and cos in this case.

And so it goes throughout the tables for all sin and cos values. And the same thing applies to the tan and cot functions as well as the sec and csc functions. Another way to remember this is to keep in mind that the values of the functions and co-functions move in exactly opposite directions numerically (we'll see why in the next chapter) when angles increase from 0° to 90° or decrease from 90° to 0°. Thus the sin increases in value from 0 at 0° to 1 at 90° whereas the cos decreases in value from 1 at 0° to 0 at 90°. Obviously, therefore, these function values pass each other at the halfway point, namely, 45°. After 45° we find our cosine values in the sin column, reading up the sin column from the bottom (as the angle increases) instead of down from the top. Conversely, we find our sine values in the cos column, again reading up from the bottom. All of which will become more apparent and more familiar to you as you continue to use the tables.

Here are a few exercises to help you get started. Find the following missing values. (Note: The trig tables do not include the sec and csc functions.)

Value from tables

(a) sin 40° = cos _____° = _____

(b) tan 25° = cot _____° = _____

(c) sec 80° = csc _____° = (not shown)

(d) csc 18° = sec _____° = (not shown)

(e) cos 45° = sin _____° = _____

(f) cot 1° = tan _____° = _____

- - - - - - - - - - - - - - -

(a) 50°, .64279; (b) 65°, .46631; (c) 10°; (d) 72°;
(e) 45°, .70711; (f) 89°, 57.290

FUNCTIONS OF 30°, 45°, AND 60° ANGLES

20. You may recall that in Chapter 3, we discussed some of the properties of the 30°—60°—90° triangle (frame 34) and the 45°—45°—90° triangle

(frame 35). We mentioned also that the unique properties of these special right triangles would be useful in trigonometry. So let's take another look at them and find out why.

Probably the main reason they are important in trigonometry is that they occur frequently in problems usually solved by trigonometric methods. It is therefore important to find the values of the trigonometric functions of these angles and to memorize the results. This will be very useful later on.

To find the functions of $45°$ we draw an isosceles right triangle (half of a square). This makes angle A = angle B = $45°$. Since the relative (rather than the actual) lengths of the sides are important, we may assign any lengths we please to the sides that satisfy the condition that the right triangle be isosceles. For simplicity's sake we will choose the lengths of the short sides as unity, that is, $a = 1$ and $b = 1$. Then $c = \sqrt{a^2 + b^2} = \sqrt{2}$, and we get

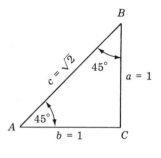

$$\sin 45° = \frac{1}{\sqrt{2}} \qquad \csc 45° = \sqrt{2}$$

$$\cos 45° = \frac{1}{\sqrt{2}} \qquad \sec 45° = \sqrt{2}$$

$$\tan 45° = 1 \qquad \cot 45° = 1$$

Now without looking up above (or, better still, after covering the upper half of the page), see if you can draw a $45°$—$45°$—$90°$ triangle and show the correct values for the sides, that is, do what we just did above.

- - - - - - - - - - - - - - - -

Check your work with the figure above.

21. To find the functions of $30°$ and $60°$, we draw an equilateral triangle, ABD, and drop the perpendicular BC from B to AD. Now if we consider just the triangle ABC, we have a triangle in which A = $60°$, angle ABC = $30°$, and the angle at C is $90°$.

Again taking the smallest side as unity,

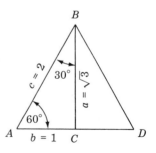

that is $b = 1$, we get $c = AB = AD = 2AC = 2$, hence $a = \sqrt{c^2 - b^2} = \sqrt{4 - 1} = \sqrt{3}$. Therefore,

$$\sin 60° = \frac{\sqrt{3}}{2} \qquad\qquad \csc 60° = \frac{2}{\sqrt{3}}$$

$$\cos 60° = \frac{1}{2} \qquad\qquad \sec 60° = 2$$

$$\tan 60° = \sqrt{3} \qquad\qquad \cot 60° = \frac{1}{\sqrt{3}}$$

And similarly, from the same triangle,

$$\sin 30° = \frac{1}{2} \qquad\qquad \csc 30° = 2$$

$$\cos 30° = \frac{\sqrt{3}}{2} \qquad\qquad \sec 30° = \frac{2}{\sqrt{3}}$$

$$\tan 30° = \frac{1}{\sqrt{3}} \qquad\qquad \cot 30° = \sqrt{3}$$

Summarizing the above we have:

Angle	30°	45°	60°
sin	$\frac{1}{2} = .500$	$\frac{1}{\sqrt{2}} = .707$	$\frac{\sqrt{3}}{2} = .866$
cos	$\frac{\sqrt{3}}{2} = .866$	$\frac{1}{\sqrt{2}} = .707$	$\frac{1}{2} = .500$
tan	$\frac{1}{\sqrt{3}} = .577$	1	$\sqrt{3} = 1.732$

Now cover up the above and try deriving our summary table of values. Start by drawing an equilateral (and therefore equiangular) triangle and dropping a perpendicular from the apex to the base to form your 30°—60°—90° triangle. Then determine the lengths of the sides and function values.

- - - - - - - - - - - - - - - - -

Check your work with our development of these values above. If you still are not clear why you are learning to find these common values quickly, be patient. You will have lots of use for them soon.

VECTORS

22. Another useful concept we should introduce at this point is that of
vectors. You may have encountered this term if you have studied
physics. Any physical quantity, such as a force or velocity, that has
both direction and magnitude is called a
vector quantity, and may be represented
by a directed line segment (arrow) called
a *vector*. The *direction* of the vector is
that of the given quantity, and the *length*
of the vector represents the magnitude of
the quantity. Thus, in the figure at the
right, the vector *AB* represents an airplane
that is traveling northeast (compass
direction of 45°) at 200 mph (miles per hour).

   In the next figure a motor boat having
a speed of 15 mph in still water is headed
directly across a river whose current is
5 mph. The boat's speed and direction are
represented by the vector *AB*, while the
current's direction and velocity (to the
same scale) are represented by the vector
CD. (Note that *AB* is three times as long
as *CD*.)

   In this third example the vector *AB*
represents a force of 20 lbs making an angle
of 35° with the positive direction on the
X-axis, and vector *CD* represents a force
of 30 lbs at 150° with the positive direction on the *X*-axis, both vectors being
drawn to the same scale (that is,
$\frac{1}{4}$ " = 5 lbs in each case).

   It is highly important in working with vectors that you use a consistent scale to represent magnitude, especially if you are going to
combine (add) vectors, as we will be doing shortly. Thus, if you decide
to let a distance of $\frac{1}{8}$ inch represent 1 lb of
force, then a 10-lb force would be shown
by an arrow $1\frac{1}{4}$ inch long, as shown at the
right.

   Choose some convenient scale and represent the following quantities.
(*Note:* Although your scale must be consistent within any one problem,
it does not, of course, have to be the same for *all* problems, since a scale
suitable to represent a force of 20 lbs would
hardly be appropriate to show speeds of
the order of 200 mph, for example.)

Remember, compass directions are measured clockwise from North.

(a)   A force of 35 lbs exerted in a direction 135° east of (measured clockwise from) compass north.

(b)   A force of 200 lbs acting directly downward.

(c)   An airplane flying due west at a speed of 300 mph.

(d)   An automobile traveling in a southwest direction (i.e., in a compass direction of 225°) at a speed of 50 mph.

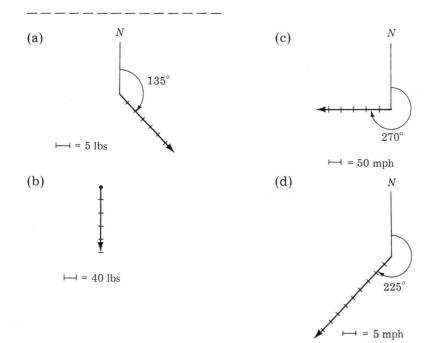

23.   Vectors can be added, thus making them an extremely valuable tool for determining aircraft position and groundspeed as a resultant of airspeed and wind velocity (dead-reckoning navigation), and similarly for

finding course, speed, and position of surface vessels (affected by ocean currents), analyzing bridge structures and designing bridges, and solving many other types of problems in applied mechanics. If you study physics, you may some day find yourself taking a course in analytic and vector mechanics, where you will discover many of the useful applications of vectors.

Here we are primarily interested in the application of vectors in mathematics, not in physical quantities. But in either case we need to know how to add them.

Let's consider first what we mean by the addition of vectors. Essentially it is this:

> The *resultant* or *vector sum* of a number of vectors, all lying in the same plane, is that vector which would produce the same effect as that produced by all of the original vectors acting together.

If two vectors α (alpha) and β (beta) have the same direction, their resultant is a vector, R, whose magnitude is equal to the sum of the magnitudes of the two vectors and whose direction is the same as that of the two vectors. Thus, as shown at the right, since α has a magnitude of 75 and β a magnitude of 125, and both are pointing due east, their resultant, R, is a vector pointing in the same direction and whose magnitude is 200, the sum of α and β. On the other hand, if two vectors have opposite directions, their resultant is a vector, R, whose magnitude is the difference (greater magnitude minus smaller magnitude) of the two vectors and whose direction is that of the vector having the greater magnitude. Thus, $\beta - \alpha = R = 50$.

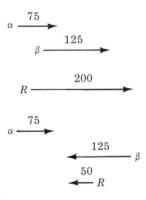

What would be the resultant of the following pairs of vectors?

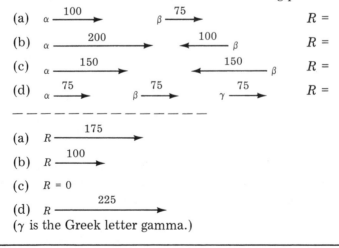

(a) $R =$

(b) $R =$

(c) $R =$

(d) $R =$

- - - - - - - - - - - - - - - -

(a) R 175

(b) R 100

(c) $R = 0$

(d) R 225

(γ is the Greek letter gamma.)

24. In the last frame we considered only the situation where two vectors were in a direct line with one another, that is, either had the same direction or opposite directions. In all other cases — where two vectors form an angle with each other — the magnitude and direction of their resultant must be obtained by one of two other methods. The first of these methods, known as the *parallelogram method*, is as follows.

Place the tail ends of both vectors at any point, *O*, in their plane and complete the parallelogram having these vectors as adjacent sides. The directed diagonal issuing from *O* is the resultant or vector sum of the two given vectors.

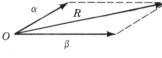

Thus, in the figure at the right, the vector *R* is the resultant of the two vectors α and β.

The other method of finding the resultant of two vectors that are not in a line is known as the *triangle method* (or head-to-tail method). The procedure is as follows.

Choose one of the vectors and label its tail end as *O*. Place the tail end of the other vector at the arrow end of the first vector. The resultant is the line segment closing the triangle and directed from *O*.

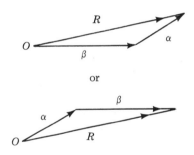

or

Thus, in the figure at the right, *R* is the resultant of the vectors α and β.

For an example of the practical application of vectors, let's return to the second example in frame 22, where we had the case of the boat moving directly across a river at a speed of 15 mph, and being acted on by a river current of 5 mph. Our objective will be to find the resultant path of the boat when acted upon by two forces: the force of the motor propelling it eastward at 15 mph, and the force of the current moving it southward at a rate of 5 mph. Using the parallelogram method of solution we would get the figure shown at the right.

Using the triangle method and laying out the vector representing the boat's speed through the water first, we get the figure shown here.

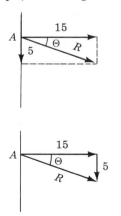

Using the triangle method and beginning with the current's vector first, we get the figure opposite. The important point is that regardless of which method we use, the resultant is the same.

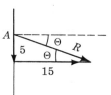

Do you have any idea how you would go about figuring out the magnitude of R or the angle (represented by the Greek letter Θ, theta) the boat's path makes with the direction in which it is headed? Try it, and then check your procedure with that shown below.

_ _ _ _ _ _ _ _ _ _ _ _ _ _ _ _

The first thing you should observe is that you have a right triangle with the two vectors as the sides and the resultant as the hypotenuse. Since you know the lengths (magnitudes) of the sides, you can solve for the length (magnitude) of the resultant by the Pythagorean Theorem. Thus, $R = \sqrt{15^2 + 5^2} = \sqrt{250} = 15.81$ mph.

The angle Θ, between the resultant path of the boat and the direction in which it is pointed, being one of the acute angles of the vector right triangle, can be found by using the tangent function since we know the lengths (magnitudes) of both sides. Thus, $\tan \Theta = \dfrac{5}{15} = 0.333 = 18° 26'$.

Therefore, the boat moves downstream in a line making an angle of $18° 26'$ with the direction in which it is pointed, at a speed of 15.81 mph.

25. Another interesting and very useful aspect of vectors is that they can be resolved into components lying along two coordinate axes at right angles to one another. This is essentially the converse of what we have been doing. That is, instead of finding the resultant of two vectors, we are resolving a resultant vector into two orthogonal (mutually perpendicular) vectors that could have produced it.

For example, the components of R in the problem above are (1) 5 mph in the direction of the current, and (2) 15 mph in a direction perpendicular to the current.

Another example is the figure at the right wherein the force F has the horizontal component $F_h = F \cos 30°$ and the vertical component $F_v = F \sin 30°$. Notice that F is the vector sum or resultant of F_h and F_v. The components of F (that is, F_h and F_v) were found by "projecting" the line F onto

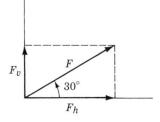

two orthogonal (perpendicular) axes. To do this we must, of course, know what angle the slanted vector F makes with the horizontal axis, which is given as $30°$ in this case.

As you can see from the illustration on the previous page, to project a line such as the vector F onto two mutually perpendicular axes, we treat it as though it were the diagonal of a rectangle, and complete that rectangle by drawing dotted lines from the tip of the arrow parallel to the two axes. The points at which the dotted lines intersect the axes mark the ends of the component vectors.

Try this in the following problem. In the figure at the right we will use X and Y as the orthogonal axes and F as the vector whose x and y components we wish to find. Note that F makes an angle of $30°$ with the X-axis. If F = 20 lbs, find the magnitude of F_x and F_y. (Hint: Use the sin function to find F_y and the cos function to find F_x.

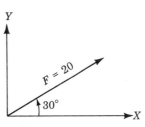

Also, use your table in frame 21 to get the sin and cos values for $30°$ if you wish. This is an example of their usefulness.)

$$\sin 30° = \frac{F_y}{20} \text{ or } F_y = 20 \sin 30°$$
$$= 20(.500) = 10 \text{ lbs}$$
$$\cos 30° = \frac{F_x}{20} \text{ or } F_x = 20 \cos 30°$$
$$= 20(.866)$$
$$= 17.32 \text{ lbs}$$

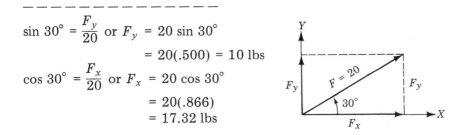

ANGULAR AND CIRCULAR MEASUREMENT

26. Since at this point we're interested only in introducing you to the concept of vectors, we won't go into further details regarding their application. However, we will meet them again later on so keep in mind what you have learned here.

The last topic we will discuss in this chapter is that of angular and circular measurement. Since we have been using degrees for angular measurement up until now in the book, it may appear to you that it is rather late to introduce the subject. But this is not so. You doubtless were familiar with the use of degrees to indicate the size of angles from your own experience and therefore willing to accept the concept intuitively. But now we need to examine the subject somewhat more explicitly, and to learn another system of circular measurement.

In frame 8, Chapter 1, we discussed the circle for the first time and defined *degree* as being 1/360th of a circle. We also defined such terms as radius, circumference, arc, and central angle. A *central angle* is, as you will recall, an angle formed by two radii, for example, the angle AOC in the figure at the right. An *arc* of a circle, on the other hand, is the curved line between two points on a circle — the curved line between A and C, for example, designated as AC.

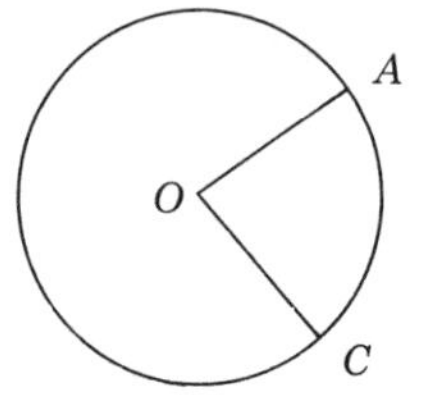

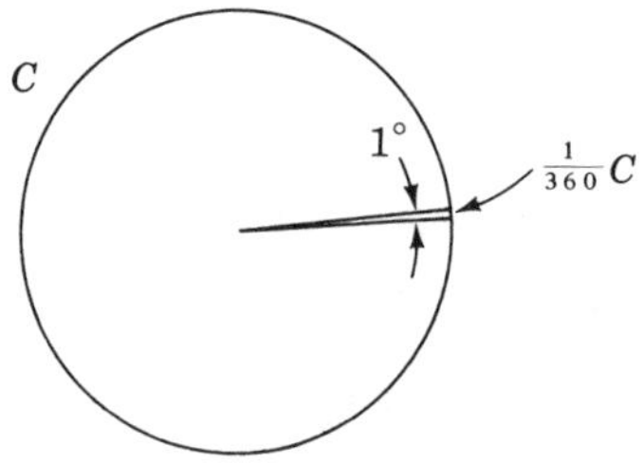

"One degree" can be defined in terms of a central angle of a circle that cuts off a definite arc length, namely, 1/360th of a circle. Thus, a central angle is one degree $(1°)$ if its arc length is $\frac{1}{360}$ of the circle, as shown at the right. A *minute* $(')$ is $\frac{1}{60}$ of a degree. A *second* $('')$ is $\frac{1}{60}$ of a minute.

Example: $\frac{1}{4}(36° 24') = 9° 06'$

Try a few more of these.

(a) $\frac{1}{2}(127° 24') = $ _____
   (Convert $1°$ to $60'$ before dividing so you'll have an even number of degrees.)

(b) $\frac{1}{4}(48° 36') = $ _____

(c) $\frac{1}{2}(81° 15') = $ _____
   (Convert $1°$ to $60'$ and $1'$ to $60''$ so you'll have an even number of degrees and minutes.)

(d) $\frac{1}{3}(42° 42') = $ _____

(e) $\frac{1}{4}(74° 29' 20'') = $ _____
   (This, too, will require some converting.)

– – – – – – – – – – – – – – –

(a) $\frac{1}{2}(127° 24') = \frac{1}{2}(126° 84') = 63° 42'$
(b) $\frac{1}{4}(48° 36') = 12° 09'$
(c) $\frac{1}{2}(81° 15') = \frac{1}{2}(80° 75') = \frac{1}{2}(80° 74' 60'') = 40° 37' 30''$
(d) $\frac{1}{3}(42° 42') = 14° 14'$
(e) $\frac{1}{4}(74° 29' 20'') = \frac{1}{4}(72° 148' 80'') = 18° 37' 20''$

27. In addition to the unit of angular measure, the degree, there also is a
unit of circular measure. This unit, known as a *radian* (rad), is defined
as follows:

A radian is the measure of the central
angle subtended by an arc of a circle
equal to the radius of the circle.

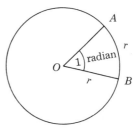

Thus, as shown at the right, the arc length
equal to the radius measures a central
angle of one radian, or, angle AOB =
1 radian. This unit of circular measurement, which was introduced early in the
last century, now is used to a certain extent in practical work and is
universally used in the higher branches of mathematics; hence it is one
with which you need to be familiar.

Since the circumference of a circle is equal to $2\pi r$ (2π times the
radius), and subtends an angle of $360°$, then 2π radians = $360°$ and

$$1 \text{ radian} = \frac{180°}{\pi} = 57.296° = 57°17'45'' \text{ (using a value of 3.1416 for } \pi).$$

And, 1 degree = $\frac{\pi}{180}$ radian = 0.01745 rad, approximately. We therefore emerge with the following rules for conversion:

(1) *To convert radians to degrees,* multiply the number of radians by
57.296 or divide them by .01745.

(2) *To convert degrees to radians,* multiply the number of degrees by
0.01745 or divide them by 57.296.

Since this is a straightforward matter of multiplication or division, we
won't ask you to do any of it now. What we will ask you to do is to
make a note of this page so you can look up these conversion factors
when you need them! What is more important at this point is that you
get accustomed to expressing angles in circular measure, as follows.

$360° = 2\pi$ radians	$60° = \frac{\pi}{3}$ radians
$180° = \pi$ radians	$30° = \frac{\pi}{6}$ radians
$90° = \frac{\pi}{2}$ radians	$45° = \frac{\pi}{4}$ radians
$270° = \frac{3\pi}{2}$ radians	$15° = \frac{\pi}{12}$ radians

Also, when writing the trigonometric functions of angles expressed in
circular measure it is customary to omit the word "radians," as shown
following.

sin (π radians) is written simply sin π and equals sin $180°$

tan ($\frac{\pi}{2}$ radians) is written simply tan $\frac{\pi}{2}$ and equals tan $90°$

cot ($\frac{3\pi}{4}$ radians) is written simply cot $\frac{3\pi}{4}$ and equals cot $135°$

cos ($\frac{5\pi}{6}$ radians) is written simply cos $\frac{5\pi}{6}$ and equals cos $150°$

csc (1 radian) is written simply csc 1 and equals csc $57.29°$

sec ($\frac{1}{2}$ radian) is written simply sec $\frac{1}{2}$ and equals sec $28.65°$

With the above in mind, write the following trigonometric functions in radian measurement (in terms of π).

Example: sin $45°$. $\frac{360°}{45°} = 8$, hence $45° = \frac{1}{8}$ of $360° = \frac{1}{8}(2\pi) = \frac{\pi}{4}$.

Therefore, sin $45°$ = sin $\frac{\pi}{4}$.

Use this procedure below.

(a)   cos $15°$ = _____

(b)   tan $30°$ = _____

(c)   sin $60°$ = _____

— — — — — — — — — — — — — —

(a) cos $\frac{\pi}{12}$;   (b) tan $\frac{\pi}{6}$;   (c) sin $\frac{\pi}{3}$

Now it's time for you to check up on yourself. When you have completed the following Self-Test, be sure to review any parts of this chapter you find you are having difficulty remembering or using.

SELF-TEST

1. Numerical trigonometry is the branch of mathematics that deals with the relationships existing between the sides and angles of triangles. (True, False)   (frame 1)

2. Plane trigonometry concerns itself with the study of plane triangles. (True, False)   (frame 1)

3. A *plane* triangle is one whose sides are straight lines lying in the same plane. (True, False)   (frame 1)

4. Which of the traingles at the right is a

   right triangle? _____

   (frame 1)

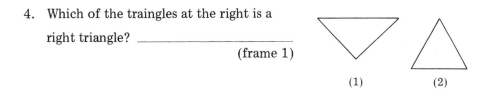

   (1)   (2)

5. The size of an angle in a right triangle depends upon the ratio existing between any two sides of the triangle. (True, False)   (frame 2)

6. The ratio of one number to another number is the result of dividing the first number by the second. This division must be performed and shown as a decimal fraction. (True, False)   (frame 3)

7. The hypotenuse of a right triangle is the side opposite the right angle. (True, False)   (frame 5)

8. The sine of an angle = ─────────────────────── .   (frame 5)

9. The names of the six trigonometric fuctions are: _____

   _____ .

   (frame 6)

10. The abbreviations for the six trigonometric functions named above are:

   _____ .

   (frame 6)

11. Referring to the triangle at the right, express each of the six functions (ratios) in terms of angle A.

   _____ A = ─────   _____ A = ─────

   _____ A = ─────   _____ A = ─────

   _____ A = ─────   _____ A = ─────

   (frame 9)

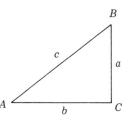

12. The three primary trigonometric functions are the:

_____ .

(frame 10)

13. Their reciprocals are the: _____ .

(frame 10)

14. The reciprocals of the three primary functions are known as the
     "secondary functions." (True, False)   (frame 10)

15. The reciprocal can always be used in place of the primary function.
     (True, False)   (frame 10)

16. Use the tables of natural trigonometric functions to find the following
     values.

     (a) $\sin 24°55'$ = _____   (c) $\tan 65°01'$ = _____

     (b) $\cos 36°18'$ = _____   (d) $\cot 84°43'$ = _____

     (frames 12, 13)

17. In the triangle shown at the right, which
     function would you select to solve for:

     (a) side b? _____

     (b) side c? _____

     (c) angle A? _____

     (frames 14, 15)

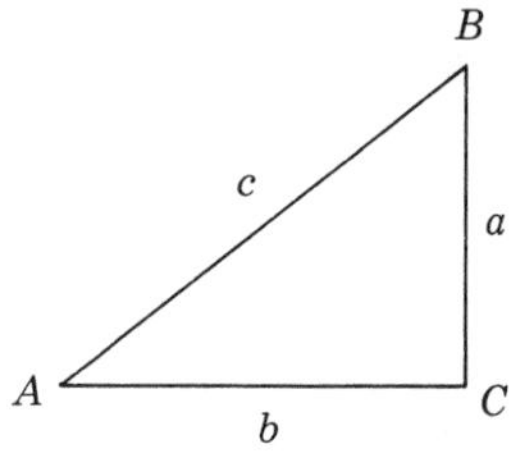

     $B = 49°22'$
     $a = 18$ ft.

18. Solve the above triangle for side c. Side c = _____

     (frame 15)

19. What is the size of angle A in the triangle in problem 17 above?

     A = _____   (from Geometry, Chapter 2, frame 17)

20. What is the length of side b in problem 17? Side b = _____

     (frame 15)

21. To find the width of a river a surveyor set up his transit at C on one bank and sighted across to a point B on the opposite bank. Then turning through an angle of $90°$ he laid off a distance $CA = 225$ ft. Finally, setting the transit at A, he measured $\angle CAB = 48°20'$. Find the width of the river.   (frames 17, 18)

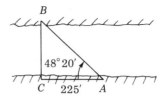

22. In the figure at the right the line AD crosses a swamp. In order to locate a point on this line a surveyor turned through an angle of $51°16'$ at A and measured 1,585 feet to a point C. He then turned through an angle of $90°$ at C and ran a line CB to B. If B is on the line AD, how far must he measure from C to reach B?

                (frames 17, 18)

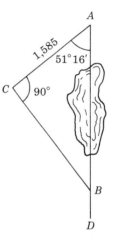

23. $\cos 50° = \sin$ _____ $° =$ _____   (frame 19)

24. Draw a $45°-45°-90°$ triangle and show the correct values for the sides.
                (frame 20)

25. Draw a $30°-60°-90°$ triangle and show the correct values for the sides.
                (frame 21)

26. Using any convenient scale draw vectors representing the following:

   (a) A force of 50 lbs exerted in a direction of due east.

(b)   A velocity of 150 mph directed
in a compass direction of 270°.

(c)   A bicycle traveling north at a
velocity of 10 mph.
(frame 22)

27. Draw the vector arrow representing the resultant, and indicate its magnitude, for the following pairs of vectors.

(a)   α $\xrightarrow{\quad 60 \quad}$ $\xleftarrow{\quad 30 \quad}$ β $R =$

(b) $\xleftarrow{\quad 30 \quad}$ α $\xleftarrow{\quad 30 \quad}$ β $R =$

(c)   α $\xrightarrow{\quad 40 \quad}$ $\xleftarrow{\quad\quad 70 \quad\quad}$ β $R =$

(frame 23)

28. Given the two vectors shown at the right,
find their resultant by both the parallelogram
and the triangle methods.   (frame 24)

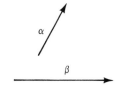

29. Given the vector shown at the right,
find its x and y (orthogonal)
components.   (frame 25)

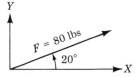

30. One degree is equal to _____ of the circumference of a circle.
(frame 26)

31. Draw a simple circular diagram showing an angular/circular measure of
one radian.
(frame 27)

32. Write the following trigonometric functions in radian measurement (in terms of π).

    (a) $\cot 45° =$ _____

    (b) $\sin 30° =$ _____

    (c) $\cos 60° =$ _____

(frame 27)

Answers to Self-Test

1. True
2. True
3. True
4. Triangle (1). Use the square corner of a sheet of paper to check this.
5. True
6. False. The division need not be performed; it can simply be expressed as an ordinary fraction.
7. True
8. $\dfrac{\text{opposite side}}{\text{hypotenuse}}$
9. sine, cosine, tangent, cosecant, secant, cotangent
10. sin, cos, tan, csc, sec, cot
11. $\sin A = \dfrac{a}{c}$, $\cos A = \dfrac{b}{c}$, $\tan A = \dfrac{a}{b}$, $\csc A = \dfrac{c}{a}$, $\sec A = \dfrac{c}{b}$, $\cot A = \dfrac{b}{a}$
12. sin, cos, tan
13. csc, sec, cot
14. True
15. True
16. (a) 0.42130;   (b) 0.80593;   (c) 2.14610;   (d) 0.09247
17. (a) tan or cot;   (b) cos;   (c) none — simply subtract angle B from 90°.
18. 27.6 ft (use cos function)
19. 40° 38′ (subtracting $\angle B$ from 90°)
20. 21.0 ft (use tan function)
21. $CB = AC \tan \angle CAB = 225 \tan 48° 20' = 225(1.1237) = 253$ ft
22. $CB = AC \tan 51° 16' = 1{,}585(1.2467) = 1{,}976$ ft
23. 40° = 0.64279
24. See frame 20. (Sides are 1; hypotenuse is $\sqrt{2}$.)
25. See frame 21. (Sides are 1, 2; hypotenuse is $\sqrt{3}$.)
26. (a)   (b) N (c)

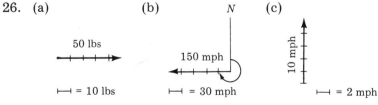

27. (a) $R = \xrightarrow{\ 30\ }$ (b) $\xleftarrow{\ 60\ } = R$ (c) $\xleftarrow{\ 30\ } = R$

28.

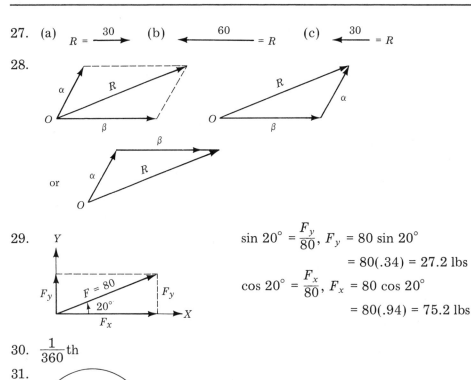

or

29.

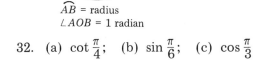

$\sin 20° = \dfrac{F_y}{80}$, $F_y = 80 \sin 20°$

$= 80(.34) = 27.2$ lbs

$\cos 20° = \dfrac{F_x}{80}$, $F_x = 80 \cos 20°$

$= 80(.94) = 75.2$ lbs

30. $\dfrac{1}{360}$th

31.

$\overgroup{AB}$ = radius
$\angle AOB$ = 1 radian

32. (a) $\cot \dfrac{\pi}{4}$;   (b) $\sin \dfrac{\pi}{6}$;   (c) $\cos \dfrac{\pi}{3}$

CHAPTER SIX

Trigonometric Analysis

In the last chapter we considered some of the general properties of plane triangles as well as the specific application of certain properties of right triangles that allow us to determine the lengths of their sides and the sizes of their angles. By now you should be generally familiar with the six trigonometric functions, what they mean and how we can use them to solve problems containing right triangles.

We also discussed two special triangles — the 30°–60°–90° and the 45°–45°–90° triangles — and learned an easy way to find the proportional lengths of their sides, hence the values of their trigonometric functions. The use of degrees for angular measure was reviewed (from geometry) and the concept of radian measurement introduced. Our treatment of vectors, though brief, should have served to acquaint you with a highly useful means for combining quantities having direction and magnitude, to find their resultant. And, conversely, of resolving a vector into its two orthogonal components, taken along any pair of selected axes.

Most of what we have covered thus far in our study of trigonometry relates to what is generally termed *numerical trigonometry*, that is, it is primarily concerned with finding number values — lengths of sides of triangles, sizes of angles, the use of the tables of natural trigonometric functions, the addition and subtraction of vectors, and so on.

However, you may recall that in the introduction to Chapter 3 we mentioned that the study of trigonometry is not limited to its application to triangles, nor to just right triangles. Not only are there many applications to oblique triangles (ones that don't contain a right angle), but there also are many purely mathematical (non-triangular) applications of the basic trigonometric concepts.

In this chapter, therefore, we are going to consider a number of new and interesting aspects and applications of trigonometry. Specifically, when you have finished this chapter you should be familiar with and able to use:

- the trigonometric functions of standard-position angles, including directed angles, the trigonometric (circular) functions of a general angle, the algebraic signs of the functions, and the line definition of the trigonometric functions;

- the relations between the various trigonometric functions or trigonometric identities;

- trigonometric analysis;

- trigonometric equations;

- graphical representation of the trigonometric functions, including periodicity of the functions, graphs of the functions by use of the unit circle, sine wave analysis, inverse functions, and reduction of trig functions to acute angle functions;

- the oblique triangles, including law of sines, law of cosines, trigonometric functions for half-angles and double angles, and formulas for the areas of oblique triangles.

TRIGONOMETRIC FUNCTIONS OF STANDARD-POSITION ANGLES

1. The notion of an angle, as presented in our study of geometry — and trigonometry so far — has been a rather intuitive concept. The kinds of angles we worked with in the last chapter are generally termed *reference angles* and all lie between $0°$ and $90°$. Now we will develop a precise definition of *angle*. We will be working with *standard-position* or *directed angles* that can be either positive or negative and of any size (such as $120°$, $460°$, or $-187°$).

Standard-position angles are angles that are generated on the coordinate system. Thus an angle is said to be *in standard position* when its vertex is at the origin and its *initial side* coincides with the positive X-axis, as shown at the right.

An *angle* is considered to be generated by a line (the *terminal side*) that revolves about the vertex (finally coinciding with the initial side). In the figure, therefore, OA is the initial side and OB_1, OB_2, and OB_3 represent successive positions of the terminal side.

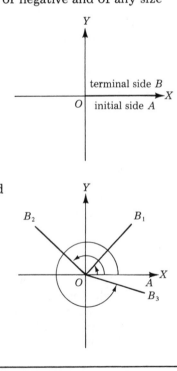

Angles generated by revolving the generating line (OB) counterclockwise are considered *positive*. Angles formed by revolving the generating line in a clockwise direction are considered *negative*. Thus, angle AOB_1 is positive, whereas angle AOB_2 is negative.

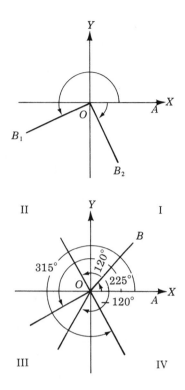

Any angle is said to be *in the first quadrant* or a *first quadrant angle* if, when in standard position, its terminal side falls in quadrant I. Thus, angle AOB is a first quadrant angle. In fact any positive angle lying between $0°$ and $90°$, or any negative angle lying between $270°$ and $360°$, is a first quadrant angle. Similarly, $120°$ is a second quadrant angle, $-120°$ is a third quadrant angle, $225°$ is a third quadrant angle, and $315°$ is a fourth quadrant angle.

Indicate which quadrant each of the following angles is in:

(a) $330°$ _____ quadrant

(b) $260°$ _____ quadrant

(c) $-45°$ _____ quadrant

(d) $110°$ _____ quadrant

(e) $-185°$ _____ quadrant

(f) $95°$ _____ quadrant

- - - - - - - - - - - - - - -

(a) fourth;   (b) third;   (c) fourth;   (d) second;   (e) second
(f) second

2. Keep in mind that there are 90° in each of the four quadrants. This fact gives us four basic angles that can be used as points of reference: 90°, 180°, 270°, and 360°, as shown in the figures at the right.

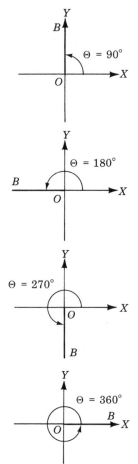

    Since the initial side is the same for every angle in standard position, we will not draw initial sides from now on. Also, for the present we will designate the angle of rotation by the Greek letter Θ (theta).

    You can see that a 360° angle involves one rotation through all four quadrants. Hence for a 360° angle (Θ = 360°), the terminal side is identical with the initial side. Two standard-position angles of the same size would, of course, have coincident terminal sides. These are, therefore, called *coterminal angles*. For example, 30° and −330°, 10° and 370° are pairs of coterminal angles. There are an unlimited number of angles that are coterminal with any given angle. The angles 0°, 90°, 180°, and 270°, and all angles coterminal with them are called *quadrantal angles*.

Quadrantal angles

Indicate whether the following angles are coterminal, quadrantal, or both, or neither.

(a)   150° and −210°   _____

(b)   180° and 0°   _____

(c)   −90° and −270°   _____

(d)   −100° and 260°   _____

(e)   −180° and 180°   _____

(f)   160° and 250°   _____

(g)   −45° and 315°   _____

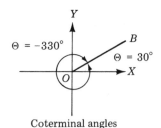

Coterminal angles

--- --- --- --- --- --- --- --- --- --- ---

(a) coterminal;   (b) quadrantal;   (c) quadrantal;   (d) coterminal; (e) coterminal and quadrantal;   (f) neither;   (g) coterminal

3. In Chapter 5 the six trigonometric functions were defined only for acute angles, that is, angles between $0°$ and $90°$. Now, however, to express the basic concepts in more general terms, we will formulate a new set of definitions of the functions that will apply to any angle whatever and which also will agree with the definitions already given for acute angles.

    So, let's select some angle, Θ, whose vertex is at O, the origin of our coordinate system, and whose initial side lies along the X-axis — that is, an angle in standard position. Now if we select a point, P, lying on the terminal side, we can define its coordinates as (x,y) and represent its distance from O by the letter r. Dropping a perpendicular from P to the X-axis gives us a triangle whose sides are the abscissa of the point P (x), its ordinate (y), and its distance from O (r).

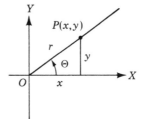

The six trigonometric functions are then defined in terms of the ordinate, abscissa, and distance of P from O as follows:

$$\sin \Theta = \frac{\text{ordinate}}{\text{distance}} = \frac{y}{r} \qquad\qquad \cot \Theta = \frac{\text{abscissa}}{\text{ordinate}} = \frac{x}{y}$$

$$\cos \Theta = \frac{\text{abscissa}}{\text{distance}} = \frac{x}{r} \qquad\qquad \sec \Theta = \frac{\text{distance}}{\text{abscissa}} = \frac{r}{x}$$

$$\tan \Theta = \frac{\text{ordinate}}{\text{abscissa}} = \frac{y}{x} \qquad\qquad \csc \Theta = \frac{\text{distance}}{\text{ordinate}} = \frac{r}{y}$$

Drawing our angle, Θ, in each of the other three quadrants we get these additional versions of the standard-position angle and their resulting triangles.

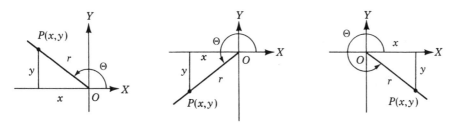

Notice that in each case x and y are the sides of the triangle and r is the length of the hypotenuse. And since the trig ratios of these $90°$ to $360°$ angles are identical to the trig ratios of their reference $(0° - 90°)$ angles, we can use our ordinary trig tables to determine their values. However, the values of the trig functions in the first quadrant and those in the other three quadrants may differ in one important respect.

Do you know what it is? _____

– – – – – – – – – – – – – – – –

Their signs may differ.

SIGNS OF THE TRIGONOMETRIC FUNCTIONS

4. As you learned in algebra, the abscissa is
    positive in quadrants I and IV and negative
    in quadrants II and III. Similarly the
    ordinate is positive in quadrants I and II
    and negative in quadrants III and IV.
    Applying this information to our six trig
    functions shown in frame 3, and keeping in
    mind that the hypotenuse, r, is always
    considered positive, we arrive at the follow-
    ing algebraic signs for the functions in the
    four quadrants.

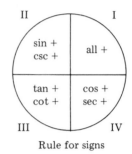

Rule for signs

> Quadrant I — all functions positive
> Quadrant II — sin and csc positive; others negative
> Quadrant III — tan and cot positive; others negative
> Quadrant IV — cos and sec positive; others negative

The sin and csc, cos and sec, tan and cot are bound to have the same
sign since they are reciprocals of one another. It will be easier to memorize, therefore, if you simply remember
that: all functions are positive in the first
quadrant, the sin is positive in the second
quadrant, the tan in the third, and the cos
in the fourth quadrant. And so are their
reciprocals; all others are negative.

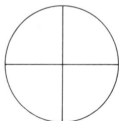

Just for practice write in the names of the
positive trig functions in the figure at the right.

- - - - - - - - - - - - - - - -

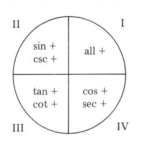

GRAPHS OF THE TRIGONOMETRIC FUNCTIONS

5. We are gradually working our way toward looking at some graphs of the
    trigonometric functions. And if you're wondering if it's really worth
    the trouble, be assured that it is, for one of the first useful things you

are going to learn from it is how the values in the trigonometric tables were derived.

To get into this we need first to examine what are referred to as *line representations* of the trig functions. This will show you how line lengths are used to represent the values of the various trig ratios. We start by drawing what is known as a "unit circle," that is, a circle with a radius of one (unity). Then we add Θ, which can be any given angle in standard position, and drop a perpendicular from P, the point where the terminal side of Θ cuts the circle, to the point M on the X-axis. This gives us the right triangle OMP, whose hypotenuse is one. Then,

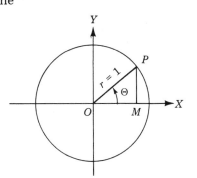

$$\sin \Theta = \frac{MP}{OP}$$

or, since OP = 1,

$$\sin \Theta = MP.$$

Thus the value of the sin is represented by the length of the line MP. It is apparent, therefore, that as the size of Θ increases from 0° to 90° the length of MP (and hence the value of the sin) *increases* from 0 to 1.

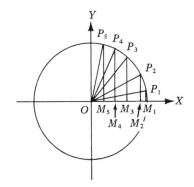

What do you think will happen to the values of sin Θ as Θ moves through the second quadrant? _____

_____

– – – – – – – – – – – – – – – – –

It will *decrease* from a value of 1 (at 90°) to 0 (at 180°).

6. Similarly, as the terminal side moves into the third quadrant (Θ increase from 180° toward 270°), the *absolute* value of sin Θ will again increase from 0 at 180° to 1 at 270°. And, as you would suspect, it will decrease in the fourth quadrant from a value of 1 at 270° to 0 at 360°, thus completing the full 360° cycle.

Now let's look at what happens to the cosine. Here is our unit circle and right triangle again in the figure at the right. What would you say is the value of cos Θ when Θ = 0°?

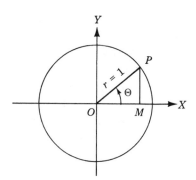

_____

_ _ _ _ _ _ _ _ _ _ _ _ _ _

Hopefully you recognized that $\cos \Theta = \dfrac{OM}{OP}$ but that since $OP = 1$, $\cos \Theta = OM$, hence $\cos 0° = OM = 1$ (radius), and that as Θ approaches 90° the value of cos Θ approaches 0. Thus the cosine varies in value from 1 to 0, just oppositely to the sine.

7. Although we will not do so here, it is possible to show how all the other functions change in value as the terminal side of the standard position angle rotates through the four quadrants (as Θ increases from 0° to 360°). However, the figure at the right shows the lengths that can be used to represent the numerical values of the six trig functions. Summarizing them we get:

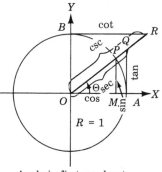

Angle in first quadrant

$$\sin \Theta = \frac{MP}{OP} = MP \qquad\qquad \cot \Theta = \frac{OM}{MP} = \frac{BR}{OB} = BR$$

$$\cos \Theta = \frac{OM}{OP} = OM \qquad\qquad \sec \Theta = \frac{OP}{OM} = \frac{OQ}{OA} = OQ$$

$$\tan \Theta = \frac{MP}{OM} = \frac{AQ}{OA} = AQ \qquad\qquad \csc \Theta = \frac{OP}{MP} = \frac{OR}{OB} = OR$$

Hence as P moves counterclockwise about the unit circle, starting at A, Θ ($\angle XOP$) varies continuously from 0° to 360° and the function values vary as shown following. (Remember, we read the symbol ∞ as "infinity" or "without limits.")

As Θ increases from	0° to 90°	90° to 180°	180° to 270°	270° to 360°
sin Θ	increases from 0 to 1	decreases from 1 to 0	decreases from 0 to −1	increases from −1 to 0
cos Θ	decreases from 1 to 0	decreases from 0 to −1	increases from −1 to 0	increases from 0 to −1
tan Θ	increases from 0 to + ∞	increases from −∞ to 0	increases from 0 to + ∞	increases from −∞ to 0
cot Θ	decreases from + ∞ to 0	decreases from 0 to − ∞	decreases from + ∞ to 0	decreases from 0 to − ∞
sec Θ	increases from 1 to + ∞	increases from − ∞ to −1	decreases from −1 to − ∞	decreases from + ∞ to 1
csc Θ	decreases from + ∞ to 1	increases from 1 to + ∞	increases from − ∞ to −1	decreases from −1 to − ∞

Looking at the above table, see if you can fill in the missing information below.

(a)   The sine and cosine can take on values only between _____ and _____ inclusive.

(b)   The tangent and cotangent can take on _____ values.

(c)   The secant and cosecant can take on any values whatever except those lying between _____ and _____ .

_ _ _ _ _ _ _ _ _ _ _ _ _ _ _

(a)   −1 and +1;   (b) any;   (c) −1 and +1

8.   Now if the sin values for angles lying between 0° and 90° do in fact increase from 0 to 1, then this is what our tables of natural trig functions should tell us.

Let's see if they do. Turn to the trig tables in the Appendix. What values do you find for:

(a)   the sin of 0°? _____   (d)   the sin of 60°? _____

(b)   the sin of 30°? _____   (e)   the sin of 90°? _____

(c)   the sin of 45°? _____

_ _ _ _ _ _ _ _ _ _ _ _ _

(a) .00000;   (b) .50000;   (c) .70711;   (d) .86603;   (e) 1.0000

9. The results in frame 8 seem to confirm our findings in frame 7, don't
    they? Let's see if the tables confirm our predictions for the cosine.
    Turn to the trig tables in the Appendix again and check the values
    shown there for the following.

(a) $\cos 0° =$ _____   (d) $\cos 60° =$ _____

(b) $\cos 30° =$ _____   (e) $\cos 75° =$ _____

(c) $\cos 45° =$ _____   (f) $\cos 90° =$ _____

_ _ _ _ _ _ _ _ _ _ _ _ _

(a) 1.0000;   (b) .86603;   (c) .70711;   (d) .50000;   (e) .25882;
(f) .00000

10. Again we seem to have confirmation of the function values. Now just
     so you'll have some assurance about what happens to the tangent, check
     the following values in the trig tables.

(a) $\tan 0° =$ _____   (d) $\tan 89° 59' =$ _____

(b) $\tan 45° =$ _____   (e) $\tan 90° =$ _____

(c) $\tan 89° =$ _____

_ _ _ _ _ _ _ _ _ _ _ _ _

(a) .00000;   (b) 1.0000;   (c) 57.290;   (d) 3437.7;   (e) ∞

PERIODICITY AND THE SINE WAVE

11. What we have just learned is going to help us to graph some of the
     trigonometric functions. For example, we are going to see how to go
     about graphing the sine ratio, generally known as a *sine wave*. The
     graph of the sine ratio from 0° to 360° is the *basic cycle* of the
     sine wave.

     Sine waves are an important graphical model in basic science and
     technology. For example, the graphs of such diverse phenomena as
     alternating currents, radio and television waves, and the vibration of a

spring are related to the sine wave in one way or another. When the graph of a phenomenon is a sine wave, we say that the phenomenon is *sinusoidal* (pronouned sigh-nah-soy-dal).

In order to find the graph of a trigonometric function such as the sine, we assume values for the angle. The circular measures (that is, measure of the angle in radians) of these angles are then taken as abscissas, and the corresponding values of the function (found in the trig tables) are taken as the ordinates of points on the graph.

Example: Plot the graph of sin x.

Solution: Let $y = \sin x$. It is easier to use degree measure of an angle when looking up its function but necessary to use circular measure when plotting. So the first thing we need is a table of values that includes all these necessary elements.

x		y	x		y
0°	0	0	210°	$\dfrac{7\pi}{6}$	−.50
30°	$\dfrac{\pi}{6}$	.50	240°	$\dfrac{4\pi}{3}$	−.86
60°	$\dfrac{\pi}{3}$	.86	270°	$\dfrac{3\pi}{2}$	−1.00
90°	$\dfrac{\pi}{2}$	1.00	300°	$\dfrac{5\pi}{3}$	−.86
120°	$\dfrac{2\pi}{3}$	.86	330°	$\dfrac{11\pi}{6}$	−.50
150°	$\dfrac{5\pi}{6}$	.50	360°	2π	0
180°	π	0			

In plotting the points we must use the circular measure of the angles for abscissas since we are dealing with circular functions. The most convenient way of doing this is to lay off distances $\pi = 3.1416$ to the right of the origin and then divide each of these into six equal parts. The ordinate values will, of course, range between 0 and 1 (+ and −).

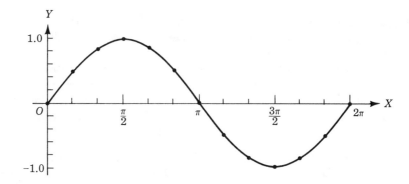

The final step is to draw a smooth curve through the plotted points. The resultant curve is a graph of sin x for values of x between 0 and 2π. This is the sine wave or *sine curve* we referred to at the beginning of this frame. Had we computed points for negative values of x we would have gotten the continuation of this curve to the left of the Y-axis.

Prepare a table of values for the cosine of x, similar to the one we prepared for $y = \sin x$, and plot the cosine curve. Use a separate sheet of paper for your work.

– – – – – – – – – – – – – – – –

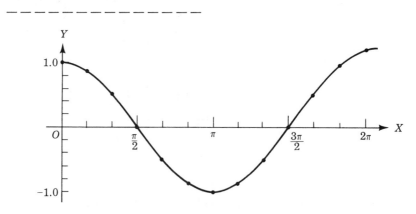

12. You will note from the graphs of the sine and cosine curves that the points at which the two functions reach their maximum is 90°, (or $\frac{\pi}{2}$ radians) apart. One way of expressing this situation is to say that the two curves are *90° out of phase.* In the study of alternating current electricity where current flow is essentially sinusoidal in nature, and thus can be represented by curves such as the ones above, the matter of the phase relationship between voltage and current is very important. Therefore if you get into the field of electricity or various other aspects of physics, you will find these concepts very useful.

Notice also from the graph of sin x in frame 11 that as the angle increased from 0 to 2π radians, the sine first increased from 0 to 1, then decreased from 1 to -1, and finally increased from -1 to 0. Had we continued to plot the curve as the angle increased from 2π radians to 4π radians, you would have seen that the sine went through the same series of changes, and so on. Thus the sine goes through all its changes while the angle changes 2π radians in value. This fact is expressed by saying that the *period of the sine is 2π*.

Similarly the cosine, secant, or cosecant passes through all its changes while the angle changes 2π radians, as shown by the table in frame 7 and also the figure at the right.

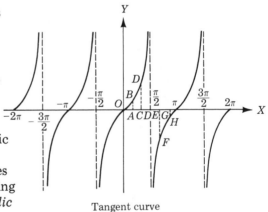

Secant curve

The tangent or cotangent, however, passes through all its changes while the angle changes by π radians, also shown in the table in frame 7 and the figure at the right.

Thus the *period* of the sine, cosine, secant, or cosecant is *2π radians*, while the *period* of the tangent or cotangent is π radians. As each trigonometric function again and again passes through the same series of values (the angle increasing or decreasing uniformly), we call them *periodic functions*.

Tangent curve

Check yourself on a few key points by answering the following questions about what we have just covered.

(a)   The period of the sine function is _____ radians.

(b)   The term "sinusoidal" means _____ .

(c)   The angular difference between the points at which the sine wave and cosine wave reach their maximum point is _____ radians.

(d)   The period of the tangent is _____ radians.

(e)   The term applied to a trigonometric function that passes repeatedly through the same series of values as the angle increases or decreases uniformly is _____ .

_ _ _ _ _ _ _ _ _ _ _ _ _ _ _

(a) 2π;   (b) like a sine wave;   (c) $\frac{\pi}{2}$;   (d) π;   (e) periodic function

INVERSE FUNCTIONS

13.   It is time we said a word about *inverse trigonometric functions*. To do so, let's go back a little bit just to make sure you are clear on where we are starting from.

The value of a trigonometric function of an angle is a function of the value of the angle. Conversely, the value of the angle is a function of the value of the function. Thus, if an angle is given, the sine of the angle can be found. Or, if the sine is given, the angle can be expressed.

It is often convenient to represent an angle by the value of one of its functions. Thus, instead of saying that an angle is $30°$, we can say what amounts to the same thing: that *it is the least positive angle whose sine is $\frac{1}{2}$*. We then consider the angle as a function of its sine, and the angle is said to be an inverse trigonometric function, and is denoted as

$$\text{arc sin } \frac{1}{2}, \text{ or } \sin^{-1} \frac{1}{2}.$$

Either of the above expressions should be read as "the angle whose sine is $\frac{1}{2}$."

In the second method of expressing this relationship it is important that you understand that the -1 is not an algebraic exponent, but is merely part of the mathematical symbol denoting an inverse trigonometric function. For example, $\tan^{-1} a$ is not the same thing at all as $(\tan a)^{-1}$, which means the reciprocal of the tangent. That is,

$$(\tan a)^{-1} = \frac{1}{\tan a}.$$

But it is because of the possibility of this confusion occurring that the expression *arc sin* is used more frequently than sin^{-1}.

Thus, the inverse of $\sin 30° = \frac{1}{2}$ would be arc sin $\frac{1}{2} = 30°$ or $\frac{\pi}{6}$.

How would you write the inverse of: $\cos 60° = \frac{1}{2}$? _____

How would you read your answer aloud? _____

— — — — — — — — — — — — — — —

arc cos $\frac{1}{2} = \frac{\pi}{3}$;   The angle whose cosine is $\frac{1}{2}$.

14.   There is another important difference between a trigonometric function and its inverse, other than the way they are written. For example,
the trigonometric function $x = \sin y$ defines a unique value of x for
each given angle y. Thus, if $y = 30°$, $x = \frac{1}{2}$. But in the inverse when
the value of x is given, the equation may have no solution or many
solutions. If for instance $x = 2$, then there is no solution since the sine
of an angle never exceeds 1. On the other hand, if $x = \frac{1}{2}$, then there
are many solutions, such as $y = 30°$, $150°$, $390°$, and so on. We can
say, therefore, that:

> *The trigonometric functions are single-valued, and the inverse
> trigonometric functions are many-valued.*

Incidentally, referring to our example
$x = \sin y$, if we wish to express y as a
function of x, we write: $y = $ arc sin x.

The inverse trigonometric functions
can, of course, be graphed, just as the
trigonometric functions can. Thus, the
graph of $y = $ arc sin x is the graph of
$x = \sin y$ and differs from the graph of
$y = \sin x$ (see frame 11) only in that
the roles of x and y are interchanged.
That is, the graph of $y = $ arc sin x is a
sine curve drawn on the Y-axis instead
of on the X-axis, as shown at the right.

The smallest value numerically of
an inverse trigonometric function is
termed its *principal value*. For example, if

   tan $x = 1$,

then

   $x = \frac{\pi}{4} = 45°$

is the *principal value* of x. And the
general value of x is

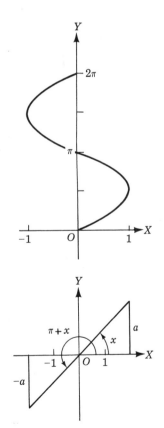

$$x = \text{arc tan } 1 = n\pi + \frac{\pi}{4},$$

where n represents zero or any positive or negative value.

What is the principal value of $\cos x = \frac{1}{2}$? $x =$ _____

- - - - - - - - - - - - - -

$60°$ or $\frac{\pi}{3}$

RELATIONS BETWEEN THE TRIGONOMETRIC FUNCTIONS

15. We will not work with some of these concepts (such as inverse functions)
to any great degree. It is enough for now that you have been introduced
to them and hence will recognize and be prepared to use them when
they occur later on in your study.

  Earlier we stated that there are many purely mathématical applications of the basic trigonometric concepts. We are going to consider
some of these now. Once again, to save space, we are not going into the
derivation of proof of the formulas stated (they are available in most
standard textbooks). It is important, however, that you learn these
formulas! If not now, then certainly before you begin the study of
calculus. You can, of course, look them up when you need them, but
having them at your mental fingertips will save you an endless amount
of time.

  In Chapter 5 you learned about the three primary trig functions
(sin, cos, and tan) and their reciporcals (csc, sec, and cot). Another
way of stating these reciprocal relationships is as follows.

$$\sin x = \frac{1}{\csc x} \qquad\qquad \csc x = \frac{1}{\sin x}$$

$$\cos x = \frac{1}{\sec x} \qquad\qquad \sec x = \frac{1}{\cos x}$$

$$\tan x = \frac{1}{\cot x} \qquad\qquad \cot x = \frac{1}{\tan x}$$

Now, making use of the unit circle
shown at the right (and which you
first saw in frame 7 of this chapter),
we can derive five more very
important relations between the
functions. These are,

$$\tan x = \frac{\sin x}{\cos x}$$

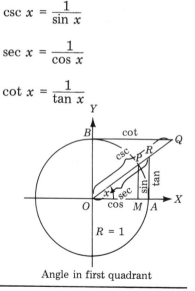

Angle in first quadrant

$$\cot x = \frac{\cos x}{\sin x}$$

$$\sin^2 x + \cos^2 x = 1$$

$$1 + \tan^2 x = \sec^2 x$$

$$1 + \cot^2 x = \csc^2 x$$

While in the figure shown the angle x has been taken in the first quadrant, the results hold true for any angle whatever. Based on the above formulas, and grouping them according to the specific function involved, we get the following formulas wherein each of the functions is expressed explicitly in terms of other functions.

(1) $\sin x = \dfrac{1}{\csc x}$

(2) $\sin x = \pm\sqrt{1 - \cos^2 x}$

(3) $\cos x = \dfrac{1}{\sec x}$

(4) $\cos x = \pm\sqrt{1 - \sin^2 x}$

(5) $\tan x = \dfrac{1}{\cot x}$

(6) $\tan x = \pm\sqrt{\sec^2 x - 1}$

(7) $\tan x = \dfrac{\sin x}{\cos x} = \dfrac{\sin x}{\pm\sqrt{1 - \sin^2 x}} = \dfrac{\pm\sqrt{1 - \cos^2 x}}{\cos x}$

(8) $\csc x = \dfrac{1}{\sin x}$

(9) $\csc x = \pm\sqrt{1 + \cot^2 x}$

(10) $\sec x = \dfrac{1}{\cos x}$

(11) $\sec x = \pm\sqrt{1 + \tan^2 x}$

(12) $\cot x = \dfrac{1}{\tan x}$

(13) $\cot x = \pm\sqrt{\csc^2 x - 1}$

(14) $\cot x = \dfrac{\cos x}{\sin x} = \dfrac{\cos x}{\pm\sqrt{1 - \cos^2 x}} = \dfrac{\pm\sqrt{1 - \sin^2 x}}{\sin x}$

By means of the above formulas it is possible to find any function in terms of the other five functions.

Example: Find $\sin x$ in terms of each of the other five functions of x.

(a) $\sin x = \dfrac{1}{\csc x}$ from (1)

(b) $\sin x = \pm\sqrt{1 - \cos^2 x}$ from (2)

(c) $\sin x = \dfrac{1}{\pm\sqrt{1 + \cot^2 x}}$ substitute (9) in (a)

(d) $\sin x = \pm\sqrt{1 - \dfrac{1}{\sec^2 x}} = \dfrac{\pm\sqrt{\sec^2 x - 1}}{\sec x}$ substitute (3) in (b)

(e) $\sin x = \dfrac{1}{\pm\sqrt{1 + \dfrac{1}{\tan^2 x}}} = \dfrac{\tan x}{\pm\sqrt{\tan^2 x + 1}}$ substitute (12) in (c)

Now it's your turn. Don't be afraid of it; the practice will give you confidence in your ability to work with these functions. Find $\cos x$ in terms of each of the other five functions.

- - - - - - - - - - - - -

(a) $\cos x = \dfrac{1}{\sec x}$ from (3)

(b) $\cos x = \pm\sqrt{1 - \sin^2 x}$ from (4)

(c) $\cos x = \dfrac{1}{\pm\sqrt{1 + \tan^2 x}}$ substitute (11) in (a)

(d) $\cos x = \pm\sqrt{1 - \dfrac{1}{\csc^2 x}} = \dfrac{\pm\sqrt{\csc^2 x - 1}}{\csc x}$ substitute (1) in (b)

(e) $\cos x = \dfrac{1}{\pm\sqrt{1 + \dfrac{1}{\cot^2 x}}} = \dfrac{\cot x}{\pm\sqrt{\cot^2 x + 1}}$ substitute (5) in (c)

TRIGONOMETRIC ANALYSIS

16.   We also have formulas (although we will not attempt to prove them here) that enable us to find the trigonometric functions of two angles. The principal formulas are as follows.

Addition Formulas

(15) $\sin(x + y) = \sin x \cos y + \cos x \sin y$

(16) $\cos(x + y) = \cos x \cos y - \sin x \sin y$

(17) $\tan(x + y) = \dfrac{\tan x + \tan y}{1 - \tan x \tan y}$

Subtraction Formulas

(18) $\sin(x - y) = \sin x \cos y - \cos x \sin y$

(19) $\cos(x - y) = \cos x \cos y + \sin x \sin y$

(20) $\tan(x - y) = \dfrac{\tan x - \tan y}{1 + \tan x \tan y}$

Let's see how we can apply these formulas.

Example 1:   Find $\sin 75°$ using the functions of $45°$ and $30°$.
Since $75° = 45° + 30°$ we get, from (15),
$\sin 75° = \sin(45° + 30°) = \sin 45° \cos 30° + \cos 45° \sin 30°$

$$= \frac{1}{\sqrt{2}} \cdot \frac{\sqrt{3}}{2} + \frac{1}{\sqrt{2}} \cdot \frac{1}{2}$$

$$= \frac{\sqrt{3} + 1}{2\sqrt{2}}$$

Example 2:   Find $\cos 15°$ using the functions of $45°$ and $30°$.
Since $15° = 45° - 30°$, we get, from (19),
$\cos 15° = \cos(45° - 30°) = \cos 45° \cos 30° + \sin 45° \sin 30°$

$$= \frac{1}{\sqrt{2}} \cdot \frac{\sqrt{3}}{2} + \frac{1}{\sqrt{2}} \cdot \frac{1}{2}$$

$$= \frac{\sqrt{3} + 1}{2\sqrt{2}}$$

Example 3:   Find $\tan 15°$ using the functions of $60°$ and $45°$.
Since $15° = 60° - 45°$, we get, from (20)

$$\tan 15° = \tan(60° - 45°) = \frac{\tan 60° - \tan 45°}{1 + \tan 60° \tan 45°}$$

$$= \frac{\sqrt{3} - 1}{1 + \sqrt{3}} = 2 - \sqrt{3}$$

Apply the formulas similarly in working out the following problems.
(Try to work out your $30°$–$45°$–$60°$ function values as you learned to
do in the last chapter. If you get stuck, refer to frame 21, Chapter 5.)

(a)   Find $\cos 15°$, taking $15° = 60° - 45°$. _____

_____

(b)   Show that $\sin 15° = \dfrac{\sqrt{3} - 1}{2\sqrt{2}}$, using the functions of $45°$ and $30°$.

_____

(c)   Verify: $\sin(45° - x) = \dfrac{\cos x - \sin x}{2}$ (Let $x = 45°$, $y = x$.)

_____

(d)   Find $\tan 15°$, taking $15° = 45° - 30°$. _____

_____

(e)   Find $\tan 75°$ from the functions of $45°$ and $30°$. _____

_____

- - - - - - - - - - - - - - -

(a) $\cos 15° = \cos(60° - 45°) = \cos 60° \cos 45° + \sin 60° \sin 45°$

$$= \frac{1}{2} \cdot \frac{1}{\sqrt{2}} + \frac{\sqrt{3}}{2} \cdot \frac{1}{\sqrt{2}}$$

$$= \frac{1}{2\sqrt{2}} + \frac{\sqrt{3}}{2\sqrt{2}} = \frac{1 + \sqrt{3}}{2\sqrt{2}}$$

(b) $\sin 15° = \sin(45° - 30°) = \sin 45° \cos 30° - \cos 45° \sin 30°$

$$= \frac{1}{\sqrt{2}} \cdot \frac{\sqrt{3}}{2} - \frac{1}{\sqrt{2}} \cdot \frac{1}{2}$$

$$= \frac{\sqrt{3}}{2\sqrt{2}} - \frac{1}{2\sqrt{2}} = \frac{\sqrt{3} - 1}{2\sqrt{2}}$$

(c) $\sin (45° - x) = \sin 45° \cos x - \cos 45° \sin x$

$$= \frac{1}{\sqrt{2}} \cdot \cos x - \frac{1}{\sqrt{2}} \cdot \sin x$$

$$= \frac{\cos x - \sin x}{\sqrt{2}}$$

(d) $\tan 15° = \tan(45° - 30°) = \dfrac{\tan 45° - \tan 30°}{1 + \tan 45° \tan 30°}$

$$= \frac{1 - \dfrac{1}{\sqrt{3}}}{1 + 1(\dfrac{1}{\sqrt{3}})} = \frac{\sqrt{3} - 1}{\sqrt{3} + 1}$$

$$= \frac{(\sqrt{3} - 1)}{(\sqrt{3} + 1)} \cdot \frac{(\sqrt{3} - 1)}{(\sqrt{3} - 1)} = \frac{3 - 2\sqrt{3} + 1}{3 - 1} *$$

$$= \frac{4 - 2\sqrt{3}}{2} = 2 - \sqrt{3}$$

*Multiplying both numerator and denominator by $(\sqrt{3} - 1)$ in order to rationalize the denominator.

(e) $\tan 75° = \tan(45° + 30°) = \dfrac{\tan 45° + \tan 30°}{1 - \tan 45° \tan 30°}$

$$= \frac{1 + \dfrac{1}{\sqrt{3}}}{1 - 1(\dfrac{1}{\sqrt{3}})} = \frac{\sqrt{3} + 1}{\sqrt{3} - 1} = 2 + \sqrt{3}$$

17. Having looked at the addition and subtraction formulas for functions of two angles, let's go on now to the formulas for double angles and half angles.

Double-Angle Formulas

(21) $\sin 2x = 2 \sin x \cos x$

(22) $\cos 2x = \cos^2 x - \sin^2 x = 1 - 2 \sin^2 x = 2 \cos^2 x - 1$

(23) $\tan 2x = \dfrac{2 \tan x}{1 - \tan^2 x}$

Half-Angle Formulas

(24) $\sin\frac{1}{2}x = \pm \sqrt{\dfrac{1 - \cos x}{2}}$

(25) $\cos\frac{1}{2}x = \pm \sqrt{\dfrac{1 + \cos x}{2}}$

(26) $\tan\frac{1}{2}x = \pm \sqrt{\dfrac{1 - \cos x}{1 + \cos x}} = \dfrac{\sin x}{1 + \cos x} = \dfrac{1 - \cos x}{\sin x}$ (from frame 15)

(*Note:* Once again, if you are interested in seeing the proofs of any of these formulas, consult any standard textbook.)

Example 1: Given $\sin x = \dfrac{2}{\sqrt{5}}$, x lying in the second quadrant, find $\sin 2x$, $\cos 2x$, $\tan 2x$.

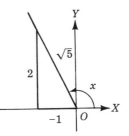

Solution: Since $\sin x = \dfrac{2}{\sqrt{5}}$ and x lies in the second quadrant, we get, using the figure at the right, the following values for the sin, cos, and tan:

$$\sin x = \frac{2}{\sqrt{5}}, \quad \cos x = -\frac{1}{\sqrt{5}}, \quad \tan x = -2.$$

(cos and tan must be negative in the second quadrant.) Substituting the sin and cos values in (21) we get,

$$\sin 2x = 2 \sin x \cos x = 2 \cdot \frac{2}{\sqrt{5}} \left(-\frac{1}{\sqrt{5}}\right) = -\frac{4}{5}$$

With the above as a guide, suppose *you* try finding $\cos 2x$.

‒ ‒ ‒ ‒ ‒ ‒ ‒ ‒ ‒ ‒ ‒ ‒ ‒ ‒ ‒ ‒

Using formula (22),

$$\cos 2x = \cos^2 x - \sin^2 x = \left(-\frac{1}{\sqrt{5}}\right)^2 - \left(\frac{2}{\sqrt{5}}\right)^2 = \frac{1}{5} - \frac{4}{5} = -\frac{3}{5}$$

18. We still haven't found the tangent in the last problem, so let's do so now — together. Stating our formula (23),

$$\tan 2x = \frac{2 \tan x}{1 - \tan^2 x}$$

and substituting the value of tan x, (-2),

$$= \frac{2 \cdot -2}{1 - (-2)^2} = -\frac{4}{3}$$

Now it's time to apply the half-angle formulas. Let's start with the sine.

Example: Given $\cos 45° = \frac{1}{\sqrt{2}}$, find $\sin 22\frac{1}{2}°$.

Solution: From (24), $\sin \frac{x}{2} = \pm \sqrt{\frac{1 - \cos x}{2}}$

If we let $x = 45°$, then $\frac{x}{2} = 22\frac{1}{2}°$, and we get

$$\sin 22\frac{1}{2}° = \sqrt{\frac{1 - \frac{1}{\sqrt{2}}}{2}} = \frac{1}{2}\sqrt{2 - \sqrt{2}}$$

Use formulas (25) and (26) to find the cosine and tangent of $22\frac{1}{2}°$.

- - - - - - - - - - - - - - -

from (25)

$$\cos \frac{1}{2}x = \pm \sqrt{\frac{1 + \cos x}{2}} = \sqrt{\frac{1 + \cos 45°}{2}}$$

$$= \sqrt{\frac{1 + \frac{1}{\sqrt{2}}}{2}} = \sqrt{\frac{\sqrt{2} + 1}{2\sqrt{2}}}$$

or rationalizing the denominator,

$$= \sqrt{\frac{\sqrt{2} + 1}{2\sqrt{2}} \cdot \frac{\sqrt{2}}{\sqrt{2}}} = \frac{1}{2}\sqrt{2 + \sqrt{2}}$$

from (26)

$$\tan \tfrac{1}{2}x = \frac{1 - \cos x}{\sin x} = \frac{1 - \dfrac{1}{\sqrt{2}}}{\dfrac{1}{\sqrt{2}}} = \sqrt{2} - 1$$

TRIGONOMETRIC EQUATIONS

19. We are not going into the subject of trigonometric equations in any
great depth, so you need not be alarmed. Nevertheless, you should not
leave any introduction to trigonometry without being aware that there
are such things as trigonometric equations and what their solution
provides.

Trigonometric equations are simply equations involving trigonometric functions of unknown angles. Basically, they are of two types:
identical and conditional.

Identical equations, or *identities*, are so termed if they are satisfied
by all values of the unknown angles for which the functions are defined.

Conditional equations, or simply *equations*, are trigonometric
equations that are satisfied only by particular values of the unknown
angles.

A *solution* of a trigonometric equation, such as $\sin x = 0$, is a value
of the angle x that satisfies the equation. In this respect they are
similar to the linear and quadratic equations you studied in algebra.
Only now we are seeking an *angular value* as a solution rather than a
numerical value.

Incidentally, the solutions of $\sin x = 0$ are $x = 0$ and $x = \pi$. You
know this already from what you have learned about the value of the
sine function, namely, that it is zero only when the angle is $0°$ or $180°$
(π). (*Note:* We will confine our discussion to angles between $0°$ and

$360°$, that is, 0 and 2π.) When the angle is $\frac{\pi}{2}$ ($90°$), its sine is 1; when

the angle is $\frac{3\pi}{2}$ ($270°$) its value is -1. (Refer to frame 7 if you need to

review the function values.)

Like some of the situations you encountered in algebra (such as
factoring quadratic expressions, solving word problems, etc.) there are
various approaches — but no set procedure — for solving trigonometric
equations. Here are a few suggested approaches:

(1) The equation may be factorable. Thus, given the equation
$\sin x - 2 \sin x \cos x = 0$, by factoring, we get $\sin x(1 - 2 \cos x) = 0$, then setting each factor equal to zero, $\sin x = 0$, hence
$x = 0, \pi$; and $1 - 2 \cos x = 0$ or $\cos x = \frac{1}{2}$, hence $x = \frac{\pi}{3}, \frac{5\pi}{3}$.

(2)   The various functions occurring in the equation may be expressed in terms of a single function. Thus, in the equation $2 \tan^2 x + \sec^2 x = 2$, replacing $\sec^2 x$ by $1 + \tan^2 x$ (from frame 15), we have $2 \tan^2 x + (1 + \tan^2 x) = 2$, or $3 \tan^2 x = 1$, and $\tan x = \pm\dfrac{1}{\sqrt{3}}$. From $\tan x = \dfrac{1}{\sqrt{3}}$, $x = \dfrac{\pi}{6}$ and $\dfrac{7\pi}{6}$. From $\tan x = -\dfrac{1}{\sqrt{3}}$, $x = \dfrac{5\pi}{6}$ and $\dfrac{11\pi}{6}$.

(3)   Sometimes it's possible to simply take the square root of both members of the equation. For example, in the equation $\sin^2 x = 1$, taking the square root we get $\sin x = \pm 1$, hence $x = \dfrac{\pi}{2}, \dfrac{3\pi}{2}$.

Try using whichever of the above approaches seems to apply best in solving the following problems. (*Note:* You may find it helpful to draw a diagram such as that shown in frame 17 as an aid to visualizing your angle solution values; also the table in frame 11 will assist you in converting from degrees to radian measure.) Show only solution values $\leqslant \dfrac{\pi}{2}$.

(a) $\tan^2 x = 1$

(b) $\cos^2 x = \dfrac{1}{4}$

(c) $2 \sin^2 x + 3 \cos x = 0$

(d) $2 \sin^2 x + \sqrt{3} \cos x + 1 = 0$

— — — — — — — — — — — — — —

(a)   Taking the square root of both members, $\tan x = \pm 1$; we are looking for the angle (x) whose tangent is 1. And since we only want first quadrant values (i.e., solution values equal to or less than $\dfrac{\pi}{2}$), we ignore the minus sign. Even without the minus sign we could still have two angle values for a function value of 1 because the tangent is positive in both the first and third quadrants. However, we're only interested in the reference angle (less than $90°$), and our table in frame 21 of Chapter 5 (or your recollection) will tell you this is an angle of $45°$, or $\frac{\pi}{4}$. The answer, then, is $x = \frac{\pi}{4}$.

(b)   Again taking the square root of both sides we get $\cos x = \frac{1}{2}$ (which we could write as $x = \text{arc } \cos \frac{1}{2}$, that is, x equals the angle whose cosine is $\frac{1}{2}$). And the angle whose cosine is $\frac{1}{2}$ is, of course, $60°$. Therefore, $x = \frac{\pi}{3}$.

(c)   The $\sin^2 x$ in this equation is a clue that we might use the relationship $\sin^2 x + \cos^2 x = 1$, or, rearranging terms, $\sin^2 x = 1 - \cos^2 x$. This would give us an equation in terms of just one function, namely, the cos. Thus, substituting, $2(1 - \cos^2 x) + 3 \cos x = 0$, or $2 - 2 \cos^2 x + 3 \cos x = 0$, from which $2 \cos^2 x - 3 \cos x - 2 = 0$, and factoring, $(2 \cos x + 1)(\cos x - 2) = 0$, or $2 \cos x = -1$ and $\cos x = 2$. (This result can't be used since no cosine is greater than 1.) Thus, $\cos x = -\frac{1}{2}$, and $x = \frac{\pi}{3}$ is the value of the reference angle.

The minus sign tells us that the terminal side of the standard angle actually would lie in quadrants II and III, since the cosine is negative in those two quadrants.

(d)   Here again it looks as though we should try substitution. Since $\sin^2 x = 1 - \cos^2 x$ we get $2 - 2 \cos^2 x + \sqrt{3} \cos x + 1 = 0$, or $2 \cos^2 x - \sqrt{3} \cos x - 3 = 0$, and since this is a quadratic in $\cos x$, factoring we get $(2 \cos x + \sqrt{3})(\cos x - \sqrt{3})$ or $\cos x = -\frac{\sqrt{3}}{2}$, from which $x = \frac{\pi}{6}$. ($\cos x = \sqrt{3}$ can't be used since cosine value is never greater than 1.)

The minus sign tells us that the terminal side actually would lie in quadrants II and III since the cosine is negative in those two quadrants.

SOLUTION OF OBLIQUE TRIANGLES

20.   As you are well aware by now, one of the principal uses of trigonometry is in the solution of triangles. That is, given three elements of a triangle (sides and angles), at least one of which is a side, the other elements may be found. In Chapter 5 we developed some unique relationships between the sides and angles of right triangles that enabled us to solve for the missing parts fairly readily. Now, however, we are going to

concern ourselves, not with the right triangle, but with the oblique triangle. In doing so we will make use of some of the concepts and trigonometric functions that evolved from our work with right triangles.

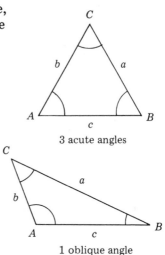

3 acute angles

1 oblique angle

An *oblique* triangle simply is one that does not contain a right angle. Such a triangle contains either three acute angles or two acute angles and one obtuse angle (greater than 90° but less than 180°).

As you learned from our study of plane geometry in Chapters 1 through 4, when three parts (not all angles) are known, the triangle is uniquely determined. The four cases of oblique triangles are:

Case 1 — given one side and two angles.
Case 2 — given two sides and the angle opposite one of them.
Case 3 — given two sides and the included angle.
Case 4 — given the three sides.

In other words, given the parts indicated in each of the cases above, we could construct the triangle by geometric methods. The parts not given could then be found by direct measurement with a scale of some kind (such as a ruler) and a protractor (to measure the angles). But the results would be rather rough. You will learn in this section, however, how to find them with great accuracy by trigonometric methods.

Before we go on let's make sure you are clear about what an oblique triangle is. An oblique triangle:

(a)   may contain a right angle. (True, False)

(b)   may contain an obtuse angle. (True, False)

(c)   must contain all acute angles. (True, False)

(d)   must either contain an obtuse angle or else all acute angles.
(True, False)

— — — — — — — — — — — —

(a) False — it would then be a right triangle, not an oblique triangle;
(b) True;   (c) False — it may contain all acute angles, but not necessarily;   (d) True

21.   Two important facts or geometrical properties common to all triangles that you should bear in mind are these (they should be familiar to you from geometry):

The sum of the three angles equals $180°$.

The greater side lies opposite the greater angle, and conversely.

Right triangles can, as we know, be solved directly by means either of the Pythagorean Theorem or by means of the three primary trigonometric ratios (or their reciprocals). But since these methods apply *directly* only to right triangles they cannot be used to solve oblique triangles — directly. They can, however, be used *indirectly* to solve oblique triangles. And we will look first into how this may be done.

Since we must have right triangles in order to apply right triangle methods, in working with oblique triangles the basic procedure is to drop a perpendicular from one vertex to the opposite side, thus dividing the oblique triangle into two right triangles, as shown in the figure at the right. We will see that solutions by this method require a two-step process.

Example: In the oblique triangle ABC, if $\angle A = 50°$, $\angle B = 30°$, and side $b = 10''$, find side a.
Solution: Drawing the altitude h divides $\triangle ABC$ into two right triangles, $\triangle ACD$ and $\triangle BCD$.

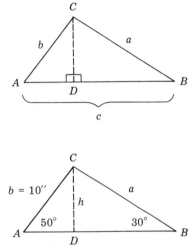

In $\triangle ACD$, $\sin A = \dfrac{h}{b}$, or $h = b \sin A$
$$h = 10 \sin 50°$$
$$= 10(0.766)$$
$$= 7.66''$$

Now, knowing the value of h we can use it to solve for side a in triangle BCD.

In $\triangle BCD$, $\sin \angle B = \dfrac{h}{a}$, or $a = \dfrac{h}{\sin B}$
$$a = \frac{7.66}{\sin 30°}$$
$$= \frac{7.66}{0.50} = 15.3''$$

Apply this two-step approach in solving the following problem. In oblique triangle ACB, altitude h is drawn and its length is given as 10 ft. Also, angle ACD is $45°$ and angle BCD is $60°$. Find the length of side c. (Hint: since

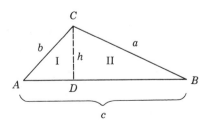

side $c = AD + DB$, find AD from $\triangle I$ and DB from $\triangle II$, then add them to find side c.)

- - - - - - - - - - - -

side $c = AD + DB$

In $\triangle I$, $\tan \angle ACD = \dfrac{AD}{h}$ or $AD = h \tan \angle ACD$

$$= 10 \tan 45°$$
$$= 10(1)$$
$$= 10 \text{ ft}$$

In $\triangle II$, $\tan \angle BCD = \dfrac{DB}{h}$ or $BD = h \tan \angle BCD$

$$= 10 \tan 60°$$
$$= 10(1.732)$$
$$= 17.32 \text{ ft}$$

Therefore, side $c = 10 + 17.32 = 27.32$ ft

22. As you can see, it is often, though not always, possible to use the trigonometric functions in the conventional way to solve oblique triangles, if those triangles can be divided into two right triangles in some convenient way. However, as we indicated earlier, there are some methods of solving oblique triangles directly, and we will consider two of them.

   The first, known as the *law of sines*, states that:

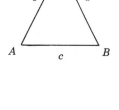

> *The sides of a triangle are proportional to the sines of the opposite angles.*

Thus, $\dfrac{a}{\sin A} = \dfrac{b}{\sin B} = \dfrac{c}{\sin C}$. The following relations (or their reciprocals) also can be obtained readily from the above.

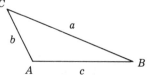

$$\frac{a}{b} = \frac{\sin A}{\sin B}, \quad \frac{b}{c} = \frac{\sin B}{\sin C}, \quad \frac{c}{a} = \frac{\sin C}{\sin A}$$

The second law, known as the *law of cosines*, states that:

> *In any triangle the square of any side is equal to the sum of the squares of the other two sides minus twice the product of these sides and the cosine of their included angle.*

Thus, $a^2 = b^2 + c^2 - 2bc \cos A$
$\qquad b^2 = c^2 + a^2 - 2ca \cos B$
$\qquad c^2 = a^2 + b^2 - 2ab \cos C$

Solving the above three equations for the cosines of the angles gives us this additional set of expressions for the cosine law.

$$\cos A = \frac{b^2 + c^2 - a^2}{2bc}$$

$$\cos B = \frac{a^2 + c^2 - b^2}{2ac}$$

$$\cos C = \frac{a^2 + b^2 - c^2}{2ab}$$

These formulas are useful in finding the angles of a triangle when its three sides are given. The first group expressed in terms of the squares of the sides of the triangle, can be used for finding the third side of a triangle when two sides and the included angle are given. The other angles can then be found either by the law of sines or by the latter three formulas.

Now let's see how we are going to apply these two laws. To do so we will employ them, as appropriate, to each of the four cases we mentioned in frame 20.

Case 1 — given one side and two angles.

Example: Suppose a, B, and C are given. Thus, $a = 7.07$, $B = 30°$, and $C = 105°$.

To find A we use $A = 180° - (B + C)$
$\qquad\qquad = 180° - (30° + 105°)$
$\qquad\qquad = 180° - 135°$
$\qquad\qquad = 45°$

To find b we use $\dfrac{b}{a} = \dfrac{\sin B}{\sin A}$ or $b = \dfrac{a \sin B}{\sin A}$

$$= \frac{7.07(\sin 30°)}{\sin 45°}$$

$$= \frac{7.07(0.5000)}{.707}$$

$$= 5$$

To find c we use $\dfrac{c}{a} = \dfrac{\sin C}{\sin A}$ or $c = \dfrac{a \sin C}{\sin A}$

$$= \frac{7.07[\sin(180° - 105°) \text{ or } 75°]}{\sin 45°}$$

$$= \frac{7.07(0.966)}{.707}$$

$$= 9.66$$

Now, you solve this practice problem.

Given: $a = 50$, $A = 65°$, $b = 40°$.
Find: C, b, and c.

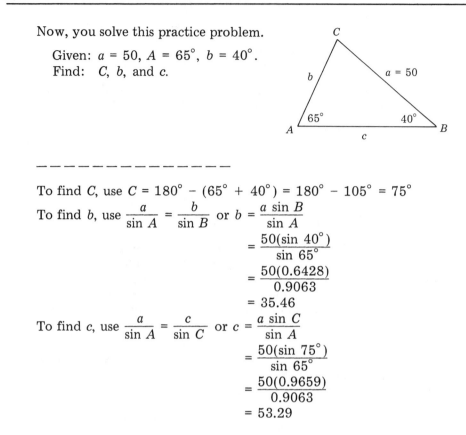

––––––––––––––––––

To find C, use $C = 180° - (65° + 40°) = 180° - 105° = 75°$

To find b, use $\dfrac{a}{\sin A} = \dfrac{b}{\sin B}$ or $b = \dfrac{a \sin B}{\sin A}$

$$= \frac{50(\sin 40°)}{\sin 65°}$$

$$= \frac{50(0.6428)}{0.9063}$$

$$= 35.46$$

To find c, use $\dfrac{a}{\sin A} = \dfrac{c}{\sin C}$ or $c = \dfrac{a \sin C}{\sin A}$

$$= \frac{50(\sin 75°)}{\sin 65°}$$

$$= \frac{50(0.9659)}{0.9063}$$

$$= 53.29$$

23. Now let's consider the next case.

Case 2 — given two sides and the angle opposite one of them.

The solution of the triangle in this case depends upon the law of sines. However, there is a built-in ambiguity in the solution that we need to examine.

The difficulty is that, when given two sides and the angle opposite one of them, we must first find the unknown angle that lies opposite one of the given sides. But when an angle is determined by its sine, it can have either one of two values which are supplements of each other. Hence either value of the angle may be taken unless one is excluded by the conditions of the problem.

Let's see what this means. In the triangle at the right, a and b are the given sides and A (opposite the side a) is the given angle. If $a > b$, then we know from geometry that $A > B$, and B must be acute regardless of the value of A, since a triangle can have only one

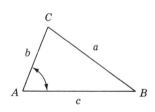

obtuse angle. Therefore, there is one, and only one, triangle that will satisfy the given conditions, and there is no ambiguity here. On the other hand, if $a = b$, then, again from geometry, both A and B must be acute, and the triangle is isosceles.

Now consider the triangle at the right. If $a < b$, then, from geometry, $A < B$ and A must be acute in order that the triangle be possible. But when A is acute it is evident that the two triangles ACB and ACB' both will satisfy the given conditions, provided that a is greater than the perpendicular CP. That is, provided $a > b \sin A$.

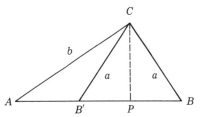

The angles ABC and $AB'C$ are supplementary (since angles $B'BC$ and $BB'C$ are equal). They are in fact the supplementary angles obtained (using the law of sines) from the formula

$$\sin B = \frac{b \sin A}{a}.$$

If $a = b \sin A$ (that is, CP), then $\sin B = 1$, $B = 90°$, and the triangle is a right triangle. If, however, $a < b \sin A$ (that is, less than CP), then $\sin B > 1$ and the triangle is impossible.

You will probably be relieved to know that all of the foregoing can be summarized as follows.

Two solutions: If A is acute and the value of a lies between b and $b \sin A$.

No solution: If A is acute and $a < b \sin A$, or if A is obtuse and $a < b$ or $a = b$.

One solution: In all other cases.

The number of solutions usually can be determined by inspection on constructing the triangle. When in doubt, find the value of $b \sin A$ and test as above.

Since you will be applying the law of sines primarily to the "all other cases" type of oblique triangle, let's consider a single solution situation.

Example: Given $a = 40$, $b = 30$, $A = 75°$. Find the remaining parts.
Solution: Since $a > b$ and A is acute we know there is only one solution. By the law of sines, then, $\dfrac{a}{\sin A} = \dfrac{b}{\sin B}$, or

$\sin B = \dfrac{b \sin A}{a} = \dfrac{30(0.9659)}{40} = 0.7244$, and $B = 46°\,25'$.

Therefore, $C = 180° - (A + B) = 180° - 121°\,25' = 58°\,35'$

To get c, using the law of sines, $\dfrac{c}{\sin C} = \dfrac{a}{\sin A}$, or

$c = \dfrac{a \sin C}{\sin A} = \dfrac{40(0.8535)}{0.9659} = 35.3$

Use the law of sines to solve the following problem. Before starting, check (using the summary above) to see how many solutions you should expect. Problem: Solve the triangle when a = 119, b = 97, and $A = 50°$. That is, find the missing parts.

-- -- -- -- -- -- -- -- -- --

Since $a > b$ and A is acute, there is only one solution.

To find B we use $\dfrac{a}{\sin A} = \dfrac{b}{\sin B}$, or $\sin B = \dfrac{b \sin A}{a} = \dfrac{97(\sin 50°)}{119} =$

$\dfrac{97(.766)}{119} = 0.62438$. Therefore $B = 38°38'$.

Hence $C = 180° - (50° + 38°38') = 91°22'$.

To find c we use $\dfrac{c}{\sin C} = \dfrac{a}{\sin A}$, or $c = \dfrac{a \sin C}{\sin A} =$

$\dfrac{119(\sin 91°22' = 88°38')}{\sin 50°} = \dfrac{119(0.99972)}{0.766} = 155.3$.

24. Having considered the application of the law of sines to Cases 1 and 2, we will go on now to Cases 3 and 4, which involve the application of the law of cosines.

   Case 3 — given two sides and the included angle.

Example: Given a = 132, b = 224, and $C = 28°40'$, solve for the other parts of the oblique triangle.

To find c we use $c^2 = a^2 + b^2 - 2ab \cos C$

$$= (132)^2 + (224)^2 - 2(132)(224) \cos 28°40'$$
$$= (132)^2 + (224)^2 - 2(132)(224)(0.8774)$$
$$= 15,714$$

or c = 125 (taking the square root of both sides, to the nearest three figures).

For A we use $\sin A = \dfrac{a \sin C}{c}$

$$= \dfrac{132 \sin 28°40'}{125}$$
$$= \dfrac{132(0.4797)}{125}$$
$$= 0.5066$$

and $A = 30°30'$.

For B we use $\sin B = \dfrac{b \sin C}{c}$

$$= \frac{224 \sin 28° \; 40'}{125}$$

$$= \frac{224(0.4797)}{125}$$

$$= 0.8596$$

or $B = 120° 40'$.

Now try this practice problem. Given an oblique triangle with side $a = 30$, side $b = 54$, and angle $C = 46°$, find the remaining parts.

_ _ _ _ _ _ _ _ _ _ _ _ _ _ _ _

Your procedure should follow that shown in the example above.
$A = 33° 06'$, $B = 100° 54'$, and $c = 39.56$.

25. Now we come to the fourth and last case. Again we will use the cosine law to solve the triangle.

    Case 4 — given the three sides.

Example: Solve the triangle ABC, given $a = 30.3$, $b = 40.4$, and $c = 62.6$.

For A we use $\cos A = \dfrac{b^2 + c^2 - a^2}{2bc}$

$$= \frac{(40.4)^2 + (62.6)^2 - (30.3)^2}{2(40.4)(62.6)}$$

$$= 0.9159, \text{ and } A = 23° 40'.$$

For B we use $\cos B = \dfrac{c^2 + a^2 - b^2}{2ca}$

$$= \frac{(62.6)^2 + (30.3)^2 - (40.4)^2}{2(62.6)(30.3)}$$

$$= 0.8448, \text{ and } B = 32° 20'.$$

And for C, $\cos C = \dfrac{a^2 + b^2 - c^2}{2ab}$

$$= \frac{(30.3)^2 + (40.4)^2 - (62.6)^2}{2(30.3)(40.4)}$$

$$= -0.5590, \text{ and } C = 124° 00'.$$

Check: $A + B + C = 180°$.

Try this practice problem. Given the following sides of the oblique triangle *ABC*, use the cosine law to find its angles: $a = 24.5$, $b = 18.6$, and $c = 26.4$.

— — — — — — — — — — — —

$A = 63°\,10'$, $B = 42°\,40'$, $C = 74°\,10'$.

26. Here are a few additional problems that will give you practice in working with the sine and cosine laws. You will have to decide in each instance which case is involved and then apply the correct law. Refer to frame 20 if you need a summary of the four cases. In fact, it would probably help if you kept a copy of it in front of you while you work these problems.

   Use a separate sheet of paper for your computations, as you solve each of the following oblique triangles *ABC*, given:

(a) $a = 125$, $A = 54°\,40'$, $B = 65°\,10'$

(b) $b = 321$, $A = 75°\,20'$, $C = 38°\,30'$

(c) $b = 215$, $c = 150$, $B = 42°\,40'$

(d) $a = 512$, $b = 426$, $A = 48°\,50'$

(e) $b = 120$, $c = 270$, $A = 118°\,40'$

(f) $a = 6.34$, $b = 7.30$, $c = 9.98$

(g)   Find the horizontal distance from a point *A* to an inaccessible point *B* on the opposite bank of a river. *AC*, which is any convenient horizontal distance, is given as 283 feet, angle *CAB* = 38°, and angle *ACB* = 66°18'. Solve triangle *ABC* for side *AB*.

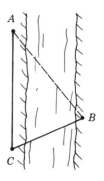

— — — — — — — — — — — — —

(a) $b = 139$, $c = 133$, $C = 60°\,10'$ (Case 1)
(b) $a = 339$, $c = 218$, $B = 66°\,10'$ (Case 1)
(c) $a = 300$, $A = 109°\,10'$, $C = 28°\,10'$ (Case 2)
(d) $c = 680$, $B = 38°\,50'$, $C = 92°\,20'$ (Case 2)
(e) $a = 234$, $B = 17°\,50'$, $C = 43°\,30'$ (Case 3)
(f) $A = 39°\,20'$, $B = 46°\,50'$, $C = 93°\,50'$ (Case 4)
(g) $AB = 267.4$ ft (Case 1)

It's time once again for you to check up on yourself and find out how much you have retained from this chapter. Before taking the Self-Test that follows, you will find it helpful to review quickly the topics we have covered. Also, don't hesitate to look up any of the formulas you need during the test. You would be quite exceptional if you had memorized them all at this point.

SELF-TEST

1. Indicate which quadrant each of the following angles is in — that is, in which quadrant its terminal side falls.

    (a) $170°$ _____   (d) $185°$ _____

    (b) $350°$ _____   (e) $-5°$ _____

    (c) $95°$ _____   (f) $-95°$ _____

                                                        (frame 1)

2. Indicate whether the following angles are coterminal, quadrantal, or both.

    (a) $-90°$ and $90°$ _____

    (b) $0°$ and $360°$ _____

    (c) $30°$ and $-330°$ _____   (frame 2)

3. Draw the standard position angle for $\Theta = 300°$; show the reference angle (angle between the terminal side and the X-axis) and the coordinates of a point P at a distance r from the center O, located on the terminal side.   (frame 3)

4. Draw a diagram showing the signs of the six trigonometric functions in all four quadrants.   (frame 4)

5. Draw a unit circle with a standard position angle in the first quadrant; show the point P where the terminal side cuts the circle, and the perpendicular from P to the X-axis. Indicate the sides of the resulting triangle that represent the sin and cos function values.   (frames 5, 6)

6. Complete the following.

   (a) The cosine value increases from 0 at _____° to 1 at _____°.

   (b) The sine value increases from 0 at _____° to 1 at _____°.

   (c) The tangent value increases from 0 at _____° to _____ at 90°.

   (frame 7)

7. Using the table of natural trigonometric functions, find the following values.

   (a) $\sin 15°$ = _____

   (b) $\cos 65°$ = _____

   (c) $\tan 80°$ = _____   (frames 8, 9, 10)

8. Plot the graph of $y = \sin(x + \frac{\pi}{2})$ at 30° (that is, $\frac{\pi}{6}$) intervals for values of x between 0 and 2π. Use a separate sheet of paper for your figures.

   (frame 11)

9. Complete the following.

   (a) The period of the sine function is _____ radians.

   (b) The period of the tangent is _____ radians.

   (c) The period of the cosine is _____ radians.   (frame 12)

10. Show two ways of writing the inverse function of $\tan 45° = 1$.

    _____

    (frame 13)

11. The principal value of $\sin x = 1$ is x = _____ or _____ .

    (frame 14)

12. Frame 15 listed five important relations between the trigonometric functions and an additional 14 formulas derived from these. On a separate sheet of paper, write down as many of these as you can recall or derive and give yourself one point for each one that is correct.

    (frame 15)

13. Apply the addition and subtraction formulas for the trigonometric functions of two angles to solve the following.

   (a) Find $\sin 15°$, taking $15° = 60° - 45°$.

   (b) Find $\cos 75°$, taking $75° = 45° + 30°$. (frame 16)

14. Given $\tan x = 2$, x lying in the third quadrant, find $\sin 2x$, $\cos 2x$, $\tan 2x$. (Start by drawing a diagram of the reference triangle in the third quadrant and finding the values for the sin, cos, and tan of x.) (frame 17)

15. Find sine, cosine, and tangent of $15°$, given $\cos 30° = \dfrac{\sqrt{3}}{2}$. (Use the half-angle formulas.) (frame 18)

16. Solve the following equation: $\sec^2 x = \dfrac{4}{3}$. (Remember, the secant is the reciprocal of the cosine.)

   _____

   (frame 19)

17. Which of the triangles shown below are/is *not* oblique? _____

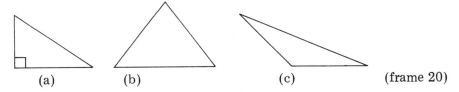

   (a)   (b)   (c)   (frame 20)

18. Use the two-step method of solving an oblique triangle to find the length of side a in triangle ABC.
   Given: $A = 45°$, $B = 27°$.
   side b = 12.0 ft.

   (frame 21)

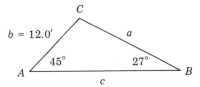

19. Given $c = 25$, $A = 35°$, and $B = 68°$ in the oblique triangle ABC, use the law of sines to find a, b, and C.

   (frame 22)

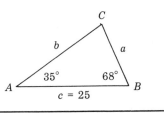

20. In oblique triangle ABC, if $c = 628$, $b = 480$, and $C = 55°10'$, find the missing parts.

(frame 23)

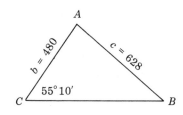

21. In the oblique triangle ABC, given $a = 20$, $c = 26$, $B = 40°$, use the cosine law to find side b.   (frame 24)

22. Given the following sides of the oblique triangle ABC, use the cosine law to find its angles: $a = 5.10$, $b = 4.60$, $c = 4.90$. (Express the angles to the nearest whole degree.)   (frame 25)

Answers to Self-Test

1. (a) second quadrant;   (b) fourth quadrant;   (c) second quadrant;
   (d) third quadrant;   (e) fourth quadrant;   (f) third quadrant
2. (a) quadrantal;   (b) quadrantal and coterminal;   (c) coterminal
3.

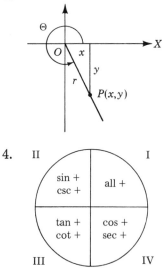

4.

II	I
sin + csc +	all +
tan + cot +	cos + sec +
III	IV

5.

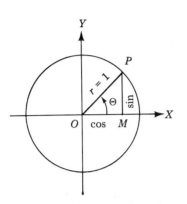

6. (a) $90°, 0°$;   (b) $0°, 90°$;   (c) $0°, \infty$ (that is, without limit)

7. (a) 0.25882;   (b) 0.42262;   (c) 5.6713

8. $y = \sin (x + \dfrac{\pi}{2})$

As you might expect, this curve looks very much like the $y = \cos x$ curve, since the sin x and cos x curves are exactly $90°\left(\dfrac{\pi}{2}\right)$ "out of phase" with one another.

x		$x + \dfrac{\pi}{2}$		y
$0°$	0	$90°$	$\pi/2$	1.00
$30°$	$\pi/6$	$120°$	$2\pi/3$	$.86$
$60°$	$\pi/3$	$150°$	$5\pi/6$	$.50$
$90°$	$\pi/2$	$180°$	π	0
$120°$	$2\pi/3$	$210°$	$7\pi/6$	$-.50$
$150°$	$5\pi/6$	$240°$	$4\pi/3$	$-.86$
$180°$	π	$270°$	$3\pi/2$	-1.00
$210°$	$7\pi/6$	$300°$	$5\pi/3$	$-.86$
$240°$	$4\pi/3$	$330°$	$11\pi/6$	$-.50$
$270°$	$3\pi/2$	$360°$	2π	0
$300°$	$5\pi/3$	$30°$	$\pi/6$	$.50$
$330°$	$11\pi/6$	$60°$	$\pi/3$	$.86$
$360°$	2π	$90°$	$\pi/2$	1.00

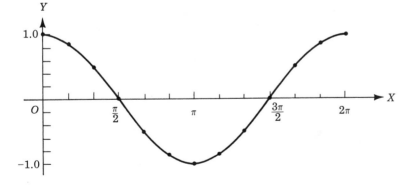

9. (a) 2π;   (b) π;   (c) 2π

10. arc sin $1 = 45°$ or $\sin^{-1} = 45°$

11. $90°$ or $\dfrac{\pi}{2}$

12. Refer to frame 15 to check your answers.

13. (a) $\sin 15° = \sin(60° - 45°)$

$= (\sin 60°)(\cos 45°) - (\cos 60°)(\sin 45°)$

$= \left(\dfrac{\sqrt{3}}{2}\right)\left(\dfrac{1}{\sqrt{2}}\right) - \left(\dfrac{1}{2}\right)\left(\dfrac{1}{\sqrt{2}}\right)$

$= \dfrac{\sqrt{3}}{2\sqrt{2}} - \dfrac{1}{2\sqrt{2}}$

$= \dfrac{\sqrt{3} - 1}{2\sqrt{2}}$

(b) $\cos 75° = \cos(45° + 30°)$

$= (\cos 45°)(\cos 30°) - (\sin 45°)(\sin 30°)$

$= \left(\dfrac{1}{\sqrt{2}}\right)\left(\dfrac{\sqrt{3}}{2}\right) - \left(\dfrac{1}{\sqrt{2}}\right)\left(\dfrac{1}{2}\right)$

$= \dfrac{\sqrt{3}}{2\sqrt{2}} - \dfrac{1}{2\sqrt{2}}$

$= \dfrac{\sqrt{3} - 1}{2\sqrt{2}}$

14. Since $\tan x = 2$ and x lies in the third quadrant, we can draw the figure shown at the right, from which $\sin x = \dfrac{-2}{\sqrt{5}}$, $\cos x = \dfrac{-1}{\sqrt{5}}$, and $\tan x = 2$. Substituting the sin and cos values in formula (21) we get: $\sin 2x = 2 \sin x \cos x$

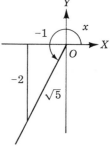

$= 2\left(-\dfrac{2}{\sqrt{5}}\right)\left(-\dfrac{1}{\sqrt{5}}\right)$

$= \dfrac{4}{5}$

and from formula (22): $\cos 2x = \cos^2 x - \sin^2 x$

$= \left(-\dfrac{1}{\sqrt{5}}\right)^2 - \left(-\dfrac{2}{\sqrt{5}}\right)^2$

$= \left(\dfrac{1}{5}\right) - \left(\dfrac{4}{5}\right)$

$= -\dfrac{3}{5}$

Finally, from formula (23): $\tan 2x = \dfrac{2 \tan x}{1 - \tan^2 x}$

$= \dfrac{2 \cdot 2}{1 - 2^2}$

$= -\dfrac{4}{3}$

15. Since $\cos 30° = \dfrac{\sqrt{3}}{2}$, then from formula (24),

$$\sin 15° = \sqrt{\dfrac{1 - \dfrac{\sqrt{3}}{2}}{2}} = \sqrt{\dfrac{2 - \sqrt{3}}{4}} = \dfrac{1}{2}\sqrt{2 - \sqrt{3}}$$

And from (25),

$$\cos 15° = \sqrt{\dfrac{1 + \dfrac{\sqrt{3}}{2}}{2}} = \sqrt{\dfrac{2 + \sqrt{3}}{4}} = \dfrac{1}{2}\sqrt{2 + \sqrt{3}}$$

Finally, from (26),

$$\tan 15° = \sqrt{\dfrac{1 - \dfrac{\sqrt{3}}{2}}{1 + \dfrac{\sqrt{3}}{2}}} = \sqrt{\dfrac{2 - \sqrt{3}}{2 + \sqrt{3}}}$$

16. Since $\sec^2 x = \dfrac{4}{3}$, then $\sec x = \dfrac{2}{\sqrt{3}}$ or $\cos x = \dfrac{\sqrt{3}}{2}$, and $x = 30°$ or $\dfrac{\pi}{6}$.

17. Triangle (a)

18. First drop a perpendicular from C to side c (call it h).

Then, $\sin A = \dfrac{h}{b}$ or $h = b \sin A$

$$= 12(0.707)$$
$$= 8.48.$$

Hence $\sin B = \dfrac{h}{a}$ or $a = \dfrac{h}{\sin B}$

$$= \dfrac{8.48}{\sin 27°}$$
$$= \dfrac{8.48}{0.454}$$
$$= 18.68$$

19. To find C: $C = 180° - (A + B) = 180° - 103° = 77°$.

To find a: $a = \dfrac{c \sin A}{\sin C} = \dfrac{25 \sin 35°}{\sin 77°} = \dfrac{25(0.5736)}{0.9744} = 15$

To find b: $b = \dfrac{c \sin B}{\sin C} = \dfrac{25 \sin 68°}{\sin 77°} = \dfrac{25(0.9272)}{0.9744} = 24$

20. Since C is acute and $c > b$, there is only one solution.

For B: $\sin B = \dfrac{b \sin C}{c}$

$$= \dfrac{480 \sin 55°10'}{628}$$
$$= \dfrac{480(0.8208)}{628}$$

$$= 0.6274, \text{ and } B = 38°50'.$$

For A: $A = 180° - (B + C) = 86°00'$.

For a: $a = \dfrac{b \sin A}{\sin B}$

$= \dfrac{480 \sin 86°}{\sin 38°50'}$

$= \dfrac{480(0.9976)}{0.6271}$

$= 764.$

21. Using the known values we get:

$b^2 = c^2 + a^2 - 2ca \cos B = 26^2 - 20^2 - 2(20)(26) \cos 40°$

$= 676 - 400 - 1040(0.766)$

$= 279 \text{ or } 280, \text{ and } b = 16.7.$

22. For C: $\cos C = \dfrac{a^2 + b^2 - c^2}{2ab}$

$= \dfrac{(5.1)^2 + (4.6)^2 - (4.9)^2}{2(5.1)(4.6)}$

$= 0.503, \text{ from which } C = 60°.$

For B: $\cos B = \dfrac{a^2 + c^2 - b^2}{2ac}$

$= \dfrac{(5.1)^2 + (4.9)^2 - (4.6)^2}{2(5.1)(4.9)}$

$= 0.576, \text{ from which } B = 55°.$

For A: $A = 180° - (B + C) = 180° - (60° + 55°) = 65°.$

This concludes our exploration into the subject of trigonometry, but you will encounter many of these concepts later on. So don't erase them from your mind! Now, however, it is time for us to move on to the very interesting subject of analytic — or coordinate — geometry.

CHAPTER SEVEN

Analytic Geometry

Analytic geometry — or coordinate geometry, as it is sometimes called — is the study of geometry by means of the analytical methods of algebra. It will, therefore, provide you an opportunity to discover how these two branches of mathematics are connected. You will learn how to express geometric figures and the facts about such figures in algebraic terms and how to obtain results from equations rather than from the figures themselves.

The Euclidean plane geometry you studied in the first two chapters of this book is called *synthetic*, meaning "put together" or "combined from related parts." The name is appropriate since the method of synthetic geometry is to put geometric facts together, rather like building blocks. Thus its primary definitions, axioms, and postulates are foundations, and its long sequences of theorems, constructions, and corollaries are superstructures. To reach any one of its higher propositions we are required to follow a step-by-step path of reasoning, all the way up from the base. This approach provides excellent training in logical, mathematical reasoning. It acquaints the learner with a great many fundamental facts that are useful in themselves and indispensable to further study.

For more advanced mathematical applications, however, the synthetic method in geometry has certain practical disadvantages. One is that it requires you to keep constantly in mind a very large number of previously demonstrated propositions. Another is that it often requires elaborate constructions and indirect methods of deduction through many intermediate steps.

The type of geometry we are going to study now, in preparation for calculus, is called *analytic*, which means, literally, "loosening up" or disentangling. Again, the name is appropriate since the method of analytic geometry is to separate out the essential elements in each new problem by stating them in the form of equations, and then resolving the geometric question by solving these equations algebraically.

An obvious advantage in this procedure is that to solve most practical problems you need keep in mind only a few basic formulas. The greatest advantage of the analytic method, however, is that it is more direct, quicker, and more powerful!

When you have reached the end of this chapter you will be familiar with, and be able to use, such concepts as:

- determining the properties of lines and curves by means of equations;

- rectangular Cartesian coordinates to define the positions of points, lines, and curves;

- finding the distance between two points, the division point of a line segment, the inclination and slope of a line, and the angle between two lines;

- the locus of an equation, infinite extent of a curve, intersections of curves, and translation of axes;

- the equation of a line, the point-slope form of the line equation, its slope-intercept form, two-point form, intercept form, and general form.

BASIC DEFINITIONS AND THEOREMS

1. You should be generally familiar already with the use of the rectangular, or Cartesian, coordinate system from your study of algebra. You used this system to aid you in the graphic solution of linear and quadratic equations, including the solution of pairs of linear equations. And of course we used Cartesian coordinates in our study of trigonometry, so we will not need to go into another complete explanation here.

   However, just to make sure you haven't forgotten how to locate points on a rectangular coordinate system, get a piece of graph (cross-section or quadrille) paper, draw a pair of X- and Y-axes, establish some convenient scale, and locate the following points.

   (a)   (4, 2)   (c)   (−4, 4)
   (b)   (−3, −3)   (d)   (3, −4)

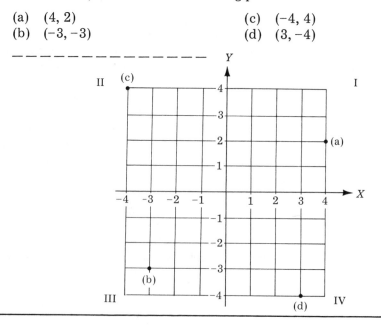

2. We didn't really need a coordinate system in our study of geometry because the properties of a given geometric configuration usually found in Euclidean plane geometry do not in any way depend upon a related coordinate system. Sometimes, however, the introduction of a coordinate system helps to simplify the work of proving a theorem, especially if the axes are chosen properly. And in our study of trigonometric analysis the selection of a coordinate system in which the initial side of the standard-position angle lay along the positive X-axis and the vertex of the angle was located at the intersection of the two axes, proved very useful indeed.

However, the use of a coordinate system is essential to the study of analytic geometry; it is the method which connects the distances of a point from two intersecting lines (the axes) by means of an equation. Without a coordinate system, then, we would have no analytic geometry.

In the figure at the right, the plane is divided into four quadrants, (I, II, III, and IV) by the two perpendicular lines (axes X and Y) intersecting at O. The arrowheads at the right end of the X-axis and at the top of the Y-axis indicate the positive *direction* of these axes. The distance from the Y-axis is called the *x-coordinate* or *abscissa* of the point, that from the X-axis the *y-coordinate* or *ordinate* of the point, and the two distances taken together and enclosed in parentheses (x,y), the *coordinates* of the point. (x,y) is an *ordered pair*. Sound familiar?

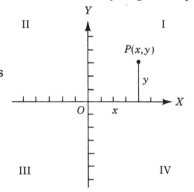

The abscissa is always written first. The origin O corresponds to the zero of our real number system. Points to the right of the Y-axis have positive abscissas, those to the left, negative. Likewise, points above the X-axis have positive ordinates, those below, negative. Thus the points $(4,2)$, $(-3,-3)$, $(-4,4)$, and $(3,-4)$ in the last problem are in the first, third, second, and fourth quadrants respectively.

A concept already employed in our coordinate system is that of the directed line segment. A line segment to which a positive or negative direction has been assigned is called a *directed line segment*. Thus, if AB (in the figure at the right) represents the length of the segment from A to B, then BA will represent the length of the segment measured in the opposite direction. That is, $BA = -AB$, or $AB + BA = 0$.

We used this idea in setting up our coordinate systems because, by definition, an abscissa has positive or negative direction according to whether it is measured to the right or left of the Y-axis. Also, an ordinate is positive when measured up from the X-axis and negative when measured down.

If we now agree that *any* line drawn parallel to one of the coordinate axes is to have the same direction as that axis, we can derive a relationship that will be of great importance in what follows. Shown at the right are two arrangements of three points, A, B, and C on a line parallel to the X-axis. In this figure,

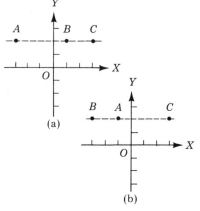

$$AB + BC = AC \qquad (1)$$

in both cases. Do you know why? Think back about what you learned about number lines and the number system and see if you can formulate an answer. Then compare your answer with the one below.

— — — — — — — — — — — — —

In (a), the segments AB and BC have the same sign and their sum is the positive number AC. In (b), AB and BC are different in sign, but BC is the greater, and again the sum is AC. There are four other possible arrangements of A, B, and C and you might enjoy checking for yourself that the given relation is valid in these cases also. Also, we could revolve the line through $90°$, the points then lying on a line parallel to the Y-axis, and our equation would still be true.

3. In finding the distance between two points, such as P_1 and P_2, whose coordinates are (x_1, y_1) and (x_2, y_2) respectively, there are two cases to consider. The first is when the given points are on a line parallel to one of the coordinate axes, and the second, when this is not the case. When P_1 and P_2 are on a line parallel to the X-axis, as shown at right, we know

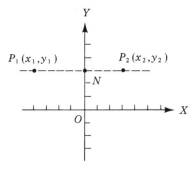

that $y_1 = y_2$, and therefore relation (1), established in the preceding frame, shows us that the distance from P_1 to P_2 is
$P_1 P_2 = P_1 N + NP_2 = NP_2 - NP_1$ (where N is the Y-intercept) for all positions of P_1 and P_2. But $NP_2 = x_2$ and $NP_1 = x_1$. Therefore $P_1 P_2 = x_2 - x_1$. Similarly, if the points are on a line parallel to the Y-axis, $P_1 P_2 = y_2 - y_1$.

We can state these results as follows:

The distance between two points on a line parallel to the X-axis is the abscissa of the terminal point minus the abscissa of the initial point. If the points are on a line parallel to the Y-axis, the distance between them is the ordinate of the terminal point minus the ordinate of the initial point.

Let's apply what we have just learned. Here are the relationships again: $P_1 P_2$ (the directed distance) $= x_2$ (abscissa of terminal point) $-$ x_1 (abscissa of initial point).

Example: Find the directed distance from P_1 $(-5,2)$ to P_2 $(3,2)$. The fact that the y-coordinate (2) is the same for both points tells us that the points lie on a line parallel to the X-axis. Therefore, $P_1 P_2 = x_2 - x_1 = 3 - (-5) = 3 + 5 = 8.$

Now try these problems. Find the directed distance from:

(a) $(-3,3)$ to $(3,3)$ _____

(b) $(0,4)$ to $(4,4)$ _____

(c) $(4,2)$ to $(-2,2)$ _____

(d) $(6,5)$ to $(2,5)$ _____

- - - - - - - - - - - - - - - - -

(a) 6;   (b) 4;   (c) −6;   (d) −4. Distances between points on a line parallel to the Y-axis would, of course, be found in the same way, using the ordinate values of the points.

4. The figure at the right represents the second, and general, case where the points P_1 and P_2 may be located anywhere in the plane.

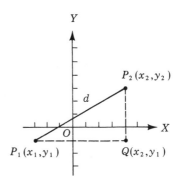

   To find the distance, d, between the two points, we draw a line through P_1 parallel to the X-axis, and a second line through P_2 parallel to the Y-axis. These lines meet at the point Q whose coordinates are (x_2, y_1). Using our results from the last frame we have $P_1 Q = x_2 - x_1$, and $QP_2 = y_2 - y_1$.
From which, using the Pythagorean theorem, we get

$$d^2 = (P_1 Q)^2 + (QP_2)^2 = (x_2 - x_1)^2 + (y_2 - y_1)^2$$

$$d = \sqrt{(x_2 - x_1)^2 + (y_2 - y_1)^2} \qquad (2)$$

And since we are interested only in the numerical value of the distance, the radical is taken with the positive sign.

Notice that since $(x_2 - x_1)^2$ and $(y_2 - y_1)^2$ are always positive (because they are squared), either (x_1,y_1) or (x_2,y_2) may be taken as the initial point when using this formula to find the distance between points.

Example: Find the distance between the points $(3,-8)$ and $(-6,4)$. Using equation (2) we get $d = \sqrt{(3 + 6)^2 + (-8 - 4)^2} = \sqrt{81 + 144} = \sqrt{225} = 15$.

Try this problem. Find the distance between the points $(2,8)$ and $(5,-3)$.

— — — — — — — — — — — — — —

$$d = \sqrt{(2 - 5)^2 + (8 + 3)} = \sqrt{9 + 121} = \sqrt{130}$$

5. The coordinates of the point dividing a line segment $P_1 P_2$ in the ratio r_1/r_2 can be found as follows.
    In the figure at the right P_1 is the initial point and P_2 the terminal point, with coordinates as shown. P_0, with coordinates (x_0,y_0) is the point on the line joining P_1 and P_2 such that

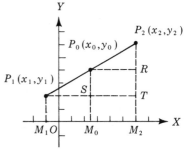

$$\frac{P_1 P_0}{P_0 P_2} = \frac{r_1}{r_2}.$$

We then draw in the segments $M_1 P_1$, $M_0 P_0$ and $M_2 P_2$, and through P_1 and P_0 draw the lines $P_1 ST$ and $P_0 R$ parallel to the X-axis. As a result we have $P_1 S = x_0 - x_1$, $P_0 R = x_2 - x_0$, $SP_0 = y_0 - y_1$, and $RP_2 = y_2 - y_0$. Since triangles $P_1 SP_0$ and $P_0 RP_2$ are similar, we can write

$$\frac{P_1 S}{P_0 R} = \frac{P_1 P_0}{P_0 P_2}, \text{ or } \frac{x_0 - x_1}{x_2 - x_0} = \frac{r_1}{r_2}.$$

Or, solving for x_0,

$$x_0 = \frac{x_1 r_2 + x_2 r_1}{r_1 + r_2}. \tag{3}$$

Similarly we can get

$$y_0 = \frac{y_1 r_2 + y_2 r_1}{r_1 + r_2}. \tag{4}$$

In the case considered above, P_0 lies between P_1 and P_2. That is, $P_1 P_0$ and $P_0 P_2$ have the same sign. If P_0 did not lie between P_1 and P_2, but fell upon $P_1 P_2$ extended, thus dividing it *externally*, $P_1 P_0$ and $P_0 P_2$ would differ in sign and the ratio $r_1 r_2$ would be negative. When the division point P_0 is the midpoint of the segment $P_1 P_2$ and therefore $r_1 = r_2$, equations (3) and (4) reduce to

$$x_0 = \frac{x_1 + x_2}{2}, \text{ and } y_0 = \frac{y_1 + y_2}{2}. \tag{5}$$

Now let's try using these formulas.

Example 1: Find the coordinates of the point that is two-thirds of the way from $(-3,5)$ to $(6,-4)$.
P_1, then, is $(-3,5)$ and P_2 is $(6,-4)$. Therefore, if P_0 is the desired point,

$$\frac{P_1 P_0}{P_0 P_2} = \frac{r_1}{r_2} = \frac{2 \text{ (i.e., } \frac{2}{3} \text{ the distance from } P_1 \text{ to } P_2)}{1 \text{ (i.e., } \frac{1}{3} \text{ the distance from } P_1 \text{ to } P_2)}$$

Thus, from equation (3)

$$x_0 = \frac{(-3)(1) + (6)(2)}{2 + 1} = 3$$

and (4),

$$y_0 = \frac{(5)(1) + (-4)(2)}{2 + 1} = -1.$$

Hence the coordinates of the point P_0 are $(3,-1)$.

Example 2: Find the coordinates of the midpoint of the segment joining $(2,6)$ and $(8,-4)$.
From equations (5), $x_0 = \dfrac{2 + 8}{2} = 5$, and $y_0 = \dfrac{6 - 4}{2} = 1$. Therefore the coordinates of the midpoint are $(5,1)$.

Here are some practice problems that will help you learn to use these equations for finding the coordinates of the division point of a line segment.

(a) Find the coordinates of the midpoint of the line segment joining $(-5,8)$ and $(2,-4)$.

(b) Find the coordinates of the point that is three-fifths of the way from $(-4,-2)$ to $(4,4)$.

(c)   Find the coordinates of the point that is three-fourths of the distance from (6,–2) to (2,6).

— — — — — — — — — — — — — —

(a) $(-\frac{3}{2}, 2)$;   (b) $(\frac{4}{5}, 1\frac{3}{5})$;   (c) $(3,4)$

6. You probably are familiar with the terms "slope" and "inclination" in a general sense. We speak of a road having a steep slope or a high angle of inclination. Highway engineers also use the word "grade" in referring to the angle a road makes with the horizontal. For example, if a road rises six feet in each 100 feet of horizontal distance, it is said to be a 6% grade. Thus, percentage is one method of measuring slope. In analytic geometry we need to (and are able to) define the concepts of inclination and slope rather precisely. Thus,

> the angle, less than 180° and measured counterclockwise, which a line makes with the positive direction of the X-axis is called the *inclination* of the line. The tangent of this angle is called the *slope* of the line.

If we designate the angle by α (alpha) and the slope by the letter m, then $m = \tan \alpha$.

Notice in the figure at the right that l_1 makes an *acute* angle, α_1, with the positive direction of the X-axis. Hence $m_1 = \tan \alpha_1$ is positive, and l_1 is said to have a *positive slope*. Similarly, since α_2 is obtuse, $m_2 = \tan \alpha_2$ is negative, and l_2 is said to have a *negative slope*.

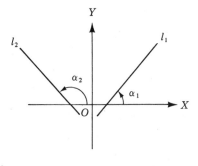

We can say, therefore, that a line that *rises* from left to right has a *positive slope*, and one that *descends* from left to right has a *negative slope*.

Since, as you will recall from Chapter 4, tan 0° = 0 and tan 90° is undefined, a line *parallel* to the X-axis has *zero slope*, and a line *perpendicular* to the X-axis has *no slope* or, if you will, *infinite slope*. The slope of a line through two points, such as P_1 and P_2 in the figure at the right, can be expressed in terms of the coordinates of those points as follows:

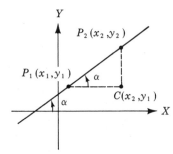

$$m = \tan \alpha = \frac{CP_2}{P_1 C} = \frac{y_2 - y_1}{x_2 - x_1} \qquad (6)$$

provided $x_1 \neq x_2$.

The relation $\dfrac{y_2 - y_1}{x_2 - x_1} = \dfrac{y_1 - y_2}{x_1 - x_2}$ is always true, regardless of whether the slope is positive or negative. We can say, therefore, that

the slope of a line not parallel to the Y-axis and passing through the points P_1 and P_2 remains the same whether the line is directed from P_1 to P_2 or from P_2 to P_1, and is equal to the difference of the ordinates divided by the corresponding difference of the abscissas.

Now let's try combining some of the concepts we have been discussing in a problem that will require us to use the proper formulas to solve for the unknown values.

Example: If the points A, B, and C, with coordinates $(-2,3)$, $(5,8)$ and $(7,-4)$ respectively, are the vertices of a triangle, find:
(a) the slope of the side AB,
(b) the length of the side BC, and
(c) the coordinates of the point two-thirds of the distance from B to the midpoint of the opposite side (AC).

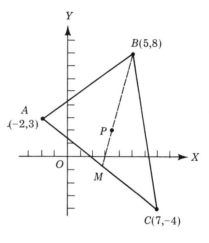

(a) From (6), the slope of the line AB is given by

$$m = \frac{y_2 - y_1}{x_2 - x_1} = \frac{8 - 3}{5 - (-2)} = \frac{5}{7}.$$

Notice that whereas this value was found by taking the direction from A to B, we know, from what we have just learned in this frame, that the slope is the same if the direction is taken from B to A. Let's verify this by trying it. Taking the direction of the line from B to A we get

$$m = \frac{y_1 - y_2}{x_1 - x_2} = \frac{3 - 8}{-2 - 5} = \frac{-5}{-7} = \frac{5}{7}.$$

(b) The length of the side BC we can find by using equation (2), thus,

$$BC = \sqrt{(x_2 - x_1)^2 + (y_2 - y_1)^2}$$
$$= \sqrt{(7 - 5)^2 + (-4 - 8)^2} = \sqrt{4 + 144} = 2\sqrt{37}.$$

(c) Before we can find the coordinates of the point two-thirds of the distance from B to the midpoint of AC, we first must find the coordinates of the midpoint of AC itself. Using equation (5) we get

$$x_0 = \frac{x_1 + x_2}{2} = \frac{-2 + 7}{2} = \frac{5}{2}, \text{ and}$$
$$y_0 = \frac{y_1 + y_2}{2} = \frac{3 - 4}{2} = -\frac{1}{2}.$$

Therefore, the coordinates of M, the midpoint of AC, are $(\frac{5}{2}, -\frac{1}{2})$. And if P is the point two-thirds of the way from B to M, then $\frac{BP}{PM} = \frac{2 \text{ (two-thirds of the distance)}}{1 \text{ (one-third of the distance)}}$. The coordinates of P can be found from equations (3) and (4) as follows:

$$x_0 = \frac{x_1 r_2 + x_2 r_1}{r_1 + r_2} = \frac{(5)(1) + (\frac{5}{2})(2)}{2 + 1} = \frac{10}{3}$$

$$y_0 = \frac{y_1 r_2 + y_2 r_1}{r_1 + r_2} = \frac{(8)(1) + (-\frac{1}{2})(2)}{2 + 1} = \frac{7}{3}$$

Thus the coordinates of point P, the point two-thirds of the distance from B to M, are $(\frac{10}{3}, \frac{7}{3})$.

In summary, then, we used equation (6) to find the slope of AB; equation (2) to find the length of side BC; equation (5) to first get the coordinates of the midpoint of AC, and then equations (3) and (4) to find the coordinates of the point P, two-thirds of the distance from B to M. So although it may have seemed to you as though we were having to write down an awful lot of letters and numbers to arrive at our solutions (and there *were* quite a few since we really combined three problems in one), the procedures themselves were quite straight-forward. Try not to be too concerned about how much writing you have to do in mathematics. The more explicitly you state things, the clearer you will be about what you're trying to do.

Here are a few practice problems for you. Do your computations on a separate sheet of paper.

(a)   Find the slope of the lines joining the following pairs of points: (3,4) and (5,9); (−3,2) and (2,−4); (1,−2) and (6,8); (2,5) and (3,−6); and (−5,−4) and (2,−3).

(b)   Find the slope and inclination of the line joining (a,b) to (c,b).

(c)   Show that the line through (1,1) and (−2,3) is parallel to the line through (3,2) and (−3,6). Draw a figure. (To prove them parallel, prove that they have the same slope.)

(d)   Prove by means of slopes that the points (0,3), (2,6), and (−2,0) lie on the same straight line.

— — — — — — — — — — — — — —

(a) $\frac{5}{2}; -\frac{6}{5}; 2; -11; \frac{1}{7}$

(b)   From (6), $m = \frac{y_2 - y_1}{x_2 - x_1} = \frac{b - b}{c - a} = 0$. And since m represents the tangent of α (which we could write as arc tan $\alpha = 0$), then $\alpha = 0°$. That is, the angle whose tangent function value is 0, is the angle $0°$.

(c) $m_1 = \dfrac{3 - 1}{-2 - 1} = -\dfrac{2}{3}$

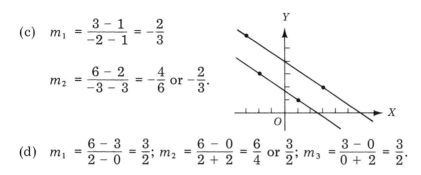

$m_2 = \dfrac{6 - 2}{-3 - 3} = -\dfrac{4}{6}$ or $-\dfrac{2}{3}$.

(d) $m_1 = \dfrac{6 - 3}{2 - 0} = \dfrac{3}{2}$; $m_2 = \dfrac{6 - 0}{2 + 2} = \dfrac{6}{4}$ or $\dfrac{3}{2}$; $m_3 = \dfrac{3 - 0}{0 + 2} = \dfrac{3}{2}$.

7. Problem (c) above brought out the fact that if two lines have the same slope, they are parallel. And in problem (d) there was the implication that if three (or more) points lie on the same line, then their slopes (taken between any two of the points) are equal. In general we can state that

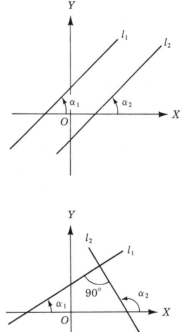

> If two lines with slopes m_1 and m_2 are parallel, their slopes are equal; and conversely.

Thus, in the figure at the right, if l_1 is parallel to l_2, then $\alpha_1 \cong \alpha_2$ and $m_1 = \tan \alpha_1$ equals $m_2 = \tan \alpha_2$. or, conversely, if $m_1 = m_2$, then $\alpha_1 \cong \alpha_2$ and the lines are parallel.
We also have this relationship:

> If two lines with slopes m_1 and m_2 are perpendicular, their slopes are negative reciprocals; and conversely.

Thus, in the figure at the right, l_1 and l_2, with slopes $m_1 = \tan \alpha_1$ and $m_2 = \tan \alpha_2$, are two lines that meet at right angles. Since each exterior angle of a triangle equals the sum of the two opposite interior angles (Chapter 2, frame 17), we can write $\alpha_2 = 90° + \alpha_1$. Thus, $\tan \alpha_2 = \tan(90° + \alpha_1) = -\cot \alpha_1 = -\dfrac{1}{\tan \alpha_1}$ and therefore

$$m_2 = -\dfrac{1}{m_1} \text{ or } m_1 m_2 = -1. \tag{7}$$

Let's look at an application of this relationship.

Example: Show that the line joining the points (5,3) and (2,−4) is perpendicular to the line joining the points (−4,2) and (3,−1).

Solution: Using equation (6) to find the slopes of the two lines we get
$$m_1 = \frac{3 - (-4)}{5 - 2} = \frac{7}{3}, \text{ and } m_2 = \frac{2 - (-1)}{-4 - 3} = -\frac{3}{7}. \text{ Hence } m_2 = -\frac{1}{m_1},$$
and the lines are perpendicular.

Now try this problem. Show that the line joining the points (3,5) and (−2,3) is perpendicular to the line joining the points (2,−1) and (−4,14).

- - - - - - - - - - - - - - - -

$$m_1 = \frac{5 - 3}{3 + 2} = \frac{2}{5}; \ m_2 = \frac{-1 - 14}{2 + 4} = -\frac{15}{6} = -\frac{5}{2}. \text{ Thus } m_2 = -\frac{1}{m_1}, \text{ and}$$
the lines are perpendicular.

8. The methods of analytic geometry also make it possible to find the angle between two intersecting lines that do *not* meet at right angles.

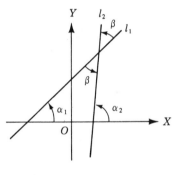

Thus, in the figure at the right, if we let l_1 and l_2 be the two lines, and β (beta) be the angle (measured counterclockwise) from l_1 to l_2, then $\alpha_2 = \alpha_1 + \beta$, or $\beta = \alpha_2 - \alpha_1$. And from this we can write (using the equation for the tangent of the difference of two angles, from frame 16 of Chapter 6),

$$\tan \beta = \tan(\alpha_2 - \alpha_1) = \frac{\tan \alpha_2 - \tan \alpha_1}{1 + \tan \alpha_1 \tan \alpha_2}.$$

But $\tan \alpha_1 = m_1$ and $\tan \alpha_2 = m_2$, hence the equation may be written

$$\tan \beta = \frac{m_2 - m_1}{1 + m_1 m_2}. \tag{8}$$

The *sign* of tan β in equation (8) tells us whether we have found the acute or the obtuse angle between the lines. If tan is positive, the angle is acute; if it is negative, the angle is obtuse.

Knowing β, the supplementary angle can be obtained by subtracting it from 180°. In this connection it is important to note that β *is measured from l_1 to l_2, hence m_2 is the slope of the line that is the terminal side of the angle.*

Except in specified cases, however, it is immaterial which line is designated as l_2 if we remember that once our choice is made, the acute angle β remains fixed.

Now let's see how to apply equation (8).

Example: Find the acute angle which the line joining the points (1,−2) and (−4,1) makes with the line joining the points (2,4) and (6,5).

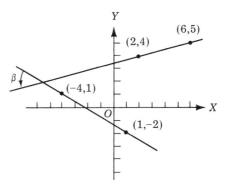

Solution: Plotting the points as shown at the right we see that m_2 is the slope of the line through the points (2,4) and (6,5). Therefore,

$$m_1 = \frac{-2 - 1}{1 + 4} = -\frac{3}{5}$$

$$m_2 = \frac{5 - 4}{6 - 2} = \frac{1}{4}$$

$$\tan = \frac{\frac{1}{4} + \frac{3}{5}}{1 - \frac{3}{20}} = \frac{17}{17} = 1,$$

and $\beta = 45°$. (If you have forgotten *why* the tangent of $45° = 1$, refer to frame 20, Chapter 5 for a quick review.)

Now suppose we hadn't plotted the points and had chosen the line through (1,−2) and (−4,1) as the terminal side of the angle. Then

$$m_1 = \frac{5 - 4}{6 - 2} = \frac{1}{4},$$

$$m_2 = \frac{-2 - 1}{1 + 4} = -\frac{3}{5},$$

$$\tan = \frac{-\frac{3}{5} - \frac{1}{4}}{1 - \frac{3}{20}} = -\frac{17}{17} = -1,$$

and $\beta = 135°$ (since the tangent is negative in the second quadrant). We can now obtain the desired angle from the relation $180° - \beta$, that is, $180° - 135° = 45°$.

Now it's your turn again. Find the acute angle which the line joining the points (−3,2) and (4,4) makes with the line joining the points (−2,−1) and (1,2). (*Note:* Be sure to plot these points and draw in the lines before attempting to solve the problem. The answer you get for tan β will not be a nice, round number this time. It will instead be a decimal fraction which you will have to look up in the table of Natural

Trigonometric Functions in order to find the value of β. Just select the nearest whole-degree value of the angle as your answer.)

— — — — — — — — — — —

l_1 is defined by the points $(-3,2)$ and $(4,4)$.
l_2 is defined by the points $(-2,-1)$ and $(1,2)$.

Therefore, $m_1 = \dfrac{4 - 2}{4 + 3} = \dfrac{2}{7}$, and $m_2 = \dfrac{2 + 1}{1 + 2} = 1$.

Hence $\tan \beta = \dfrac{1 - \dfrac{2}{7}}{1 + (\frac{2}{7})(1)} = \dfrac{\dfrac{7 - 2}{7}}{\dfrac{7 + 2}{7}} = \dfrac{5}{9} = 0.55555$

and $\beta = 29°$.

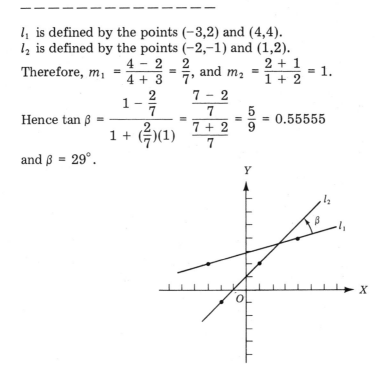

9. Here is another problem for the brave of heart. (Handle it just as you did the last problem, except that you need not plot it. It will require some persistence in handling the algebraic solution of the tangent function, particularly since it will contain a square root term. But if you stay with it, it reduces very nicely to a simple answer.) Find the acute angle between the line joining the points $(-1,3)$ and $(3,5)$ and the line joining the points $(-2,8)$ and $(-3,5\sqrt{3})$.

— — — — — — — — — — —

l_1 is defined by the points $(-1,3)$ and $(3,5)$.

l_2 is defined by the points $(-2,8)$ and $(-3,5\sqrt{3}$.

$$m_1 = \frac{5-3}{3+1} = \frac{1}{2}, \ m_2 = \frac{8-5\sqrt{3}}{-2+3} = 8 - 5\sqrt{3}.$$

Hence $\tan \beta = \dfrac{(8-5\sqrt{3}) - \dfrac{1}{2}}{2 + (8-5\sqrt{3}}$

$$= \frac{2(8-5\sqrt{3}) - 1}{2 + (8-5\sqrt{3})}$$

$$= \frac{3 - 2\sqrt{3}}{2 - \sqrt{3}} \text{ or, rationalizing the denominator,}$$

$$= -\sqrt{3},$$

from which $\beta = 120°$, hence the acute angle $= 180° - \beta = 60°$.

EQUATIONS AND LOCI

10. The term "loci" is simply the plural of the word *locus* which we worked with in Chapter 4, frames 15 and on. The two fundamental problems in analytic geometry are concerned with the concept of locus. These are, the *locus of an equation* and the *equation of a locus.* We can state these two problems as follows:

   (1) *Given an equation, find the corresponding locus and its properties.*

   (2) *Given a locus defined geometrically, find the corresponding equation.*

   You may recall (from Chapter 4) that the word locus, in Latin, means location. Thus, the locus of a point is the set of points, and only those points, that satisfy given conditions. And in your study of algebra you learned to use plotted points for the purpose of drawing the graph of a simple equation, either linear (first degree) or quadratic (second degree). Now we are going to extend these ideas a bit in order to become thoroughly familiar with the fundamental concepts of analytic geometry. Let's start by extending our definition of "locus" to read as follows:

   The locus or graph of an equation in two variables is the curve (including straight lines) that contains all of the points, and no others, whose coordinates satisfy the given equation.

And by "satisfy" we mean, as you will again recall from algebra, that they will reduce the equation to an identity (that is, the same value on both sides of the equal sign).

   While the above definition is perfectly correct, you should be aware that it is equally correct to think of a curve as the path traced by a moving point, in which case we can define it as follows:

If a variable point $P(x,y)$ moves in such a way that its coordinates must always satisfy a given equation, then the curve traced by P is called the locus of the equation: that is, the curve is the locus, or place, of all points (and no others) whose coordinates satisfy the equation.

The following examples should help clarify the concept of the locus of an equation.

Example 1: Suppose we decide to choose coordinates such that they must satisfy the equation $x = 2$. Here the value of y is not restricted and may assume any value whatever. The points of this locus will, therefore, lie on a straight line 2 units to the right of the Y-axis and parallel to it, and no points not on the line will satisfy the equation. The line is known, then, as the *locus of the equation*, and $x = 2$ is the equation of the line.

Example 2: If the values of the coordinates x and y are restricted by the equation $x - 2y + 2 = 0$, then notice that for each arbitrary choice of a value for x, the value of y is definitely determined. Thus, if we write the equation in the form $y = \frac{1}{2}x + 1$ and substitute $x = 2$, we find that $y = \frac{1}{2}(2) + 1 = 2$. The other points in the table are computed similarly, just as you did when plotting linear curves in

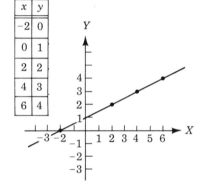

x	y
-2	0
0	1
2	2
4	3
6	4

algebra. As you can see, when we plot these points we find that instead of falling at random over the plane (that is, the surface of our coordinate system), they lie on a definite curve which appears to be a straight line, and this curve is the locus of the equation $x - 2y + 2 = 0$.

Example 3: Plot the locus of the equation $4y^2 - 9x - 18 = 0$. Solving the equation for y — just as you learned to do when working with quadratic equations in algebra — we get

$$y = \pm \frac{3}{2}\sqrt{x + 2}.$$ Assigning arbitrary values to x we get the y values shown in the table. Plotting these points and connecting them by a smooth curve gives us the locus of the equation.

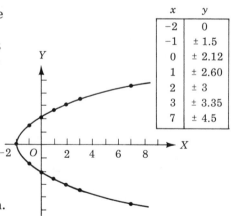

x	y
-2	0
-1	± 1.5
0	± 2.12
1	± 2.60
2	± 3
3	± 3.35
7	± 4.5

Using coordinate paper draw the locus of the following equations by plotting points.

(a) $y = x - 2$

(d) $x^2 = 4y - 12$

(b) $2y = x - 6$

(e) $y^2 + 2x - 4 = 0$

(c) $2x - 3y = 6$

- - - - - - - - - - - - - - - -

(a) $y = x - 2$

x	y
-2	-4
0	-2
2	0
4	2

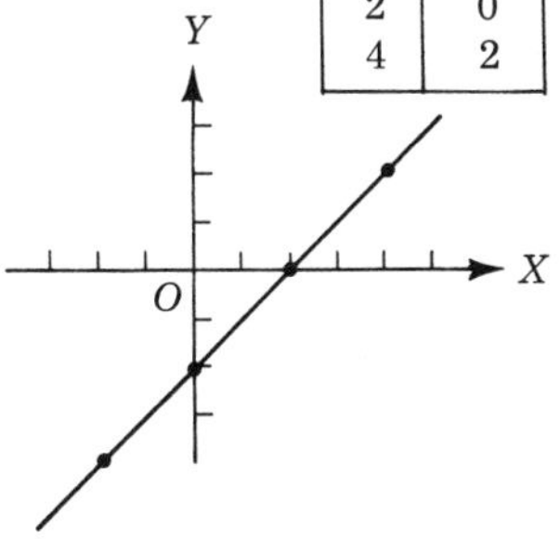

(b) $2y = x - 6$

$y = \dfrac{x - 6}{2}$

x	y
-2	-4
0	-3
2	-2
4	-1

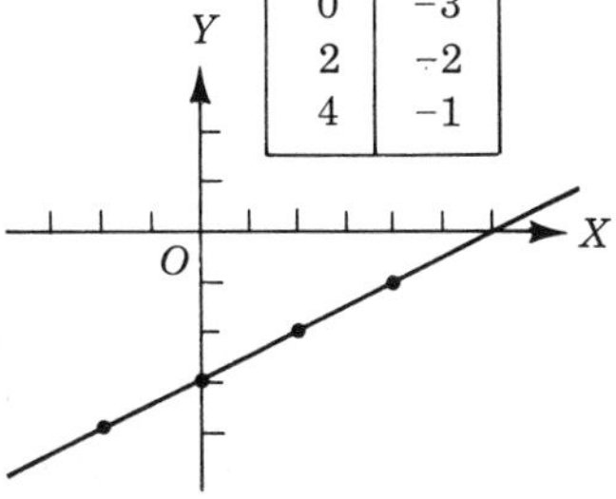

(c) $y = \dfrac{2x - 6}{3}$

x	y
-3	-4
0	-2
3	0
6	2

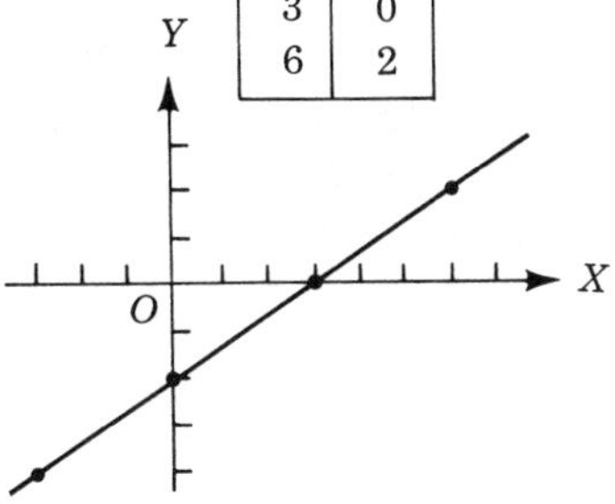

(d) $x^2 = 4y - 12$

$y = \dfrac{x^2 + 12}{4}$

x	y
-4	7
-2	4
0	3
2	4
4	7

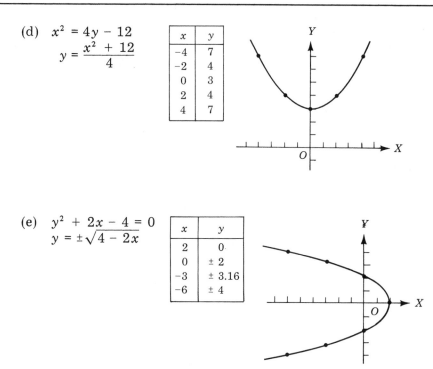

(e) $y^2 + 2x - 4 = 0$

$y = \pm\sqrt{4 - 2x}$

x	y
2	0
0	± 2
-3	± 3.16
-6	± 4

11. The graph of an equation when drawn by plotting separate points usually is an approximation since we cannot possibly plot all the points, and the position of a point cannot be drawn precisely. However, there are a few ways to check the geometric properties of a particular equation that will help verify our graphing. And since these are useful to know we will discuss them briefly.

The first property of an equation we will consider are its *intercepts*. This, again, is a term that should be familiar to you from your work in plotting from algebra. The intercepts of a curve are the directed distances from the origin to the points where the curve cuts the coordinate axes. Thus, to find the x-intercept, we let $y = 0$ in the equation of the curve and solve algebraically for x. This will give us the x-coordinate of the point where the curve cuts the X-axis. Similarly, to find the y-intercept, we substitute $x = 0$ in the equation and solve for y. This gives us the y-coordinate of the point where the curve cuts the Y-axis. Of course, in order for a curve to cut an axis the intercept on that axis must be real. That is, the equation must have real roots (i.e., not imaginary), as we learned in algebra.

Example: Examine the curve
$y^2 = 4x + 4$ for intercepts.
For $x = 0$ we get $y = \pm 2$ as the
y-intercepts.
For $y = 0$ we get $x = -1$ as the
x-intercept.

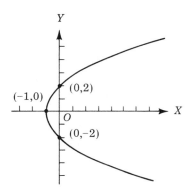

What are the x- and y-intercepts
of the equation $x^2 = 2y + 2$?

- - - - - - - - - - - - - -

When $x = 0$, $y = -1$; when $y = 0$, $x = \pm\sqrt{2}$.

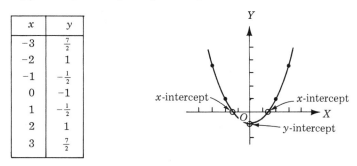

x	y
-3	$\frac{7}{2}$
-2	1
-1	$-\frac{1}{2}$
0	-1
1	$-\frac{1}{2}$
2	1
3	$\frac{7}{2}$

12. Another interesting and useful property
    often associated with curves is that of
    symmetry. We say that two points are
    symmetrical with respect to a line,
    called the *axis of symmetry,* if that line
    is the perpendicular bisector of the
    segment joining the two points. Thus,
    in the figure at the right the points P
    and P' are considered to be symmetrical
    with respect to a line (the Y-axis)
    because that line (the Y-axis) is the
    perpendicular bisector of PP', the
    segment joining the two points.

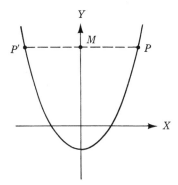

Two points are *symmetrical with respect to a third point,* called the
center of symmetry, if this third point is the midpoint of the segment
joining the two points. Thus, in the above figure, P and P' are symmetrical with respect to the midpoint M.

A curve is said to be symmetrical with respect to a line as an axis of
symmetry, or with respect to a point as a center of symmetry, *if each*

*point on the curve has a symmetrical point with respect to the axis or
center which is also on the curve.*

Thus, referring once more to the figure above, not only the two
points P and P', but the *entire curve* is symmetrical about (or with
respect to) the Y-axis, since for every point on the right side of the
curve there is a symmetrically-positioned point lying on the left side of
the curve.

Stating this a little more explicitly we can say that in order for a
curve to be symmetrical about the Y-axis (for example) to each point
of the curve in the first, or in the fourth, quadrant, there must be a
symmetrical point in the second, or in the third, quadrant which is also
on the curve. All of which leads us to the following tests for symmetry:

(1)   If an equation remains unchanged when x is replaced by $-x$, the
locus is symmetrical with respect to the Y-axis.

(2)   If an equation remains unchanged when y is replaced by $-y$, the
locus is symmetrical with respect to the X-axis.

(3)   If an equation remains unchanged when x is replaced by $-x$ and
y is replaced by $-y$ at the same time, the curve is symmetrical
with respect to the origin.

Thus, $x + y^2 = 5$, $x^2 + y = 5$, and $x^3 + y = 0$ are symmetrical with
respect to the X-axis, the Y-axis, and the origin, respectively.

In frame 11 we used as an example the equation $y^2 = 4x + 4$. Apply
the above three tests for symmetry and indicate your conclusions.

If we replace x by $-x$, we have $y^2 = -4x + 4$, which is not the same
equation as the original, hence the curve is not symmetrical about the
Y-axis. Replacing y by $-y$ gives us $(-y)^2 = y^2 = 4x + 4$ and therefore
leaves the equation unchanged, therefore the curve is symmetrical about
the X-axis. If both x and y are replaced by their negative opposites,
the equation is not the same, which tells us that the curve is not symmetrical about the origin.

13.   The third property of an equation we will discuss is that of *extent*.
When we consider an equation in two variables it is natural to ask if
there are values of one of the variables that will cause the other to
become imaginary. We might call these *excluded values* since they do
not give points on the curve. To find these values we begin by solving
the equation for y in terms of x, and for x in terms of y. If either
solution produces radicals of even order, the values of the variable that

make the expression under the radical sign negative must be excluded, since the corresponding values of the other variable will be imaginary.

Thus, in the equation $y^2 - x + 4 = 0$, solving for y gives us $y = \sqrt{x - 4}$. This is an even-order radical because it is the square root (rather than the cube root, for example, which would be of an odd order). And since values of x less than 4 would result in a negative value under the radical, such values must be excluded because corresponding values of y would be imaginary.

Shown at the right is the graph of the equation $y^2 = 4x + 4$. This is the equation we first saw in frame 11 and which we used to illustrate the concepts of intercepts and symmetry in an equation. Examine this curve for extent, using the procedure we have just discussed, and see what conclusions you can draw.

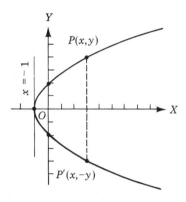

Solving the equation for y gives us $y = \pm 2\sqrt{x + 1}$, which shows that the expression under the radical is positive or zero for $x \geqslant -1$. This means that y is real for any value of $x \geqslant -1$, or that the curve lies entirely to the right of the line $x = -1$.

14. Try putting it all together now by examining the curve $9x^2 + 25y^2 = 225$ for intercepts, symmetry, and extent, and then draw the curve on graph paper. Refer to frames 11, 12, and 13 only as you need to for assistance. (*Note:* If a curve is symmetrical about both axes and the origin, it is a *closed* curve.)

(a) When $y = 0$ we have $x = \pm 5$ as the x-intercepts; when $x = 0$ we have $y = \pm 3$ as the y-intercepts.

(b) The equation remains unchanged when x is replaced by $-x$, when y is replaced by $-y$, and when both x and y are replaced by $-x$ and $-y$ at the same time. This means that the curve is symmetrical about both axes and the origin.

(c)   Solving the equation for y we have $y = \pm\dfrac{3}{5}\sqrt{25 - x^2}$. This shows us that in order for y to be real, x must not be greater than 5 or less than -5.

Similarly, $x = \pm\dfrac{5}{3}\sqrt{9 - y^2}$

shows us that only values of y from -3 to 3, inclusive, will give real values to x. These facts indicate that the curve is closed and that it lies wholly within the rectangle bounded by the lines $x = \pm 5$ and $y = \pm 3$.

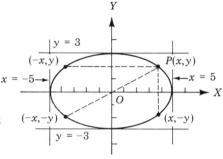

15.   Now we need to say a word about the matter of the *infinite extent of a curve*. It often happens that one of the variables of an equation becomes infinite for a finite value of the other variable. In such cases the tracing point of the curve recedes into infinity and, generally, we have two or more branches of the curve. Since it is important to know such values of the variables in discussing and graphing an equation, we're going to consider a method for finding them when they exist.

Example: Draw the graph of the equation $xy + x - 3y - 4 = 0$. Solving this equation for y in terms of x and for x in terms of y we get

(1) $y = \dfrac{4 - x}{x - 3}$ and (2) $x = \dfrac{3y + 4}{y + 1}$. In (1) observe that as x approaches 3, y becomes infinite and therefore the tracing point of the curve recedes to infinity for this value of x. Likewise in (2), as y approaches -1, x becomes infinite and the curve recedes to infinity for this value of y. By drawing the lines $x = 3$ and $y = -1$ first, and then computing a table of values, we get the curve shown at the right.

From the foregoing, then, we can state this rule for finding the infinite extent of a curve:

Solve the equation for x and, if the result is a fraction, place the denominator equal to zero and solve for y; solve the equation for y and, if the result is a fraction, place the denominator equal to zero and solve for x.

In general, the values found by equating the denominators to zero will represent lines along which the curve recedes to infinity.

Find the lines of infinite extent and plot the graph of the equation
$xy - 2y = 8$.

$x = \dfrac{2y + 8}{y}$, from which we get $y = 0$ as one of the lines.

$y = \dfrac{8}{x - 2}$, from which we get $x - 2 = 0$, or $x = 2$ as the other line.

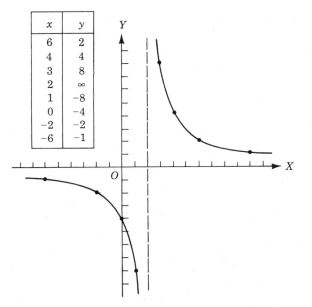

x	y
6	2
4	4
3	8
2	∞
1	-8
0	-4
-2	-2
-6	-1

16. The first fundamental problem of analytics — that of finding the locus of an equation — we have just discussed. The second problem which we mentioned in frame 10 is that of finding the equation of a locus, or curve, which is defined by means of a geometric property common to all points on the locus, and to no other points. That is, we are given the condition under which a point $P(x,y)$ moves in tracing a locus and are asked to find an equation in terms of the variables x and y that is satisfied by the coordinates of all points on the locus and by those of no other points. Such an equation is called the *equation of the locus.*

   Although there are no specific rules for finding such an equation, the following steps often prove useful:

   (1) If the coordinate axes are not determined by the statement of a given problem, choose them in such a way that the resulting equation will have a simple form. This choice of axes is permissible since the locus is independent of the axes to which it is referred.

(2) After constructing the axes, place the point $P(x,y)$, whose locus you wish to determine, in a representative position.

(3) Express the condition which P must satisfy in terms of x,y and any other constants involved in the definition of the locus. The equation thus obtained (or its simplified form) is the equation of the locus if it contains no variables except x and y and is satisfied by the coordinates of all points on the locus, and by those of no other points.

(4) Properties of the locus may be obtained by studying the equation thus obtained.

Let's see how this works.

Example 1: Find the locus of a point that is always equidistant from the extremities of the line segment joining the points $(-1,4)$ and $(2,2)$.
Here the coordinate axes are given since the points are located with reference to the X/Y coordinate system. So if we let the tracing point be $P(x,y)$, the geometric condition states that $PA = PB$. Expressing this condition in terms of coordinates we get (using equation (2) from frame 4 for the distance between two points),

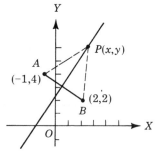

$$\sqrt{(x + 1)^2 + (y - 4)^2} = \sqrt{(x - 2)^2 + (y - 2)^2}$$

which, when simplified, becomes $6x - 4y + 9 = 0$. Plotting this equation we find the locus to be a straight line, the perpendicular bisector of the given segment. We encountered this fact first in Chapter 2, frame 15, where it was discussed as a distance principle of geometry.

Example 2: A point moves so that the sum of its distances from the points $(4,0)$ and $(-4,0)$ is 10 units. Find the equation of the locus.
Again we'll let $P(x,y)$ be the tracing point and let F and F' represent the given points. Then the geometric condition on the point P is that $PF' + PF = 10$. Then $PF' = \sqrt{(x + 4)^2 + y^2}$ and $PF = \sqrt{(x - 4)^2 + y^2}$. Hence

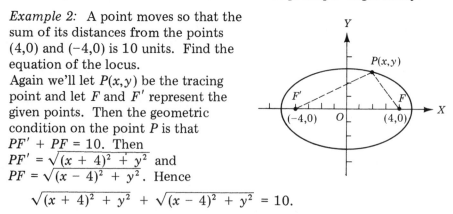

$$\sqrt{(x + 4)^2 + y^2} + \sqrt{(x - 4)^2 + y^2} = 10.$$

Transposing the second radical to the right side and then squaring both sides we get

$$x^2 + 8x + 16 + y^2$$
$$= 100 - 20\sqrt{(x - 4)^2 + y^2} + x^2 - 8x + 16 + y^2,$$

which, although it looks rather long and involved, reduces to

$$4x - 25 = -5\sqrt{(x - 4)^2 + y^2}.$$

Squaring again, and reducing, gives us the even simpler form

$$9x^2 + 25y^2 = 225.$$

By drawing the graph of this equation we find it is the symmetrical curve, or ellipse, shown above.

Apply this general approach in solving the following problems. Sketch the figure where possible and do your computations on a separate sheet of paper.

(a) A point moves so that it is always 4 units distant from the point $(-2,3)$. Find the equation of its locus.

(b) Find the equation of the locus of a point that moves so that it always is equidistant from the line $x = -2$ and the point $(2,0)$.

(c) If a point moves so that its distance from $(2,0)$ is twice its distance from $(-2,0)$, what is the equation of its locus?

– – – – – – – – – – – – – – – –

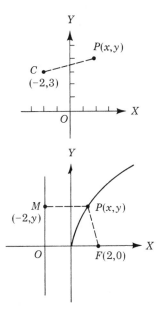

(a) Plotting the points and identifying them as shown, using formula (2) and the conditions of the problem we get $PC = \sqrt{(x + 2)^2 + (y - 3)^2} = 4$ or $(x + 2)^2 + (y - 3)^2 = 16$, and the locus equation is $x^2 + y^2 + 4x - 6y - 3 = 0$, which of course represents a circle.

(b) Sketching the situation described in the problem and identifying the parts as shown at the right we can establish that $PM = \sqrt{(x + 2)^2 + (y - y)^2}$ and $PF = \sqrt{(x - 2)^2 + y^2}$ or, since $PM = PF$, and squaring both sides, $(x + 2)^2 = (x - 2)^2 + y^2$, from which we find the equation of the locus to be $y^2 = 8x$.

(c) Letting $P(x,y)$ be the moving point, F_1 be the fixed point $(2,0)$ and F_2 the fixed point $(-2,0)$, we can define the distances as $PF_1 = \sqrt{(x - 2)^2 + y^2}$ and $PF_2 = \sqrt{(x + 2)^2 + y^2}$.

And since $PF_1 = 2(PF_2)$, then $\sqrt{(x-2)^2 + y^2} = 2\sqrt{(x+2)^2 + y^2}$, or squaring both sides gives us
$x^2 - 4x + 4 + y^2 = 4(x^2 + 4x + 4 + y^2)$,
from which we get the equation of the locus,
$3x^2 + 3y^2 + 20x + 12 = 0$.

17. In finding the equation of a curve, the coordinates of the tracing point are, of course, referred to a set of coordinate axes. If these axes are *moved*, not only will the coordinates of any fixed point change, but the equation of any fixed curve likewise will change.

Sometimes it is desirable to change the axes to which a curve is referred in order to simplify the equation of the curve. When such a change is made and the new axes are drawn parallel to the old, the transformation on the coordinates is known as a *translation*.

To obtain the relations that exist between the coordinates of a point referred to one set of axes and the coordinates of the same point referred to a second set of axes, parallel to the original set, we proceed as follows.

Let OX and OY be a set of coordinate axes, and $O'X'$ and $O'Y'$ be a second set parallel to the first. Then each point in the plane will have two sets of coordinates: (x,y) with reference to the original axes, and (x',y') with reference to the new axes. If we let (h,k) be the coordinates of the new origin with respect to the old axes, and let P be any point in the plane, then $x = SP$, $x' = AP$, $h = SA$, $y = NP$, $y' = BP$, and $k = NB$, as shown in the above figure. But $SP = SA + AP$ and $NP = NB + BP$, and therefore

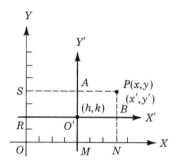

$$x = x' + h \quad \text{and} \quad y = y' + k. \tag{9}$$

These formulas, known as *translation formulas*, are true for any position of the point P, or of the axes, so long as the two sets of axes are parallel to one another.

Example 1: Transform the equation $3x - 2y + 6 = 0$ by translating the origin to the point $(2,6)$.
In this case the formulas of translation become $x = x' + 2$ and $y = y' + 6$. Substituting these values in the equation of the given line we get
$3(x' + 2) - 2(y' + 6) + 6 = 0$,
or $3x' - 2y' = 0$ as the equation of

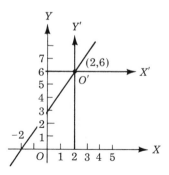

the line referred to the $O'X'$ and $O'Y'$ axes. This transformation leaves the line unaltered but, by moving the frame of reference, changes the equation of the line.

Example 2: Transform the equation $x^2 + y^2 + 6x + 4y - 3 = 0$ by translating the axes to the new origin $(-3,-2)$.
The translation formulas thus become $x = x' - 3$ and $y = y' - 2$. Substituting these values in the equation we obtain $(x' - 3)^2 + (y' - 2)^2 + 6(x' - 3) + 4(y' - 2) - 3 = 0$, or $x'^2 + y'^2 = 16$. As you can see, the transformation changes the form of the equation but not of the locus. This equation represents a circle of radius 4 and center $(0,0)$

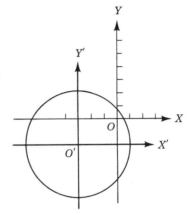

with reference to the new axes, or a circle of radius 4 and center $(-3,-2)$ with reference to the original axes.

Apply this approach in the following problems.

(a) Find the new coordinates of the points $(2,4)$, $(-2,2)$, and $(-2,0)$ if the axes are translated to a new origin at

   (1) $(4,4)$

   (2) $(2,2)$

   (3) $(0,-4)$.

(b) Find the equation of each of the following curves if the axes are translated to the new origin indicated.

   (1) $2x - 3y = 6$; $(4,1)$

   (2) $x^2 + y^2 - 6x + 4y - 12 = 0$; $(3,-2)$

   (3) $3y^2 - 12y - 7x - 2 = 0$; $(-2,2)$

— — — — — — — — — — — — —

(a) (1) $x = x' + h$ or $2 = x' + 4$, from which $x' = -2$, and $y = y' + k$ or $4 = y' + 4$, from which $y' = 0$; hence the new coordinates are $(-2,0)$. Similarly the new coordinates of $(-2,2)$ and $(-2,0)$ are $(-6,-2)$ and $(-6,-4)$ respectively.
   (2) $(0,2)$, $(-4,0)$, and $(-4,-2)$.
   (3) $(2,8)$, $(-2,6)$, and $(-2,4)$.

(b) (1) $x = x' + 4$ and $y = y' + 1$. Therefore,
    $2(x' + 4) - 3(y' + 1) = 6$, or $2x' - 3y' - 1 = 0$.
  (2) $x'^2 + y'^2 = 25$.
  (3) $3y'^2 - 7x' = 0$.

18. A very important use of the translation formulas is to simplify the equation of a given curve by some suitable choice of axes. Two methods of simplification are shown in the following example.

Example: Simplify the equation $x^2 + y^2 - 10x + 4y - 7 = 0$ by removing the first degree terms.

First Method: Substitute $x = x' + h$ and $y = y' + k$, expand the equation (that is, square the binomials and perform the indicated multiplications), and collect like terms. Thus,
$(x' + h)^2 + (y' + k)^2 - 10(x' + h) + 4(y' + k) - 7 = 0$; or
$x'^2 + 2x'h + h^2 + y'^2 + 2y'k + k^2 - 10x' - 10h + 4y' + 4k - 7 = 0$.
From which, collecting like terms, we get
$x'^2 + y'^2 + (2h - 10)x' + (2k + 4)y' + (h^2 + k^2 - 10h + 4k - 7) = 0$.
To remove the x' and y' terms it is necessary that the coefficients of these terms become zero, that is, $2h - 10 = 0$, or $h = 5$, and $2k + 4 = 0$, or $k = -2$. Substituting these values in the equation gives us $x'^2 + y'^2 = 36$ as the equation of the locus with reference to the new axes, chosen in such a way as to remove the first degree terms. The new origin, then, is the point $(5, -2)$.

Second Method: Another procedure that often can be used is that of completing the square. (If you have forgotten how to do this, you should refer to your favorite algebra text for a short review.) Thus, to complete the square of $x^2 - 10x$ we need to add $+25$. To complete the square of $y^2 + 4y$ we need to add $+4$. But if we add them to the *left* member, we also must add these values to the *right* member as well to maintain the equality. Doing so — and moving the -7 to the right side — we get $(x - 5)^2 + (y + 2)^2 = 7 + 25 + 4 = 36$. Now if we let $x - 5 = x'$ and $y + 2 = y'$, the equation becomes simply $x'^2 + y'^2 = 36$, where, again, the coordinates of the new origin (h, k) are $(5, -2)$.

The above two methods of determining the new origin give the same results, but the second method would be preferable in the present case. Incidentally, this second method should not be used with an equation that contains an xy-term.

Now you should practice using these methods. Remove the first degree terms from the following equations by translating the axes, using the first method.

(a) $4x^2 + 4y^2 + 12x - 4y - 6 = 0$

(b) $x^2 + y^2 + 10x - y + 3 = 0$

Use the second method to simplify the following equations by removing the first degree terms for translation.

(c) $x^2 + y^2 + 5x + 3y - 4 = 0$

(d) $y^2 - 8x^2 - 8y + 40 = 0$

- - - - - - - - - - - - - - -

(a) $x'^2 + y'^2 = 4$;   (b) $4x'^2 + 4y'^2 = 89$;   (c) $2x'^2 + 2y'^2 = 25$;
(d) $8x'^2 - y'^2 = 24$

THE STRAIGHT LINE

19.   We can define the equation of a straight line as an equation in x and y that is satisfied by the coordinates of every point on the line and by the coordinates of no other points. The *form* of a straight line equation will depend upon the information used to determine the line. Thus, if two points are used to determine the line the equation assumes one form. However, if one point and a direction are used, the equation will have a different form.

The essential facts about a straight line are that it is determined by two independent conditions, that its equation is of the first degree in the coordinates x and y, and that it may be expressed in several standard forms.

When a line is parallel to either axis its equation can be determined directly from a figure. Thus, as shown at the right, if the line L_1 is drawn parallel to the Y-axis and a units distant from it, then $x = a$ for every point on L_1.

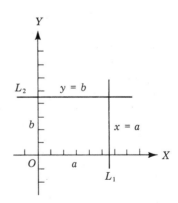

Since the equation $x = a$ is satisfied by the coordinates of every point on the line and by those of no other point, it is the equation of the line. The line lies to the right or left of the Y-axis according to whether a is positive or negative. Similarly, $y = b$ is the equation of L_2, a line parallel to the X-axis.

Now let's consider what is known as the *point-slope form* of the equation of a line. Specifically, we will seek to find the equation of a line L that passes through a fixed point $P_1(x_1,y_1)$ with a given slope m. Taking $P(x,y)$ as any other point on the line, since (x_1,y_1) and (x,y) are on the same line, we can write its slope as

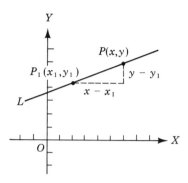

$$m = \frac{y - y_1}{x - x_1}.$$

Clearing fractions we get

$$y - y_1 = m(x - x_1). \tag{10}$$

This equation is true for any position of the point P on the line. We can, therefore, consider P as a tracing point since, as it moves, its coordinates will vary but will always satisfy the equation. This first degree equation is called the *point-slope form* of the equation of a line and should be used to write the equation of any straight line that passes through a fixed point with a given slope. If the coordinates of the given point P_1 are $(0,0)$, equation (10) becomes $y = mx$ and represents a line *through the origin with the slope m.*

Example: Find the equation of the line that passes through the point $(2,-\frac{5}{2})$ with the slope $-\frac{3}{4}$.

To draw the figure, we plot the given point $P_1(2,-\frac{5}{2})$ and then obtain a second point B by measuring from P_1 four units to the left and three units up (remember, the slope is $\frac{-3}{4}$ — that is, a ratio of 3:4,

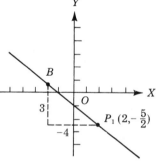

and negative). The equation of the line through B and P_1 is then, from (10), $y + \frac{5}{2} = -\frac{3}{4}(x - 2)$, which reduces to $3x + 4y + 4 = 0$.

This equation can (and should) be checked by plotting its graph from a table of values to show that the line actually satisfies the given conditions.

Write the equations of the lines that pass through the following points with the indicated slopes.

(a) $(-3,2)$, $m = \dfrac{2}{3}$ _____

(b) $(2,4)$, $m = 3$ _____

(c) $(-4,-6)$, $m = \dfrac{5}{7}$ _____

_ _ _ _ _ _ _ _ _ _ _ _ _ _

(a) $2x - 3y + 12 = 0$;   (b) $3x - y - 2 = 0$;   (c) $5x - 7y - 22 = 0$

20. Now let's consider another form of the equation of a line — the *slope-intercept* form.

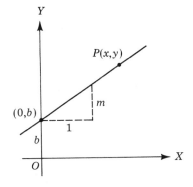

If the y-intercept of a line is b, the coordinates of the point of intersection of the line and the Y-axis are $(0,b)$. To express the equation of a line in terms of its y-intercept b and its slope m, we write the equation of the line through the point $(0,b)$ with the slope m, using (10). This gives us $y - b = m(x - 0)$, which reduces to

$$y = mx + b. \tag{11}$$

This is the slope-intercept form of the equation of a line. Note particularly the form of this equation. It not only allows us to write down the equation of a line when the y-intercept and slope are known, but it also enables us to find the slope and the y-intercept when the equation is given.

Example: Find the slope and y-intercept of the line whose equation is
$2x + 3y - 12 = 0$.
We first solve the equation for y, which changes it to the slope-intercept form
$y = -\dfrac{2}{3}x + 4$. By comparing this equation with the standard form $y = mx + b$ (11), we find that the slope is
$m = -\dfrac{2}{3}$ and the y-intercept is $b = 4$.

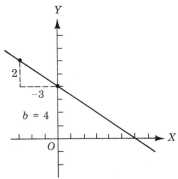

Using these two quantities the line can be easily drawn by measuring 4 units on the positive Y-axis and then constructing an angle whose tangent is $-\dfrac{2}{3}$, as shown in the figure above.

Try it and see how easy it is. Find the slopes and y-intercepts of the following lines.

(a) $3x - 5y - 10 = 0$

(b) $4x + 3y - 18 = 0$

(c) $3x + y = 7$

———————————————

(a) $3x - 5y - 10 = 0$ or $y = \frac{3}{5}x - 2$. Therefore, from $y = mx + b$,

   $m = \frac{3}{5},\ b = -2.$

(b) $m = -\frac{4}{3},\ b = 6$

(c) $m = -3,\ b = 7$

21.   To find the equation of a line determined by two points, we use a method which we developed in the last frame. First, we find the slope of the line through the two points. Then by substituting this slope and one of the points in the point-slope form, we get the required equation. Thus, if $P_1\,(x_1,y_1)$ and $P_2\,(x_2,y_2)$ are the given points, the slope of the line is

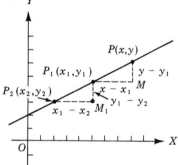

   $$m = \frac{y_1 - y_2}{x_1 - x_2}$$

and by using this slope and one of the points, say (x_1,y_1), in (10), we get the equation

   $$y - y_1 = \left(\frac{y_1 - y_2}{x_1 - x_2}\right)(x - x_1).$$

This equation can be written

   $$\frac{y - y_1}{x - x_1} = \frac{y_1 - y_2}{x_1 - x_2} \tag{12}$$

and is called the *two-point* form of the equation of a straight line.

   The above figure shows that the formula can be derived by using similar triangles. Thus, taking $P(x,y)$ as any point on the line, we can write

$$\frac{MP}{P_1 M} = \frac{M_1 P_1}{P_2 M_1}, \text{ or } \frac{y - y_1}{x - x_1} = \frac{y_1 - y_2}{x_1 - x_2}.$$

Example: Find the equation of the line determined by the points (−2,−2) and (5,2).

By finding the slope first, we can then use the point-slope equation. Thus, from (10),

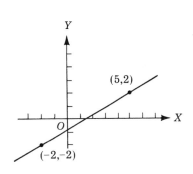

$$m = \frac{-2 - 2}{-2 - 5} = \frac{-4}{-7} = \frac{4}{7}.$$

Hence, using this slope with one of the points, say (−2,−2), we get the equation

$$y + 2 = \frac{4}{7}(x + 2), \text{ or } 4x - 7y - 6 = 0.$$

This same result can be written directly by using the two-point form. Thus, from (12),

$$\frac{y + 2}{x + 2} = \frac{-2 - 2}{-2 - 5} = \frac{-4}{-7}, \text{ or } \frac{y + 2}{x + 2} = \frac{4}{7},$$

and finally

$$4x - 7y - 6 = 0.$$

Once again, it is a good idea to test the accuracy of your work by substituting the coordinates of the given points in the final equation of the line.

Use formula (12) to write the equations of the lines determined by the following pairs of points.

(a)   (2,3) and (−3,5).

(b) $\left(\frac{3}{4}, \frac{3}{2}\right)$ and $\left(\frac{7}{2}, 3\right)$.

(c) $\left(-3\frac{1}{2}, 5\frac{1}{2}\right)$ and (4,−6).

- - - - - - - - - - - - - - -

(a)   From the given points (2,3) and (−3,5), $y_1 = 3$, $y_2 = 5$, $x_1 = 2$, and $x_2 = -3$. Hence, from (12), $\dfrac{y - 3}{x - 2} = \dfrac{3 - 5}{2 + 3} = -\dfrac{2}{5}$ or $5(y - 3) = -2(x - 2)$, from which we get $2x + 5y - 19 = 0$.

(b) $6x - 11y + 12 = 0$
(c) $23x + 15y - 2 = 0$

22. In the last frame we developed the two-point form of the equation of a line. This equation (12) was, as you will recall,

$$\frac{y - y_1}{x - x_1} = \frac{y_1 - y_2}{x_1 - x_2}.$$

Now if our two points should happen to be intercepts of the X- and Y-axes, then we can call the x-intercept a and the y-intercept b, in which case the coordinates of the points of intersection of the line and the axes are $(a,0)$ and $(0,b)$. Substituting these values in equation (12) we get

$$\frac{y - b}{x - 0} = \frac{b - 0}{0 - a}, \text{ or } \frac{y - b}{x} = -\frac{b}{a}.$$

This last equation can then be reduced to $bx + ay = ab$, or, dividing both sides by ab,

$$\frac{x}{a} + \frac{y}{b} = 1, \qquad (13)$$

which is the *intercept* form of the equation of a line.

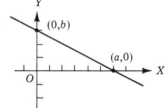

Example: Change the equation $4x - 5y - 8 = 0$ to the intercept form. By transposing the constant term to the right side and dividing the equation by it we get $\dfrac{x}{2} - \dfrac{5y}{8} = 1$. Expressing this in the intercept form we have

$$\frac{x}{2} + \frac{y}{-\dfrac{8}{5}} = 1.$$

Another method of accomplishing the same result would be to find the intercepts directly from the equation and then substitute them in the intercept form. Thus, setting x and y alternately equal to zero in the equation $4x - 5y - 8 = 0$, $a = 2$ for the x-intercept and $b = -\dfrac{8}{5}$ for the y-intercept. Substituting these values in the general intercept form gives us the same result we obtained above.

Use whichever procedure seems easiest to you to change the following equations to the intercept form.

(a) $4x + 3y - 12 = 0$

(b) $2x - 5y - 10 = 0$

(c) $3x - 7y - 6 = 0$

_ _ _ _ _ _ _ _ _ _ _ _ _

(a) $\dfrac{x}{3} + \dfrac{y}{4} = 1$;   (b) $\dfrac{x}{5} + \dfrac{y}{-2} = 1$;   (c) $\dfrac{x}{2} + \dfrac{y}{-\dfrac{6}{7}} = 1$

23.   We will conclude our discussion of the various forms for the equation of a line with a word or two about the *general equation* of a line.

The most general form of the equation of the first degree in the variables x and y is, as you no doubt have observed,

$$Ax + By + C = 0, \tag{14}$$

where A, B, and C are any constants, including zero, but with the restriction that A and B cannot be zero at the same time. Thus we have the theorem (which we will not attempt to prove here) that:

Every equation of the first degree in x and y is the equation of a straight line (and conversely).

From our general equation we can arrive at the following interesting and helpful conclusions:

(1)   If $C = 0$, the line passes through the origin.
(2)   If $B = 0$, the line is vertical.
(3)   If $A = 0$, the line is horizontal.
(4)   Otherwise, the line has the slope $m = -\dfrac{A}{B}$ and the y-intercept

$b = -\dfrac{C}{B}$.

Based on the above information, what can you conclude about these equations:

(a) $4x + 3y = 0$ _____

(b) $4x - 12 = 0$ _____

(c) $3y = 0$ _____

(d) $4x + 3y - 12 = 0$ _____

_ _ _ _ _ _ _ _ _ _ _ _ _

(a)   The line passes through the origin.
(b)   The line is vertical.

(c)   The line is horizontal.

(d)   The line has the slope $m = -\frac{4}{3}$; $b = 4$.

Since we have considered several forms of the equation of a line, let's review them briefly in order to help you fix them in your mind.

Point-Slope Form. Here we seek to find the equation of a line that passes through a fixed point with a given slope (m). From frame 19, the equation (10) is

$$y - y_1 = m(x - x_1)$$

Slope-Intercept Form. In this case we know the slope (m) and the y-intercept (b). From frame 20, the equation (11) is

$$y = mx + b.$$

Two-Point Form. We use this form of the equation when we know the coordinates of two points that the line passes through. From frame 21, the equation (12) is

$$\frac{y - y_1}{x - x_1} = \frac{y_1 - y_2}{x_1 - x_2}$$

Intercept Form. This form is used when we know, or can calculate, the x- and y-intercepts (a and b) of the line. From frame 22, the equation (13) is

$$\frac{x}{a} + \frac{y}{b} = 1$$

General Form. From frame 23, the general form of the equation of a line is

$$Ax + By + C = 0$$

Briefly, then, in this chapter we have considered the nature of analytic geometry, how we go about determining the properties of lines and curves by equations, the relation of Euclidean geometry to analytic geometry, basic definitions and theorems, equations and loci, and several forms of the equation of a straight line. Now it is time for a review test.

SELF-TEST

1.   Using graph paper draw a pair of coordinate (X, Y) axes, establish a scale, and plot the following points.

(a) $(-3,-5)$ (c) $(-2,5)$

(b) $(2,-4)$ (d) $(5,1)$ (frame 1)

2. Find the directed distance from:

    (a)   (3,2) to (7,2) _____   (c)   (−6,4) to (−2,4) _____

    (b)   (7,2) to (3,2) _____   (d)   (−2,3) to (−6,3) _____

                                     (frame 3)

3. Find the distance between the points (−2,3) and (4,−3). _____

                                     (frame 4)

4. Find the coordinates of the midpoint of the line segment joining (−3,4) and (−5,2) _____   (frame 5)

5. Find the coordinates of the point that is two-thirds of the way from (−5,−5) to (7,7). _____   (frame 5)

6. Find the slope of the line joining (−7,−3) and (1,5). _____

                                     (frame 6)

7. Show that the line through (−1,−4) and (4,2) is parallel to the line through (−3,−2) and (2,4). _____

                                   (frame 6)

8. Show that the line joining the points (5,−2) and (7,4) is perpendicular to the line joining the points (−3,4) and (9,0). _____

                                   (frame 7)

9. Find the acute angle which the line joining the points (−4,−2) and (2,3) makes with the line joining the points (−4,−1) and (4,1), to the nearest whole degree. _____   (frame 8)

10. Draw the locus of the equation $4x + y - 8 = 0$.   (frame 10)

11. Find the x- and y-intercepts of the equation $x^2 + y - 9 = 0$.

                                   (frame 11)

12. Apply the three tests for symmetry to the equation $4x^2 + y^2 - 16 = 0$ and state your conclusions. _____

_____

(frame 12)

13. Examine the equation $x^2 + y - 9 = 0$ for extent and see what conclusions you can draw. _____

_____

(frame 13)

14. Examine the curve $9x^2 - 4y^2 = 36$ for intercepts, symmetry, and extent and state your conclusions. _____

_____

(frame 14)

15. Find the lines of infinite extent and plot the graph of the equation $xy + 3x = 6$.   (frame 15)

16. Find the equation of the path traced by a point that moves in such a way as to remain equidistant from the points $(-2,4)$ and $(4,-2)$.

_____

(frame 16)

17. Find the equation of the curve $y^2 = 4x$ if the axes are translated to the new origin $(1,0)$. _____

(frame 17)

18. Transform the equation $9x^2 + 4y^2 - 54x + 32y + 1 = 0$ by translating the axes to the new origin $(3,-4)$. _____

(frame 17)

19. Remove the first degree term from the equation $9x^2 - y^2 + 2y - 10 = 0$ by translating the axes. _____

(frame 18)

20. Write the equation of the line that passes through the point $(7,-9)$ with the slope $m = 4$. _____

(frame 19)

21. Find the slope and y-intercept of the line whose equation is $2x + y = 8$. _____

(frame 20)

22. Write the equation of the line determined by the pair of points $(0,6)$ and $(-2,-3)$. _____

(frame 21)

23. Change the equation $12x + 5y + 50 = 0$ to the intercept form.

_____

(frame 22)

24. What can you conclude about the equation $3x + 2y - 6 = 0$?

_____

(frame 23)

Answers to Self-Test

1.

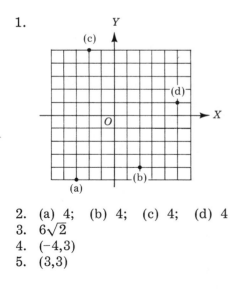

2. (a) 4;   (b) 4;   (c) 4;   (d) 4
3. $6\sqrt{2}$
4. $(-4,3)$
5. $(3,3)$

6. $m = 1$

7. $m_1 = m_2 = \dfrac{6}{5}$

8. $m_1 = 3$, $m_2 = -\dfrac{1}{3}$; since $m_1 = -\dfrac{1}{m_2}$, the slopes of the lines are negative reciprocals of one another and the lines are perpendicular.

9. $\beta = 26°$

10. $y = 8 - 4x$

x	y
0	8
1	4
2	0
3	−4
4	−8

11. When $x = 0$, $y = 9$; when $y = 0$, $x = \pm 3$.

12. Since the equation remains unchanged when x is replaced by $-x$ and y is replaced by $-y$, the curve is symmetrical with respect to the origin.

13. Solving for x gives us $x = \pm\sqrt{9 - y}$, hence values of $y > 9$ must be excluded since they make x become imaginary.

14. (a) When $y = 0$, $x = \pm 2$ as the intercepts; when $x = 0$, y becomes imaginary (i.e., the curve doesn't intersect the Y-axis).

    (b) Since the equation remains unchanged when x and y are replaced simultaneously by their negatives, the curve is symmetrical with respect to both axes and the origin.

    (c) Since $y = \pm\dfrac{3}{2}\sqrt{x^2 - 4}$, only values of $x \geqslant \pm 2$ will yield real values of y. And since $x = \pm\dfrac{2}{3}\sqrt{y^2 + 9}$, y can assume any positive or negative values and x will remain a real value.

15. $x = 0$ and $y = -3$ are the lines of infinite extent.

16. $x - y = 0$

17. $y'^2 = 4x' + 4$

18. $9x'^2 + 4y'^2 - 144 = 0$

19. $9x'^2 - y'^2 = 9$

20. $4x - y - 37 = 0$

21. $m = -2$; y-intercept is $b = 8$.

22. $9x - 2y + 12 = 0$

23. $\dfrac{x}{-25/6} + \dfrac{y}{-10} = 1$

24. The line has the slope $-\dfrac{3}{2}$ and the y-intercept is $b = 3$.

CHAPTER EIGHT
Conic Sections

The curves we will be studying in this chapter all are curves we have met briefly before in previous chapters, although they were not always identified by name.

The *conic sections* (sections of a right circular cone), or simply *conics*, consist of the circle, the parabola, the ellipse, and the hyperbola.

By definition, a *right circular cone* is the surface generated by rotating one straight line about another straight line, intersected at an oblique angle. The fixed line is the cone's *axis*. The possible positions of the generating line in its rotation are the cone's *elements*. The common intersection points of all the elements is the cone's *vertex*. And the two symmetrical parts of the generated surface on each side of the vertex are the *nappes* of the cone.

When an intersecting plane cuts through both nappes of the cone the resulting curve has two parts and is called a *hyperbola*. When the intersecting plane is at right angles to the axis the curve is a *circle*. When the intersecting plane cuts completely across one nappe at an oblique angle to the axis, the curve is an *ellipse*. When the intersecting plane is parallel to an element the curve is a *parabola*.

Although the Greeks studied conic sections for their aesthetic properties, it is more convenient to study these curves, by modern analytic methods, as loci. For by defining curves as loci we can more readily derive the equations that are their analytic equivalents.

Our overall objective in this chapter will, therefore, be to gain a working familiarity with the equations of the conics, be able to "know one when you see one," and be able

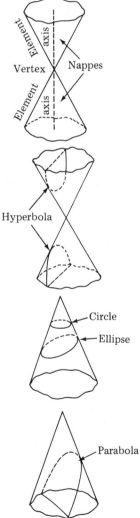

to plot and interpret these curves. A knowledge of the conics not only is important because they are encountered so frequently in every branch of engineering and science, but also because it is essential in the study of calculus.

Specifically, when you have completed this chapter you will be familiar with and able to use the following:

- definition, standard and general equation of a circle, its determination by three conditions, circles through the intersection of two given curves, and the radical axis of two circles;

- the definition and construction of a parabola, its equation and method of plotting, and other equations of the parabola;

- definition, equation, and construction of an ellipse, its special properties, and other equations of the ellipse;

- definition and equation of the hyperbola, construction, asymptotes, conjugate hyperbolas, equilateral hyperbola, and other equations of the hyperbola;

- lines associated with second degree curves, applications of the conics;

- polar coordinates.

The circle, parabola, ellipse and hyperbola all are represented by second degree (that is, non-linear) equations. We are going to approach our study of equations of the second degree from the point of view of finding *the equation of a locus.* That is, the law governing the motion of a point in a plane will be given as the definition of a curve, and from this definition we will find the algebraic expression that describes the path traced by the moving point. And, as before, since all the points, lines, etc. used to define these curves lie in the same plane, they are called *plane curves.*

THE CIRCLE

1. *A circle is the locus of a point that moves in such a way that its distance from a fixed point is always constant.* The fixed point is called the *center,* and the constant distance is, of course, the *radius* of the circle.

   To consider its equation let's look at the figure at the right. Let $C(h,k)$ be the fixed point, $P(x,y)$ the

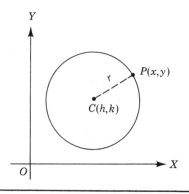

moving or tracing point, and $CP = r$ the constant distance. Thus, using the distance formula we get $\sqrt{(x - h)^2 + (y - k)^2} = r$, or

$$(x - h)^2 + (y - k)^2 = r^2. \tag{1}$$

Since this equation is satisfied by all points on the circle and by no other points, it is called the equation of a circle with center (h,k) and radius r.

If the center is at the origin, $h = k = 0$ and our equation becomes

$$x^2 + y^2 = r^2.$$

However, in order to develop a general equation for the circle we need to return to (1) and expand the binomial. Doing so and transposing r^2, this expression becomes

$$x^2 + y^2 - 2hx - 2ky + h^2 + k^2 - r^2 = 0.$$

But this is of the form

$$x^2 + y^2 + Dx + Ey + F = 0 \tag{2}$$

if we make the substitutions $D = -2h$, $E = -2k$, $F = h^2 + k^2 - r^2$. Therefore, we can say that it's always possible to write the equation of a circle in the form (2).

We also need to be able to show that the converse of the above is true. That is, that every equation of the form $x^2 + y^2 + Dx + Ey + F = 0$ represents a circle. To do so we transpose the constant term F to the right side and complete the square on each set of terms, $x^2 + Dx$ and $y^2 + Ey$, thereby obtaining

$$x^2 + Dx + \frac{D^2}{4} + y^2 + Ey + \frac{E^2}{4} = \frac{D^2}{4} + \frac{E^2}{4} - F,$$

where $\frac{D^2}{4}$ and $\frac{E^2}{4}$ have been added to the right member to preserve the equality. An equivalent (binomial) form of this equation is

$$\left(x + \frac{D}{2}\right)^2 + \left(y + \frac{E}{2}\right)^2 = \frac{1}{4}(D^2 + E^2 - 4F),$$

which is the algebraic expression of the condition that a tracing point (x,y) remain at a constant distance $\frac{1}{2}\sqrt{D^2 + E^2 - 4F}$ from a fixed point $\left(-\frac{D}{2}, -\frac{E}{2}\right)$. Hence it is the equation of a circle with center $\left(-\frac{D}{2}, -\frac{E}{2}\right)$ and radius $\frac{1}{2}\sqrt{D^2 + E^2 - 4F}$.

Since the radius is expressed as a radical, the following three cases may arise:

(1) When $D^2 + E^2 - 4F < 0$, the radius is imaginary and there is no real locus. The circle is called *imaginary*.

(2)   When $D^2 + E^2 - 4F = 0$, the radius is zero and the circle shrinks to a point, the center. In this case it is sometimes termed a *point circle.*

(3)   When $D^2 + E^2 - 4F > 0$, the radius is real and we have a real circle.

Since $ax^2 + ay^2 + bx + cy + d = 0$ (where $a \neq 0$) can be reduced to the form (2) by dividing through by a and substituting

$$D = \frac{b}{a}, \ E = \frac{c}{a}, \ F = \frac{d}{a}, \text{ we can say that:}$$

Every equation of the second degree in x and y, in which the xy term is missing and the coefficients of the x^2 and y^2 terms are the same, is the equation of a circle.

And since the symbols chosen to represent the variables are a matter of choice, the above statement is true when x and y are replaced by other letters.

Now let's see how we can apply this information.

Example: Find the center and radius of the circle
$4x^2 + 4y^2 - 12x + 4y - 26 = 0$, and draw the figure.
Solution: Dividing through by 4 in order to reduce the equation to the general form (2), we get $x^2 + y^2 - 3x + y - \dfrac{13}{2} = 0$. Hence, $D = -3$,

$E = 1$, $F = -\dfrac{13}{2}$, and $-\dfrac{D}{2} = \dfrac{3}{2}$, $-\dfrac{E}{2} = -\dfrac{1}{2}$, and

$r = \dfrac{1}{2}\sqrt{D^2 + E^2 - 4F} = \dfrac{1}{2}\sqrt{9 + 1 + 26} = 3$.

As the figure at the right shows, this is a circle with center $\left(\dfrac{3}{2}, -\dfrac{1}{2}\right)$ and $r = 3$.

This problem also can be solved directly, without the necessity of remembering the formulas for center and radius, by completing the squares on the x and y terms. After dividing through by 4, the equation can be written

$x^2 + y^2 - 3x + y = \dfrac{13}{2}$. Hence, by completing

the squares, $x^2 - 3x + \dfrac{9}{4} + y^2 + y + \dfrac{1}{4} = \dfrac{13}{2} + \dfrac{9}{4} + \dfrac{1}{4} = 9$, or

$\left(x - \dfrac{3}{2}\right)^2 + \left(y + \dfrac{1}{2}\right)^2 = 9$. By comparing this with

$(x - h)^2 + (y - k)^2 = r^2$ we see that the center is at $\left(\dfrac{3}{2}, -\dfrac{1}{2}\right)$ and the radius is $r = 3$.

Now it's your turn. Write the equations of the following circles and draw the figures on a sheet of graph paper.

(a)   Center at $(0,0)$; $r = 4$

(b)   Center at $(2,-2)$; $r = 6$

Find the center and radius of each of the following circles.

(c) $x^2 + y^2 - 2x + 4y - 11 = 0$

(d) $4x^2 + 4y^2 - 4x + 8y + 5 = 0$

———————————————

(a)   From (1): $(x - 0)^2 + (y - 0)^2 = (4)^2$ or $x^2 + y^2 = 16$
(b)   From (1): $(x - 2)^2 + (y + 2)^2 = 6^2$ or
      $x^2 - 4x + 4 + y^2 + 4y + 4 = 36$ and
      $x^2 + y^2 - 4x + 4y - 28 = 0.$
(c)   Completing the square,
      $x^2 - 2x + 1 + y^2 + 4y + 4 = 11 + 1 + 4$ or
      $(x - 1)^2 + (y + 2)^2 = 16$, and by comparison with (1), $h = 1$,
      $k = -2$, $r = 4$. Coordinates of center are, therefore, $(1,-2)$; $r = 4$.
(d)   Dividing through by 4 gives us $x^2 + y^2 - x + 2y = -\dfrac{5}{4}$, and completing the square we get
      $$x^2 - x + \frac{1}{4} + y^2 + 2y + 1 = -\frac{5}{4} + 1 + \frac{1}{4} \text{ or}$$
      $\left(x - \dfrac{1}{2}\right)^2 + (y + 1)^2 = 0$, from which the coordinates of the
      center are $\left(\dfrac{1}{2},-1\right)$ and $r = 0$.

2.   If we examine the standard forms of the equation of a circle,
     $(x - h)^2 + (y - k)^2 = r^2$, and $x^2 + y^2 + Dx + Ey + F = 0$, we see
     that each contains three arbitrary constants. Therefore, in order to
     obtain the equation of a particular circle we must be able to set up
     three independent equations from which the values of these constants —
     h, k, r or D, E, F — can be found. Such equations are the analytical
     expressions of conditions that the circle must satisfy. And since in
     general three such conditions will lead to three independent equations,
     we speak of a circle as being determined by three conditions.
         While it often is true that the given conditions determine just one
     circle — as, for instance, "three points not in the same straight line
     determine one and only one circle" — this is not always the case since it
     may happen that several circles satisfy the same conditions.

The usual method of solving problems of the type considered here is to decide which of the standard forms of the equation of a circle is to be used and then set up the three independent equations in the constants involved. And since it is easier to show than to talk about, let's look at a couple of examples.

Example 1: Find the equation of the circle through the points (1,2), (−2,1), and 2,−3).
Selecting the standard form
$x^2 + y^2 + Dx + Ey + F = 0$ to represent the circle, we reason that since each point is on the circle, then the coordinates of the given points must satisfy the equation.

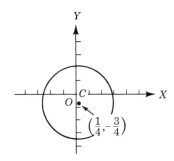

Substituting the coordinates of the three known points in the above equation, we obtain the following three equations:

$$1 + 4 + D + 2E + F = 0$$
$$4 + 1 - 2D + E + F = 0$$
$$4 + 9 + 2D - 3E + F = 0,$$

which when solved* for D, E, and F yield the values $D = -\dfrac{1}{2}$, $E = \dfrac{3}{2}$, and $F = -\dfrac{15}{2}$. Substituting these values in the general equation gives us

$$2x^2 + 2y^2 - x + 3y - 15 = 0$$

as the equation of a circle through the points (1,2), (−2,1), and (2,−3), with center at $\left(\dfrac{1}{4}, -\dfrac{3}{4}\right)$ and $r = \dfrac{1}{4}\sqrt{130}$. As usual, it is a good idea to check the accuracy of your work by substituting the coordinates of the given points in the final equation.

Example 2: Find the equation of the circle passing through the points (1,−2) and (5,3) and having its center on the line $x - y + 2 = 0$.
Choosing $(x - h)^2 + (y - k)^2 = r^2$ as the standard form, we substitute the given points and obtain

$$(1 - h)^2 + (-2 - k)^2 = r^2$$
$$(5 - h)^2 + (3 - k)^2 = r^2$$

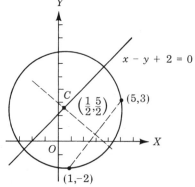

*If you have forgotten how to do this, review (from algebra) methods of solving systems of linear equations.

as two of the equations in h, k, and r. The third equation is found by substituting the coordinates of the center (h,k) in the equation of the line $x - y + 2 = 0$, since this line passes through the center. Doing so gives us

$$h - k + 2 = 0.$$

Solving the three equations simultaneously for h, k, and r we find that $h = \frac{1}{2}$, $k = \frac{5}{2}$, and $r = \frac{1}{2}\sqrt{82}$. Therefore the required equation of the circle is

$$\left(x - \frac{1}{2}\right)^2 + \left(y - \frac{5}{2}\right)^2 = \frac{82}{4}, \text{ or}$$
$$x^2 + y^2 - x - 5y - 14 = 0.$$

Problems of this kind offer you a good opportunity to exercise your ingenuity and knowledge of basic geometry. For although the solutions above are perfectly valid, it often is possible to devise a shorter and better method of attack by considering the geometry involved in a particular problem.

Try exercising your ingenuity on the following problems. Find the equations of the circles satisfying the following conditions. You might want to use a separate sheet of paper to do your computations, draw a figure for each case, and check your results.

(a) Passing through the points $(0,0)$, $(-2,-1)$, $(4,5)$

(b) Passing through the points $(-1,3)$, $(7,-1)$, $(2,9)$

(c) Having intercepts of 2 and 3 on the X- and Y-axis respectively, and passing through the origin

(d) Passing through the points $(5,0)$, $(0,-3)$ and having its center on the line $x - y = 0$

(e) Passing through the points $(-3,2)$, $(1,5)$ and having its center on the line $x = 5$.

$- - - - - - - - - - - -$

(a) Substituting the coordinates of the three points in the equation $x^2 + y^2 + Dx + Ey + F = 0$, we get the following three equations:

(1) $F = 0$
(2) $4 + 1 - 2D - E + F = 0$ or $5 - 2D - E = 0$
(3) $16 + 25 + 4D + 5E + F = 0$ or $41 + 4D + 5E = 0$

Multiplying (2) by 2 and adding (3) gives us

$$10 - 4D - 2E = 0$$
$$41 + 4D + 5E = 0$$
$$51 \qquad + 3E = 0, \text{ and } E = -\frac{51}{3} = -17$$

Substituting this value of E in equation (2) we get

$$5 - 2D - E = 0, \text{ or}$$
$$5 - 2D + 17 = 0, \text{ and } D = 11.$$

Finally, substituting the values of D, E, and F back in our original equation for a circle produces the answer, namely,
$$x^2 + y^2 + 11x - 17y = 0.$$

(b) $x^2 + y^2 - 9x - 8y + 5 = 0$

(c)   The coordinates of the three points would be (2,0), (0,3), and (0,0), hence the equation would be $x^2 + y^2 - 2x - 3y = 0$.

(d)   Substituting the values of the two points in the equation $(x - h)^2 + (y - k)^2 = r^2$ and restating the coordinates of the line gives us the three equations

$$(1) \quad 25 - 10h + h^2 + k^2 = r^2$$
$$(2) \quad h^2 + 9 + 6k + k^2 = r^2$$
$$(3) \quad h - k = 0$$

From which we find (by simultaneous solution) the values $h = 1$, $k = 1$, $r^2 = 17$. Substituting these values back in the original equation gives us our answer: $x^2 + y^2 - 2x - 2y - 15 = 0$.

(e) $x^2 + y^2 - 10x + 9y - 61 = 0$

THE PARABOLA

3.   *A parabola is the locus of a point that moves so that its distance from a fixed point is always equal to its distance from a fixed straight line.* The fixed point is called the *focus* and the fixed line the *directrix* of the parabola.

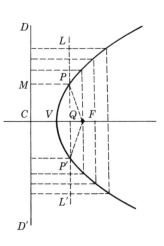

    To see what this looks like, consider the figure at the right. If we let F be the given point (focus) and DD' the given line (directrix), we can then draw a line through F perpendicular to DD' at C and let V be the midpoint of the segment CF. Since V is equidistant from C and F, it is, by definition, a point of the parabola.

To construct other points we proceed as follows: Through any point Q, lying on the line through C and F and to the right of V, we draw a line LL' parallel to the directrix DD'. Then with F as a center, describe (swing) an arc of radius CQ, intersecting LL' in P and P'. Since $FP = CQ = MP$, the point P is equidistant from the focus and the directrix, hence it lies on the parabola. Likewise P' is a point on the curve.

The line through C and F, which is seen to be a line of symmetry, is called the *axis* of the parabola. The point V, where the curve intersects its axis, is called the *vertex* of the parabola. These are a few new names for you to learn. But it will be worth the effort because you will encounter them frequently in calculus.

Just to make sure you remember the key reference elements of a parabola, complete the following.

(a)   The *focus* is _____ .

(b)   The *directrix* is _____ .

(c)   The *axis* is _____ .

(d)   The *vertex* is _____ .

- - - - - - - - - - - - - - - - - -

(a)   the fixed point of a parabola
(b)   the fixed line of a parabola
(c)   the line of symmetry of a parabola
(d)   the point where the curve intersects its axis

4.   The simplest form of the equation of a parabola is obtained by using one of the coordinate axes as the axis of the parabola and taking the vertex at the origin. Thus, in the figure at the right we let F have the coordinates $(p,0)$ and take V at O. Then the equation of the directrix is $x + p = 0$, since V is the midpoint of CF.

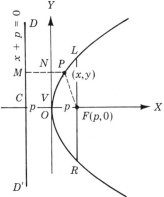

From frame 3 we know that for any point $P(x,y)$ on the curve we have, by definition, $FP = MP$ or, by squaring both sides, $(FP)^2 = (MP)^2$. However, $(FP)^2 = (x - p)^2 + y^2$, by the distance formula, and $(MP)^2 = (MN + NP)^2 = (p + x)^2$. Therefore, $(x - p)^2 + y^2 = (p + x)^2$, or simply

$$y^2 = 4px. \qquad\qquad (3)$$

This is the equation we want because, as we have just shown, it is true for every point on the curve. Equally important is the fact that it is *not* true for any other point, since for a point not on the curve $FP \neq MP$, $(FP)^2 \neq (x - p)^2 + y^2$, $(x - p)^2 + y^2 \neq (p + x)^2$, and finally, $y^2 \neq 4px$.

Looking at the equation $y^2 = 4px$ we can see that it consists of only two terms — the square of y and a constant times x. Therefore it is satisfied by $x = 0$, $y = 0$, and remains unchanged when y is replaced by $-y$. This tells us that the locus of the equation passes through the origin and is symmetrical about the X-axis.

If we reduce the equation (take the square root of both sides) to the form $y = \pm 2\sqrt{px}$, we see that p and x must be of like sign in order for y to be real and that for each value of x there are two values of y numerically equal but opposite in sign, these values of y increasing as x increases. Hence the curve opens to the right or the left according to whether p is positive or negative and extends indefinitely away from both coordinate axes.

When $x = p$, we find that $y = \pm 2p$. Therefore the length of the chord through the focus, perpendicular to the axis of the parabola, is $4p$, the coefficient of x in the equation $y^2 = 4px$. This chord is called by the fascinating name of the *latus rectum* and is shown in the figure above by the line LR.

If the focus is taken at the point $(0,p)$ on the Y-axis and the equation of the parabola derived, we get

$$x^2 = 4py, \tag{4}$$

which represents a parabola with the origin as its vertex, the Y-axis as its axis, the point $(0,p)$ as its focus, and the line $y + p$ as the equation of its directrix. It opens up or down according to whether p is positive or negative. (It would be good practice for you to derive equation (4)).

We can plot either equation (3) or (4) by computing a table of values. However, if we just want a sketch we can obtain it by drawing the curve through the vertex and the ends of the latus rectum. Let's see how this is done.

Example: Discuss the equation $y^2 = -6x$ and sketch the curve. The equation is satisfied by $(0,0)$ and remains unchanged when $-y$ is substituted for y. Hence the curve passes through the origin and is symmetrical about the X-axis. By comparing the equation with the standard form $y^2 = 4px$ we see that

$4p = -6$, or $p = -\dfrac{3}{2}$ and, therefore,

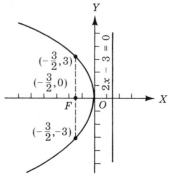

the curve has its focus at $\left(-\dfrac{3}{2},0\right)$ and opens to the left.

The equation of the directrix is $x - \dfrac{3}{2} = 0$, or $2x - 3 = 0$. When

$x = -\dfrac{3}{2}$, $y = \pm 3$, hence the coordinates of the ends of the latus rectum

are $\left(-\dfrac{3}{2}, \pm 3\right)$. The length of the latus rectum is 6 units. With these

facts known, the curve can be readily drawn, as shown above.

Use this general approach to guide you in solving the following problems. For each of these parabolas, find the coordinates of the focus and ends of the latus rectum, and the equation of the directrix. Sketch each curve.

(a) $y^2 = 8x$

(b) $y^2 = -4x$

(c) $4x^2 = 5y$

(a) Since $y^2 = 8x$, then $4p = 8$, or $p = 2$. Hence the coordinates of the focus are $(2,0)$. Substituting the x-coordinate of the focus, 2, into the equation of the curve gives us $y^2 = 8 \cdot 2 = 16$, hence $y = \pm 4$. Therefore the coordinates of the latus rectum are $(2, \pm 4)$. Substituting the value of p, 2, in the equation $x + p = 0$, we get $x + 2 = 0$ as the equation of the directrix. Your sketch of the equation should look generally like the one at the right.

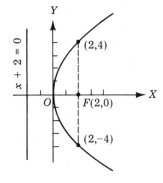

(b) $(-1,0)$; $(-1, \pm 2)$; $x - 1 = 0$.

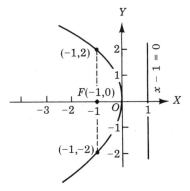

(c) $\left(0, \frac{5}{16}\right)$; $\left(\pm\frac{5}{8}, \frac{5}{16}\right)$; $16y + 5 = 0$.

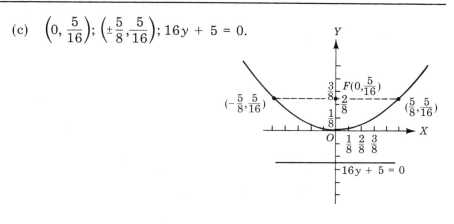

5. Now let's try an applied problem so you will get some feel for the way in which parabolas can appear and be solved in a real-life situation.

Example: A parabolic reflector is to be designed with a light source at its focus, $2\frac{1}{4}$ inches from its vertex. If the reflector is to be 10 inches deep, how broad must it be and how far will the outer rim be from the source?

Solution:

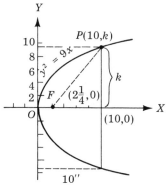

(1) Draw a diagram of the situation with the vertex of the reflector's parabolic cross-section at the origin and the focus at $(2\frac{1}{4}, 0)$ as shown at the right.

(2) The standard equation for the cross-section is, from (3), $y^2 = 4px$, but since $p = 2\frac{1}{4}$, then $y^2 = 4(\frac{9}{4})x = 9x$, hence the equation of the parabolic cross-section is $y^2 = 9x$.

(3) Since the reflector is to be 10″ deep, we can designate a point on its outer rim as $(10, k)$. Substituting these coordinate values in the equation $y^2 = 9x$ we get $k^2 = 9(10) = 90$, or $k = 9.486″$, and the total breadth is $2k = 2(9.486) = 18.972″$.

(4) To find the focal radius (that is, the distance of the point P from the focus F) since, by the definition of a parabola, the focal radius to any point on the curve is equal to the distance of the same point from the directrix, we can use the relationship

$$FP = x + p = 10 + 2\frac{1}{4} = 12.25″.$$

Apply this approach to the following problems. Use a separate sheet of paper for your computations.

(a) Compute the breadth of the parabolic reflector in the example above if it is designed to be 5 inches deep. What is the length of the focal radius to the rim of the reflector?

(b) Compute the breadth and focal radius if the reflector is designed to be 15 inches deep.

— — — — — — — — — — — — — — — —

(a) Breadth = $6\sqrt{5}$ inches;

focal radius = $7\frac{1}{4}$ inches

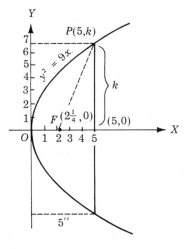

(b) Breadth = $6\sqrt{15}$ inches; focal radius = $17\frac{1}{4}$ inches.

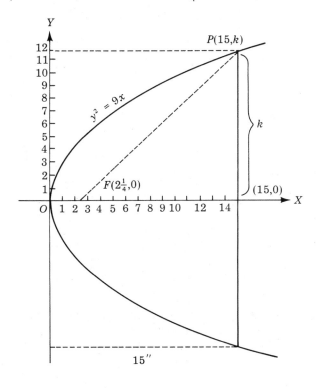

THE ELLIPSE

6. *An ellipse is the locus of a point that moves so that the sum of its distances from two fixed points is a constant.* The two fixed points are called *foci* (plural of *focus*) and the midpoint of the line segment joining them is known as the *center* of the ellipse.

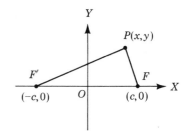

To obtain a simple form of the equation of the ellipse, let's take the foci on the X-axis and the center at the origin, as shown at the right. Then, if the distance between the foci F' and F is assumed to be $2c$ units in length, the coordinates of these points are $(-c,0)$ and $(c,0)$, respectively. Furthermore, if the sum of the distances of any point $P(x,y)$ on the ellipse from the foci is denoted by $2a$, we have by definition $F'P + FP = 2a$.

By looking at the triangle $F'PF$ we can see that $2a > 2c$ for any point not on the segment $F'F$, since the sum of two sides of a triangle is always greater than the third side. This being true, we will consider $a > c$ in the discussion that follows.

Expressing the above relation in terms of coordinates, we have

$$\sqrt{(x + c)^2 + y^2} + \sqrt{(x - c)^2 + y^2} = 2a, \qquad (5)$$

which, by transposing the second radical (although we could have used the first radical with the same results), squaring and reducing, becomes

$$a^2 - cx = a\sqrt{(x - c)^2 + y^2}.$$

Squaring again and simplifying, we get

$$(a^2 - c^2)x^2 + a^2 y^2 = a^2 (a^2 - c^2).$$

But $a^2 - c^2$ is a positive number since $a > c$, hence $a^2 > c^2$. Let's call this number b^2, where b is real, and make the substitution $b^2 = a^2 - c^2$. Our equation then becomes

$$b^2 x^2 + a^2 y^2 = a^2 b^2, \qquad (6)$$

or, dividing both sides by $a^2 b^2$,

$$\frac{x^2}{a^2} + \frac{y^2}{b^2} = 1. \qquad (7)$$

Now, what have we accomplished with all of the above? What we have done so far is to show that every point that satisfies the condition $F'P + FP = 2a$ has coordinates that satisfy equation (7). It is quite possible to prove the converse, namely, that every point whose coordinates satisfy equation (7) must also satisfy equation (5) and therefore be a point on the ellipse. But we're going to spare you that. If you are interested in seeing this proof you can find it in any good textbook on analytics.

In summary, what we have shown is that the equation of an ellipse with center at the origin, foci at $(\pm c, 0)$, and having $2a$ as a constant, is

$$\frac{x^2}{a^2} + \frac{y^2}{b^2} = 1, \text{ where } b^2 = a^2 - c^2.$$

Are you still clear as to what an ellipse is? Let's see. Complete the following definition.

An ellipse is the _____ of a point that moves so that

the _____ of its distances from two fixed points is

_____ .

---

locus; sum; a constant.

7. Now let's take a closer look at the equation we have just derived and see what more we can learn about it. The ellipse represented algebraically by equation (7) is symmetrical about both coordinate axes and the origin. How do we know this? By virtue of the tests for symmetry we learned in frame 12 of Chapter 7. Thus, the equation remains unaltered when x is replaced by $-x$, y is replaced by $-y$, and finally, when both x and y are replaced by $-x$ and $-y$ simultaneously.

Solving the equation of the ellipse for y in terms of x, and for x in terms of y, we find that $y = \pm\frac{b}{a}\sqrt{a^2 - x^2}$ and $x = \pm\frac{a}{b}\sqrt{b^2 - y^2}$.

The first of these equations shows that the only values of x that give real values of y are those for which $x^2 \leqslant a^2$. Likewise, from the second equation, values of y such that $y^2 \leqslant b^2$ are the only ones that give real values of x. Hence, as shown at the right, the curve lies between the lines $x = \pm a$ and $y = \pm b$. If $x = \pm a$ we find that $y = 0$, and if $y = \pm b$, $x = 0$.

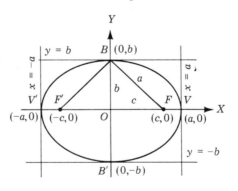

Therefore, the curve cuts the X-axis at $(\pm a, 0)$ and the Y-axis at $(0, \pm b)$.

The line segment $V'V$, of length $2a$, passing through the foci is called the *major axis*, while the chord $B'B$, of length $2b$, passing through the center perpendicular to the major axis is called, not surprisingly, the *minor axis*. The lengths a and b are called the *semi-major* and *semi-minor* axes, respectively. The end points, V' and V, of the major axis are known as the *vertices* of the ellipse.

From the Pythagorean Theorem, the relationship between the constants a, b, and c is expressed by the equation $a^2 = b^2 + c^2$. Interpreted geometrically this means that a line drawn from a focus to an end of the minor axis has the same length as the semi-major axis. The chord through either focus perpendicular to the major axis is called the *latus rectum*, a name that should sound familiar to you from our study of the parabola. Its length is found by substituting $x = c$ or $x = -c$ in the equation of the ellipse and solving for y. This gives us

$$y = \pm \frac{b}{a}\sqrt{a^2 - c^2} = \pm \frac{b^2}{a},$$ since $a^2 - c^2 = b^2$. Hence the length of the

latus rectum is $2\left(\dfrac{b^2}{a}\right)$ since it is the double ordinate (twice the length of the ordinate to the curve at that point) at a focus.

The value of the ratio $\dfrac{c}{a}$ indicates the shape of the ellipse, since for a of fixed length the curve flattens out as $c \to a$ and approaches a circle of radius a as $c \to 0$. The ratio takes values between 0 and 1 since $c < a$. It is called the *eccentricity* and is designated by the letter e, that is, $e = \dfrac{c}{a}$.

The equation of an ellipse with major axis along the Y-axis and foci at $(0, \pm c)$ is given by

$$\frac{y^2}{a^2} + \frac{x^2}{b^2} = 1. \tag{8}$$

Check your understanding and recollection by completing the following.

(a) The line segment $V'V$, of length $2a$, passing through the foci is

called the _____ .

(b) The chord $B'B$, of length $2b$, passing through the center is called

the _____ .

(c) The lengths a and b are called the _____ and

_____ axes respectively.

(d) The end points, V' and V, of the major axis are known as the

_____ of the ellipse.

(e) The chord through either focus perpendicular to the major axis is

called the _____ .

(f) The ratio $e = \dfrac{c}{a}$ is called the _____ of the ellipse.

(a) major axis;   (b) minor axis;   (c) semi-major, semi-minor;
(d) vertices;   (e) latus rectum;   (f) eccentricity

8. Just for a change of pace, let's try a little construction task. Specifically,
we're going to talk about how to construct an ellipse. So get out your
drawing compass and some graph paper.
   An ellipse can be constructed by means of points in the following
way.

(1) Lay off the major axis $V'V$ (as
shown at the right) and locate
the foci F' and F.

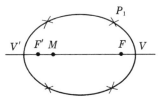

(2) Let M be any point on the line
segment $F'F$.

(3) With the foci as centers and a
radius MV, draw arcs above and below the major axis. With the
same centers and a radius MV', draw arcs intercepting those just
found. This will give four points of the ellipse; others can be
found by varying the position of M.

Now, why does this work? We can check the validity of this construction by calling one of the points P_1 and observing that $MV = F'P_1$ and
$MV' = FP_1$, hence $F'P_1 + FP_1 = MV + MV' = V'V$, which is the
length of the major axis. And since this length remains constant (that
is, $V'M + MV$ always $= V'V$), the sum of the distances from the foci
to any point on the curve is constant.
   If you are only required to make a *sketch* of the ellipse, however, it
is enough just to draw a curve through the x and y intercepts and the
extremities of the latera recta (plural of latus rectum). But let's see
how all this happens.

Example: Find the semi-major and semi-minor axes, the coordinates of
the foci and vertices, the length of the latus rectum, the eccentricity,
and sketch the ellipse $9x^2 + 4y^2 = 36$.
Solution: First, in order to reduce the equation to standard form we
will divide both sides by 36. This gives us $\dfrac{x^2}{4} + \dfrac{y^2}{9} = 1$. Since the
larger of the two numbers 9 and 4 appears in the term containing y^2,
this tells us that the major axis lies along the Y-axis. And since, from
(8), $a^2 = 9$ and $b^2 = 4$, we now know that the length of the semi-major
axis is $a = 3$ and of the semi-minor axis is $b = 2$. The coordinates of
the vertices are, therefore, $(0,\pm 3)$ and those of the ends of the minor
axis are $(\pm 2,0)$.
   From the relation $c^2 = a^2 - b^2$ (frame 7), we find that $c = 5$,
hence the coordinates of the foci are $(0,\pm\sqrt{5})$. And since we found
(again in frame 7) that the length of the latus rectum is given by the

formula $\dfrac{2b^2}{a}$, substituting the values $a = 3$ and $b = 2$ we find the length of the latus rectum in this instance to be $\dfrac{8}{3}$, hence the coordinates of the extremities are $(\pm\frac{4}{3}, \pm\sqrt{5})$. The eccentricity is $e = \dfrac{c}{a} = \dfrac{\sqrt{5}}{3}$, and the sketch of the figure is as shown at right.

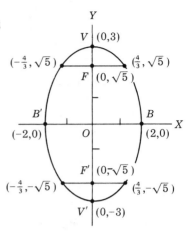

Here are a couple of problems for you to practice on. Find the semi-axes, the foci, the vertices, the latus rectum, and the eccentricity of the following ellipses, and sketch the curves.

(a) $x^2 + 2y^2 = 4$.

(b) $3x^2 + y^2 = 12$.

— — — — — — — — — — — —

(a) $x^2 + 2y^2 = 4$, or $\dfrac{x^2}{4} + \dfrac{y^2}{2} = 1$.

$a^2 = 4$, $b^2 = 2$, hence
$a = 2$, $b = \sqrt{2}$
(semi-axes).
$c^2 = a^2 - b^2 = 4 - 2 = 2$,
hence $c = \sqrt{2}$ and the
foci are $(\pm\sqrt{2}, 0)$.
From $a = 2$, the
vertices are $(\pm 2, 0)$.

$LR = \dfrac{2b^2}{a} = \dfrac{2(2)}{2} = 2$.

$e = \dfrac{c}{a} = \dfrac{\sqrt{2}}{2} = \frac{1}{2}\sqrt{2}$.

(b) $3x^2 + y^2 = 12$, or
$\dfrac{x^2}{4} + \dfrac{y^2}{12} = 1$. Since the
larger number, 12, is in
the term containing the
y^2, the major axis lies
along the Y-axis.
$a = 2\sqrt{3}$, $b = 2$
foci are $(0,\pm 2\sqrt{2})$
vertices are $(0,\pm 2\sqrt{3})$
latus rectum $= \frac{4}{3}\sqrt{3}$
(after rationalizing the
denominator)
eccentricity, $e = \frac{1}{3}\sqrt{6}$.

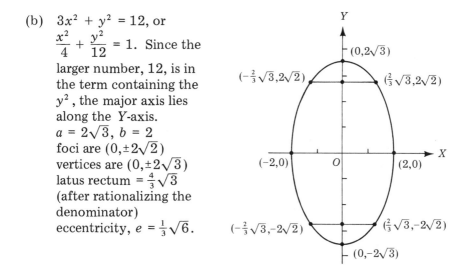

THE HYPERBOLA

9. *A hyperbola is the locus of a point that moves so that the difference of
its distances from two fixed points is a constant.* Does this definition
sound a bit like that of the ellipse? It differs from it by only one
word — *difference* rather than *sum*. You will recall that the ellipse
represents the locus of a point that moves so that the sum of its distances from two fixed points is a constant. With the hyperbola it is the
difference in these distances that is constant.

   If you have had any military experience with the LORAN (LOng
RAnge Navigation) system, you probably are aware that it is based on a
series of hyperbolic curves representing the loci of points at which the
time difference between signals received from two transmitting stations
is constant. By using maps overlaid with these hyperbolic curves and a
LORAN receiver to measure the time differential, the navigator can
plot his position (actually two such lines of position are required to
establish a "fix"). This is just another indication that the conic curves
have many useful and very practical applications.

   The two fixed points of a hyperbola
are called *foci*, just as with the ellipse,
and the midpoint of the line segment
joining them is called the *center* of the
hyperbola.

   A simple form of the equation of the
curve can be obtained by taking the foci
on the X-axis and the center at the
origin. Thus, in the figure at the right,
if the coordinates of the foci F' and F

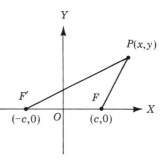

are $(-c,0)$ and $(c,0)$, respectively, and if $P(x,y)$ is any point on the hyperbola such that the difference of its distances from the foci is $2a$, we have by definition $F'P - FP = 2a$. In terms of coordinates this condition becomes

$$\sqrt{(x + c)^2 + y^2} - \sqrt{(x - c)^2 + y^2} = 2a,$$

which can be reduced to the form

$$(c^2 - a^2)x^2 - a^2 y^2 = a^2 (c^2 - a^2)$$

by following the steps used to find the equation of the ellipse in frame 6.

Since the difference of two sides of a triangle is always less than the third side, we have for triangle $F'PF$ (of the figure above), $F'P - FP < F'F$, or $2a < 2c$. Hence $a < c$ and $c^2 - a^2$ is a positive number. If we let b^2 represent this number, where b is real, and make the substitution $b^2 = c^2 - a^2$, our equation then assumes the form

$$b^2 x^2 - a^2 y^2 = a^2 b^2, \;.$$

or, dividing through by $a^2 b^2$,

$$\frac{x^2}{a^2} - \frac{y^2}{b^2} = 1 \tag{9}$$

where, as with the ellipse, $b^2 = a^2 - c^2$.

What we have shown above is that every point on the hyperbola has coordinates that satisfy equation (9). The converse is, of course, also true: that every point whose coordinates satisfy equation (9) lies on the hyperbola.

Just to make sure you understand what a hyperbola is, complete the following definition (without looking back to the beginning of this frame, if you can help it).

A hyperbola is the locus of a point that moves so that the _____

of its distances from _____ is a constant.

- - - - - - - - - - - - - - - -

difference, two fixed points

10. As in the corresponding case of the ellipse, the hyperbola that is represented algebraically by equation (9) is symmetrical about both axes and the origin.

Solving equation (9) first for y and then for x, we get

$$y = \pm\frac{b}{a}\sqrt{x^2 - a^2} \text{ and } x = \pm\frac{a}{b}\sqrt{y^2 + b^2}. \text{ From the first of these equa-}$$

tions we can see that in order for y to be real, x must take values such that $x^2 \geqslant a^2$; that is, no values of x between $x = -a$ and $x = a$ will give a point on the curve. The second equation shows that x is real for all real values of y. If $y = 0$, we find that $x = \pm a$, and if $x = 0$, y is imaginary. Hence the curve (shown at the right) cuts the X-axis at $(\pm a, 0)$ but does not intersect the Y-axis. It consists of two branches lying outside of the lines $x = \pm a$ and extending indefinitely away from both coordinate axes. The line through the foci is called the *principal axis* and the segment $V'V$, of length $2a$, is called the

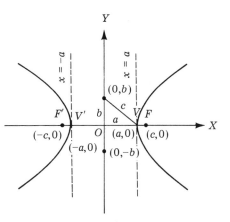

transverse axis. The line segment on the Y-axis between the points $(0, b)$ and $(0, -b)$, of length $2b$, is called the *conjugate axis*.

The lengths a and b are called the *semi-transverse axis* and *semi-conjugate axis*, respectively. The points V' and V, at the ends of the transverse axis, are called the *vertices* of the hyperbola.

From the relationship $c^2 = a^2 + b^2$ we can see that the distance from the center to a focus is the same as the distance from a vertex to an end of the conjugate axis.

The chord through a focus perpendicular to the principal axis is called (familiarly) the *latus rectum*. Its length, $\dfrac{2b^2}{a}$, is found in the same way as for the ellipse.

The eccentricity, $e = \dfrac{c}{a}$, is seen to be greater than 1 for the hyperbola since $c > a$. Its value indicates the shape of the curve.

The equation of the hyperbola with transverse axis along the Y-axis and foci at $(0, \pm c)$ is given by

$$\frac{y^2}{a^2} - \frac{x^2}{b^2} = 1. \tag{10}$$

Check your understanding and recollection of the above by completing the following (refer to the figure as necessary).

(a) The line through the foci is called the _____ .

(b) The segment $V'V$, of length $2a$, is called the _____ .

(c) The line segment on the Y-axis between the points $(0, b)$ and $(0, -b)$, of length $2b$, is called the _____ .

(d) The length a is called the _____ .

(e) The length b is called the _____ .

(f)   The points V' and V at the ends of the transverse axis are called the _____ of the hyperbola.

(g)   The chord through a focus perpendicular to the principal axis is called the _____ .

- - - - - - - - - - - - - -

(a) principal axis;   (b) transverse axis;   (c) conjugate axis;
(d) semi-transverse axis;   (e) semi-conjugate axis;   (f) vertices;
(g) latus rectum

11.   Now let's find out how we would go about constructing a hyperbola. A point-by-point construction of a hyperbola is quite similar to that of an ellipse. First, locate the foci, F' and F, and the vertices, V' and V, as shown at the right. Then, with the foci as centers and a radius MV, describe arcs above and below the principal axis. With the same centers and a radius MV', describe arcs intersecting those just drawn.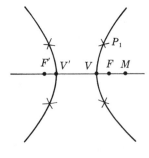

The four points thus found lie on the hyperbola. Other points may be constructed by varying M where M may coincide with F' or F, or may fall to the left of F' as well as to the right of F. We can see the construction is correct from the following argument, where P_1 represents a typical point located as described above: $MV' = F'P_1$ and $MV = FP_1$, therefore $F'P_1 - FP_1 = MV' - MV = VV'$, the length of the transverse axis, and a constant value.

Very shortly we're going to be looking at an example of how to go about finding the various elements of a hyperbola and sketch the curve, but first we need to introduce a final and very important concept, namely that of the *asymptotes* of a hyperbola.

We have seen from our discussion in frame 10 that the hyperbola $b^2 x^2 - a^2 y^2 = a^2 b^2$ consists of two branches opening outward to the right and left of the Y-axis. Now let's draw a line through the origin intersecting these branches in the points P and P', as shown at the right. If $y = mx$ is the equation of this line, we obtain, by solving it simultaneously with the equation of the hyperbola,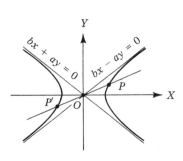

$$x = \pm \frac{ab}{\sqrt{b^2 - a^2 m^2}}$$

as the abscissas of the points of intersection of the line and the curve. As P moves to the right along the curve, or P' to the left, the numerical value of x increases without limit, and therefore, since a and b are fixed numbers, the denominator of the above fraction approaches zero.

When $b^2 = a^2 m^2 = 0$ we have $m = \pm \frac{b}{a}$, and the equation $y = mx$

becomes $y = \pm \frac{b}{a} x$, or $bx \pm ay = 0$. Hence, there are two lines,

$bx - ay = 0$ and $bx + ay = 0$, passing through the origin with slopes $\frac{b}{a}$ and $-\frac{b}{a}$, respectively, which the hyperbola gradually approaches as the numerical value of x increases.

If we define an *asymptote* of a curve as a straight line such that the perpendicular distance from the line to a point on the curve becomes and remains less than any positive value we can assign to it, as the point on the curve recedes indefinitely from the origin, then the lines $bx \pm ay = 0$ are *asymptotes* of the hyperbola $b^2 x^2 - a^2 y^2 = a^2 b^2$.

An easy way to find the equations of the asymptotes is to make the right member of the equation of the hyperbola zero and then factor. Thus, $\frac{x^2}{a^2} - \frac{y^2}{b^2} = 0$, or $b^2 x^2 - a^2 y^2 = 0$ factors into $bx - ay = 0$, the equations of the asymptotes.

If the equation of the hyperbola is $b^2 y^2 - a^2 x^2 = a^2 b^2$, showing that the foci are on the Y-axis, the equations of the asymptotes are $by \pm ax = 0$.

Asymptotes are very useful in sketching a hyperbola. Let's take the transverse axis $2a$ and the conjugate axis $2b$ as shown in the figure at the right, and construct a rectangle with its center at the center of the hyperbola and sides $2a$ and $2b$ parallel to the transverse and conjugate axes, respectively. Since the diagonals of this rectangle have slopes $\frac{b}{a}$ and $-\frac{b}{a}$,

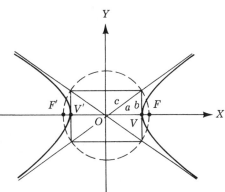

they become, when produced (extended), the asymptotes of the hyperbola. Thus to sketch a hyperbola we first draw the asymptotes and then use them as guidelines for the curve that is drawn tangent to the rectangle at a vertex and passing through the extremities of the latera recta.

Notice that a circle having the diagonals as diameters will pass through the foci of the hyperbola.

Now we will look at an example that should put it all together for you.

Example: Find the values of a, b, c, the coordinates of the foci, vertices and ends of the latera recta, the length of a latus rectum, and the equations of the asymptotes for the hyperbola $49y^2 - 16x^2 = 196$. Also sketch the curve.

Solution: Reducing the above equation to standard form by dividing through by 196 (the value of the right-hand member), we get

$$\frac{y^2}{4} - \frac{x^2}{\frac{49}{4}} = 1.$$

Hence $a = 2$, $b = \frac{7}{2}$, $c = \sqrt{a^2 + b^2} = \frac{1}{2}\sqrt{65}$, and $e = \frac{c}{a} = \frac{1}{4}\sqrt{65}$.

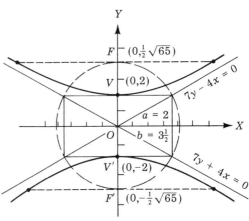

Since the term containing y is positive, we know that the transverse axis is along the Y-axis. Hence the coordinates of the desired points are: foci, $(0, \pm\frac{1}{2}\sqrt{65})$; vertices, $(0, \pm 2)$; ends of the latera recta, $(\pm\frac{49}{8}, \pm\frac{1}{2}\sqrt{65})$. The length of a latus rectum is $\frac{2b^2}{a} = \frac{49}{4}$. The equations of the asymptotes can be found by factoring $49y^2 - 16x^2 = 0$. They turn out to be $7y - 4x = 0$ and $7y + 4x = 0$. The figure above shows the sketch.

Using the above example as a guide, find the values of a, b, c, and e, the coordinates of the foci, of the vertices and of the ends of the latera recta, the length of a latus rectum, and the equations of the asymptotes for the following hyperbolas, and sketch the curves. (Remember to use equations (9) and (10) to help you determine whether the transverse axis lies along the X-axis in each case.) Use a separate sheet of paper for your work.

(a) $9x^2 - 16y^2 = 144$

(b) $81y^2 - 144x^2 = 11{,}664$

(c) $y^2 - x^2 = 16$

- - - - - - - - - - - - - -

(a)   To solve for the required values we proceed as follows.

(1)   Divide through by 144, giving us $\dfrac{x^2}{16} - \dfrac{y^2}{9} = 1$. Comparing this with (9) and noting that the term containing the x term is positive tells us that the transverse axis is along the X-axis.

(2)   Then $a^2 = 16$, or $a = 4$; $b^2 = 9$, or $b = 3$; $c = \sqrt{a^2 + b^2}$, or $c = \sqrt{4^2 + 3^2}$, from which $c = 5$; $e = \dfrac{c}{a} = \dfrac{5}{4}$.

(3)   Since $c = 5$, the coordinates of the foci are $(\pm 5,0)$.

(4)   Since $a = 4$, the coordinates of the vertices are $(\pm 4,0)$.

(5)   Length of latus rectum $= \dfrac{2b^2}{a} = \dfrac{9}{2}$.

(6)   The ends of the latera recta have as their coordinates $(\pm 5, \pm\tfrac{9}{4})$, the x-coordinate being that of the foci and the y-coordinate equal to one-half the length of the latus rectum.

(7)   By setting $9x^2 - 16y^2 = 0$ and factoring we find the equations of the asymptotes to be $3x \pm 4y = 0$.

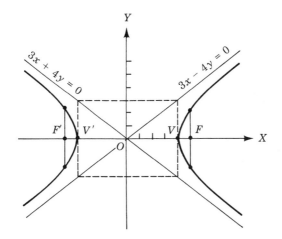

(b) $a = 12$; $b = 9$; $c = 15$;

$e = \dfrac{5}{4}$; foci $(0,\pm 15)$; vertices

$(0,\pm 12)$; length of latus

rectum $= \dfrac{27}{2}$; ends of latera

recta $(\pm \frac{27}{4}, \pm 15)$; equations

of the asymptotes are
$4x \pm 3y = 0$.

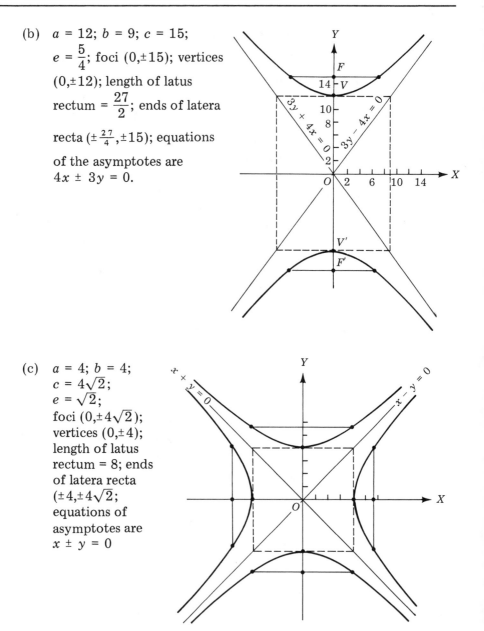

(c) $a = 4$; $b = 4$;
$c = 4\sqrt{2}$;
$e = \sqrt{2}$;
foci $(0,\pm 4\sqrt{2})$;
vertices $(0,\pm 4)$;
length of latus
rectum $= 8$; ends
of latera recta
$(\pm 4, \pm 4\sqrt{2}$;
equations of
asymptotes are
$x \pm y = 0$

12. Before leaving the subject of the hyperbola there are two special pairs of hyperbolas we should talk about. The first of these is *conjugate hyperbolas*.

Two hyperbolas are said to be conjugate when the transverse axis of each is the conjugate axis of the other.

Thus the equations $\dfrac{x^2}{a^2} - \dfrac{y^2}{b^2} = 1$ and $\dfrac{y^2}{b^2} - \dfrac{x^2}{a^2} = 1$ represent conjugate hyperbolas. Because $a^2 + b^2$ has the same value in both cases, it is

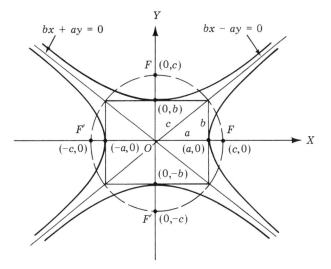

evident that the foci are equidistant from the center. Also, the two hyperbolas have common asymptotes since the left member of each equation, when set equal to zero, factors into $bx \pm ay = 0$. All this is shown in the figure above.

And since $\dfrac{y^2}{b^2} - \dfrac{x^2}{a^2} = 1$ is the same as $\dfrac{x^2}{a^2} - \dfrac{y^2}{b^2} = -1$, we can write

the equation for the conjugate of the hyperbola $\dfrac{x^2}{a^2} - \dfrac{y^2}{b^2} = 1$ by chang-

ing the sign of the constant term.

The other interesting hyperbola is the *equilateral hyperbola.*

When the transverse and conjugate axes are of the same length, the hyperbola is said to be *equilateral.*

From our study of the ellipse we know that the locus becomes a circle when the major and minor axes are equal. In the corresponding case of a hyperbola we have an equilateral hyperbola. Thus $b^2 x^2 - a^2 y^2 = a^2 b^2$ becomes $x^2 - y^2 = a^2$ when $b = a$, and the second equation represents an equilateral hyperbola with center at the origin and foci on the X-axis.

Since the asymptotes of an equilateral hyperbola meet at right angles, such a hyperbola often is called *rectangular.* Notice that the rectangle associated with a hyperbola is now a square.

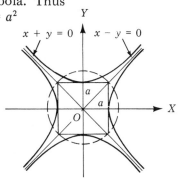

Make sure you understand the difference between conjugate and equilateral hyperbolas by completing the following definitions.

(a) Two hyperbolas are said to be conjugate when the _____ axis of each is the _____ axis of the other.

(b) When the _____ and _____ axes of a single hyperbola are _____ the hyperbola is said to be equilateral.

_ _ _ _ _ _ _ _ _ _ _ _ _ _

(a) transverse, conjugate;   (b) transverse, conjugate, of equal lengths

LINES ASSOCIATED WITH SECOND DEGREE CURVES

13. In working with curves of the second degree — the circle, ellipse, parabola, and hyperbola — we have found it useful to define and use certain lines, such as the directrix, the asymptote, and others. Now we are going to consider two more lines associated with second degree curves, namely, the secant and the tangent. The tangent is by far the more important of the two.

You may recall that in frame 1 of Chapter 2 we defined the *secant of a circle* as a line that intersects a circle at two points. Similarly, we defined the *tangent of a circle* as a line that touches the circle at only one point. From Chapter 5, frame 6, we also have our trigonometric definitions of the secant and tangent as ratios of the sides of a right triangle. However, although both definitions and applications are valid, we are mainly interested in the geometric concepts of these two lines at the moment.

Let's assume we have a curve and a secant through points P and Q on that curve, as shown at the right. As point Q moves closer to point P, the secant approaches the tangent as a limiting position, that is, the position at which the secant and tangent coincide. Stating this somewhat more formally we can say that

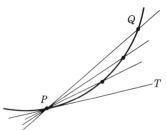

a tangent PT at a point P of a curve is defined as the limiting position of a secant PQ as Q approaches P along the curve.

This definition applies to any curve that has a tangent, and the methods of differential calculus are used to find the slope and hence the equation of PT.

Since our study is limited to second degree curves, we can use a special method rather than the general method of calculus in finding the equations of tangents. To be specific, let's suppose we wish to find the equation of the line of slope m that is tangent to the parabola $y^2 = 4px$. When a straight line $y = mx + k$ cuts the parabola $y^2 = 4px$ at the two points, P and Q, we find the coordinates of these points by solving the equations of the line and the parabola simultaneously. That is, by substituting $y = mx + k$ in the equation $y^2 = 4px$, we get $m^2 x^2 + 2mkx + k^2 = 4px$, or

$$m^2 x^2 + 2(mk - 2p)x + k^2 = 0, \tag{11}$$

which, solved for x, will give the abscissas of P and Q. The ordinates can be found by substituting these values of x back in the equation $y = mx + k$.

If the two points P and Q coincide, the line is said to be tangent to the parabola.

In this case it turns out, from what we know about quadratic equations, that the discriminant of equation (11) must be zero. (The *discriminant*, in case you have forgotten from algebra, is the quantity $b^2 - 4ac$ for the general quadratic equation $ax^2 + bx + c = 0$.) Thus, for our equation the discriminant has the value $4(mk - 2p)^2 - 4m^2 k^2$. Setting this equal to zero and solving we find that $k = \dfrac{p}{m}$, hence the equation of the tangent in terms of the slope m is given by

$$y = mx + \frac{p}{m}, \tag{12}$$

which is true for all finite values of m except $m = 0$.

Similarly, the tangents to the other second degree curves are found to be

$$y = mx \pm r\sqrt{1 + m^2}, \tag{13}$$

when the curve is the circle $x^2 + y^2 = r^2$;

$$y = mx \pm \sqrt{a^2 m^2 + b^2}, \tag{14}$$

when the curve is the ellipse $b^2 x^2 + a^2 y^2 = a^2 b^2$; and

$$y = mx \pm \sqrt{a^2 m^2 - b^2}, \tag{15}$$

when the curve is the hyperbola $b^2 x^2 - a^2 y^2 = a^2 b^2$.

Now let's see how all of this will help us in some useful way.

Example: Find the equations of the tangents to the circle $x^2 + y^2 = 16$ that have slope $-\frac{1}{2}$.

Solution: In order for the line $y = -\frac{1}{2}x + k$ to be a tangent to the given circle, the discriminant of the quadratic equation in x must be zero, so $x^2 + (-\frac{1}{2}x + k)^2 = 16$ (substituting $-\frac{1}{2}x + k$) for y, or $5x^2 - 4kx + 4k^2 - 64 = 0$. From the last equation, $a = 5$, $b = -4k$, and $c = 4k^2 - 64$. Therefore, the discriminant $b^2 - 4ac$ becomes $16k^2 - 20(4k^2 - 64) = 0$ and $k = \pm 2\sqrt{5}$. Hence $y = -\frac{1}{2}x \pm 2\sqrt{5}$ are the equations for the tangents.

You may be relieved to know that, since it is not really important at this point that you be able to solve problems by use of the secant/tangent concept, you are not going to be given any to work here. (But if you are one of those individuals who feels compelled to try out every new concept, you will have no difficulty in finding appropriate problems for the exercise of equations (11) through (15) in any good textbook on analytics). What is important for you to be aware of, however, is this notion that a tangent can be thought of as the *limiting position of a secant.* That is, when the points P and Q on our parabola coincide, the straight line connecting them (the secant) becomes tangent to the parabola. As mentioned earlier, this is a very fundamental concept in differential calculus and provides one of the classic examples of its application.

In the next chapter we will go into the matter of limits, so keep in mind what we have just discussed; it should help prepare you for a fuller investigation of the subject.

To make sure you've caught this new definition of a tangent, complete the following definition.

A tangent PT at a point P of a curve is defined as the _____

_____ of a _____ PQ as Q approaches P along the curve.

— — — — — — — — — — — — —

limiting position, secant

APPLICATIONS OF THE CONICS

14. Let's relax for a moment from the hard thinking you had to do in the last frame and reflect (*reflect* is a very appropriate word in this case) on some of the applications of the conics. Since a detailed discussion of many of the scientific applications requires a knowledge of calculus, we'll stick to a few basic uses.

The cable of a suspension bridge uniformly loaded along the horizontal hangs in the shape of a parabola. (If this same cable were supporting only its own weight, it would assume the shape of a different curve called a *catenary.*) The path of a projectile fired at an angle with the horizontal is a parabola, if air resistance is neglected. Arches of buildings and bridges often are parabolic in shape. Parabolic reflectors and reflecting telescopes make use of parabolic mirrors, these mirrors being formed by revolving a parabola about its axis. Such reflectors are highly effective since light emanating from a source placed at the focus will strike the parabolic surface and be reflected in parallel rays, giving a beam of light that can be controlled by turning the mechanism. This is the principle used in designing headlights, searchlights, and the like.

This same type of mirror is used in reflecting telescopes where the rays of light, coming from a distant source, strike the mirror in parallel lines and are collected at the focus. The design of a burning glass also is based on this property of the parabola. In this case the rays from the sun strike the convex surface of the glass and, after passing through it, are collected at the focus on the other side. In fact, it seems likely that the word *focus* was coined from this use of the parabola since the Latin meaning of the word is *hearth* or *fireplace.*

The conic sections have their application in more aspects of astronomy than simply the design of reflectors for telescopes. They are, in fact, used to describe the motion of celestial bodies such as planets, comets, and asteroids. Thus, in the middle of the sixteenth century when Copernicus completed his work on celestial orbits he concluded that planets revolve in circular orbits about the sun. (This was a departure from the Ptolemaic Theory that the sun and planets revolved around the earth.) However, the idea of uniform circular motion caused his results to be at variance with some of the known facts about planetary motion. Accordingly, about fifty years later, Kepler, after much computation, concluded that the planets move in *ellipses* with the sun at one focus. This was later confirmed by other astronomers and mathematicians, including Newton, who showed that the law of gravitation conforms to such a theory.

Ellipses are also used in architecture and bridge design. The Colosseum at Rome is in the shape of an ellipse, and many beautiful stone and concrete bridges have elliptical arches. The design of whispering galleries is based on the ellipse, where a sound from one focus may be heard at the other but is inaudible between these two points. Elliptical gears are used in such machines as power punches and planers, where a slow but powerful stroke is required.

A hyperbola, referred to by its asymptotes as axes, can be used to express Boyle's law of a perfect gas. This equation also is used in the study of economics and in locating a source of sound, as in range finding. The use of hyperbolas in position-finding (navigation) systems such as LORAN was mentioned earlier.

You will come across these and many other applications if you continue your study of science and mathematics, and you will find many ways in which to extend your knowledge of the conics when you study calculus. But now let's proceed to one final topic before leaving our brief investigation of analytic geometry and the conic sections.

POLAR COORDINATES

15.   In this chapter and the previous one we have discussed equations of the first and second degree using rectangular coordinates to show their corresponding graphs. Now we are going to introduce a new system of coordinates, called *polar*, and you will see that many equations assume a simpler form when expressed in terms of such coordinates.

> *In polar coordinates the position of a point is determined by a direction and a distance (rather than by two distances as in rectangular coordinates) and the frame of reference consists of a point and a directed line.*

Thus, in the figure at the right, let O be a fixed point, called the *origin*, or *pole*, and OX be a fixed directed line, called the *initial line*, or *polar axis*. The position of any point, P, is determined by two numbers, the angle $XOP = \Theta$, and the distance $OP = r$.

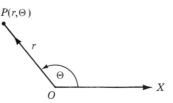

The coordinate r is called the *radius vector* and Θ the *vectorial angle*. (You may find it helpful at this point to turn to Chapter 5, frame 22, and review some of the things we learned about vectors.)

The usual convention of signs used in trigonometry applies to the vectorial angle. That is, a positive angle is generated by a counterclockwise rotation and a negative angle by a clockwise rotation of the initial side. The radius vector r is positive when it is measured from the pole along the terminal side of the angle, and negative when measured in the opposite direction.

Since the position of a point is determined by direction and distance, to plot a point, the angle is first drawn in the proper direction, thus locating the terminal side, and then the distance r is measured either along the terminal side, if positive, or along the terminal side produced through the pole, if negative.

Notice that *one* pair of polar coordinates will determine one, and only one, point in the plane, but that any given point may have an unlimited number of polar coordinates. If the angle is restricted to values between $0°$ and $360°$, any given point may be designated by *four* different pairs of polar coordinates, as shown below. Here the points $(2,120°)$, $(-2,300°)$, $(-2,-60°)$, and $(2,-240°)$ all determine the same point Q.

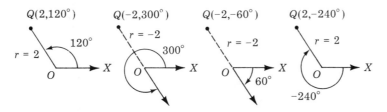

For a little practice in using polar coordinates, plot the following points, using the same pole and polar axis (you will need the aid of a protractor to measure angles and will, of course, need to establish some convenient scale).

(a) $(5,30°)$

(b) $(-3,15°)$

(c) $(8,-60°)$

(d) $(2,90°)$

(e) $(6,-180°)$

(f) $(-5,45°)$

– – – – – – – – – – – – – –

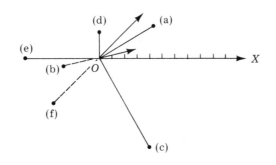

16. If r and Θ are connected by an equation, values may be assigned to Θ and corresponding values for r calculated. We then will have a table of values for points that may be plotted and joined by a curve, thus describing the locus of the equation.

Example: Construct a table of values and plot the curve $r = 2(1 - \cos \Theta)$.

Solution: Computing the table of values shown below and with the aid of a sheet of polar coordinate paper (which is nearly indispensable for plotting polar coordinates), we plot the r values for the selected angles and obtain the curve shown at the right.

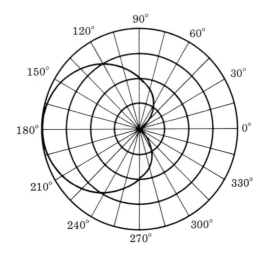

In plotting equations in polar coordinates it is helpful to be aware of the *symmetry* of the curve. Thus, if we can replace Θ by $-\Theta$ and obtain the same value of r, then the locus is said to be *symmetric with respect to the polar axis.* Or if we can replace r by $-r$ for the same value of Θ, then we say the curve is *symmetric with respect to the pole.*

Θ	$\cos \Theta$	$1 - \cos \Theta$	r
0°	1.000	0.000	0.00
30°	0.866	0.134	0.27
60°	0.500	0.500	1.00
90°	0.000	1.000	2.00
120°	−0.500	1.500	3.00
150°	−0.866	1.866	3.73
180°	−1.000	2.000	4.00

Because of the symmetry of the cosine function we do not need to compute values of Θ from 180° to 360° in the above example.

With the aid of polar coordinate paper found in the Appendix, construct a table of values and plot the curve $r = 3 \cos \Theta$. (Cos function values to two decimal places is adequate.)

- - - - - - - - - - - - - - -

Assigning values to Θ and calculating the corresponding values of r we get:

Θ	$\cos \Theta$	$r = 3 \cos \Theta$
0°	1.00	3
30°	.87	2.61
60°	.50	1.50
90°	0	0
120°	−0.50	−1.50
150°	−0.87	−2.61
180°	−1.00	−3

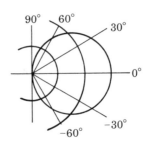

17. The simplest problem of tracing polar curves is the case in which there is only one value of r for each value of Θ, such as the equation $r = 5 \cos 2\Theta$. Such curves are called *single-valued functions.* On the other hand, the case in which r^2 is expressed as a function of Θ yields two values of r for each value of Θ and is called a *double-valued function.* An example of this would be an equation such as $r^2 = 25 \sin 2\Theta$.

There are many interesting and even beautiful curves represented by various polar equations and often having intriguing names, such as the Lemniscate of Bernoulli, Limaçon, Spiral of Archimedes, Conochoid of Nicomedes, With of Agnesi, or Cissoid of Diocles. If you are interested, look up the equations of some of these in one of the referenced text-books and have some fun plotting them.

Now, having discussed both rectangular and polar coordinates in our work thus far, we need to find some way of relating these different sets of coordinates so that we can convert from one to the other as the occasion requires.

If the pole, in our polar coordinate system, coincides with the origin in our rectangular coordinate system, and the polar axis OX is taken as the positive X-axis as shown in the figure at the right, then any point P may be considered as having rectangular coordinates (x,y) and polar coordinates (r,Θ). The relations between the two systems can be taken directly from the figure. Thus,

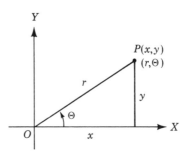

$$x = r \cos \Theta, \qquad y = r \sin \Theta;$$
$$r^2 = x^2 + y^2, \quad \text{and} \quad \Theta = \tan^{-1} \frac{y}{x}. \tag{16}$$

See Chapter 6, frame 3 if you need to review the derivation of these relationships.

By means of these relations we can transform an equation in polar coordinates to one in rectangular coordinates, and vice versa.

Example 1: Transform the equation $r = 5 \cos \Theta$ into an equation in rectangular coordinates.

Solution: Substituting $r = \sqrt{x^2 + y^2}$ and $\cos \Theta = \frac{x}{r}$ from the formulas above (16), we get

$$\sqrt{x^2 + y^2} = \frac{5x}{\sqrt{x^2 + y^2}},$$

or $x^2 + y^2 - 5x = 0$.

Example 2: Transform $(3 - 2 \cos \Theta)r = 2$ into an equation in rectangular coordinates.

Solution: Performing the indicated multiplication in the left member we can write the equation as $3r - 2r \cos \Theta = 2$. Then by substituting $r = \sqrt{x^2 + y^2}$ and $r \cos \Theta = x$, we get

$$3\sqrt{x^2 + y^2} - 2x = 2.$$

Transposing $-2x$, squaring both sides, and combining terms, we finally have $5x^2 + 9y^2 - 8x - 4 = 0$.

Example 3: Transform $r = 4 \sin 2\Theta$ into rectangular coordinates.

Solution: By using $\sin 2\Theta = 2 \sin \Theta \cos \Theta$ (the double-angle formula from frame 17, Chapter 4), we can write the equation as $r = 8 \sin \Theta \cos \Theta$. Then by substituting values of $\sin \Theta$ and $\cos \Theta$, we have $r = 8\frac{y}{r} \cdot \frac{x}{r}$, or $r^3 = 8xy$. But $r = \sqrt{x^2 + y^2}$, hence we can write

$$(x^2 + y^2)^{\frac{3}{2}} = 8xy, \text{ or } (x^2 + y^2)^3 = 64x^2 y^2.$$

Example 4: Transform $x^2 + 2y^2 = 8$ into polar coordinates.

Solution: Substituting $x = r \cos \Theta$ and $y = r \sin \Theta$ we get $r^2 \cos^2 \Theta + 2r^2 \sin^2 \Theta = 8$, or $r^2(1 + \sin^2 \Theta) = 8$.

Try a few of these transformations just for practice. Transform the following equations into rectangular coordinates.

(a) $r = 8 \sec \Theta$ (remember that $\sec \Theta = \dfrac{1}{\cos \Theta}$)

(b) $\Theta = \dfrac{\pi}{6}$

(c) $r = 3 \cos \Theta$

Transform the following equations into polar coordinates.

(d) $x + y = 0$

(e) $x^2 + y^2 = 16$

(f) $x^2 + y^2 - 4x - 4y = 0$

— — — — — — — — — — — —

(a) $x = 8$, or $x - 8 = 0$.
(b) $x - 3y = 0$.
(c) $x^2 + y^2 - 3x = 0$.
(d) $\sin \Theta + \cos \Theta = 0$.
(e) $r = \pm 4$.
(f) $r = 4(\cos \Theta + \sin \Theta)$

Now it is time for us to take a look back over what we have covered in this chapter. The following Self-Test is, as usual, intended to assist you with your review and check-up.

SELF-TEST

1. Write the equation of the circle whose center is at $(-4,2)$ and which touches the Y-axis, and draw the figure on graph paper. (You *know* the radius, r, from the fact that the abscissa of the center is -4 and the circle touches the Y-axis.) _____

   (frame 1)

2. Find the center and radius of the circle whose equation is

   $3x^2 + 3y^2 + 8x + 4y = 0.$ _____

   (frame 1)

3. Find the equation of the circle that touches both axes and passes through

   the point $(6,3)$. _____

   (frame 2)

4. Complete the following:

   (a) The fixed point of a parabola is called the _____ .

   (b) The fixed line of a parabola is called the _____ .

   (c) The line of symmetry of a parabola is called the _____ .

   (d) The point where the parabola intersects its axis is called the

   _____ .

   (frame 3)

5. Find the coordinates of the focus and ends of the latus rectum and the equation of the directrix of the curve $y^2 - 2x = 0$. Sketch the curve.
(frame 4)

6. We know that when a cable suspends a load of equal weight for equal horizontal distances it assumes a parabolic shape. The ends of such a cable on a bridge are 1000 feet apart and 100 feet above the horizontal road bed, while the center of the cable is level with the road bed. Find the height of the cable above the road bed at a distance of 300 feet from either end. (Use the sketch below to help you. What you need to do is compute the value of y in the equation $x^2 = 4py$ when $x = 500 - 300 = 200$. But in order to do this you must first find the value of $4p$ by inserting the coordinates of the end of the cable in the above equation for the parabola. Then insert this value back in the equation when $x = 200$ feet and solve for the value of y.)
(frame 5)

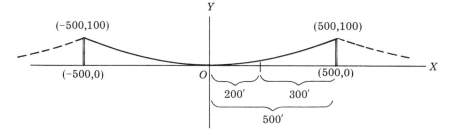

7. An ellipse is the locus of a point that moves so that the _____ of its distances from two fixed points is _____ .
(frame 6)

8. The long axis of an ellipse is called the _____ axis. The short axis is called the _____ axis. The equation $e = \dfrac{c}{a}$ represents the _____ of an ellipse.
(frame 7)

9. Find the semi-axes, the foci, the vertices, the latus rectum, and the eccentricity of the ellipse $8x^2 + 4y^2 = 32$.
(frame 8)

10. A hyperbola is the locus of a point that moves so that the _____ of its distances from _____ is a constant.

(frame 9)

11. In the figure at the right, the line through the foci (X-axis in this case) is called the

_____ .

The line segment $V'V$, of length $2a$, is called the

_____ .

The line segment on the Y-axis between the points $(0,b)$ and $(0,-b)$, of length $2b$, is called the

_____ .

(frame 10)

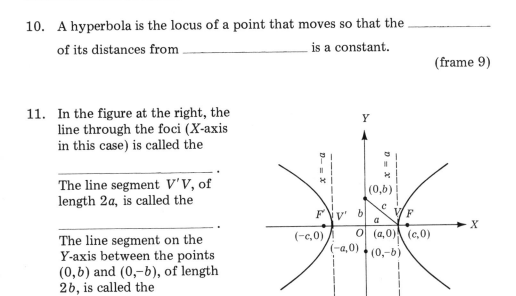

12. Find the values of a, b, c, and e, the coordinates of the foci, vertices and ends of the latera recta, the length of a latus rectum, and the equations of the asymptotes for the hyperbola $x^2 - y^2 = 64$.

(frame 11)

13. Two hyperbolas are said to be conjugate if the transverse axis of each is the conjugate axis of the other. (True, False)   (frame 12)

14. A hyperbola is said to be equilateral if the transverse and conjugate axes are of the same _____ .   (frame 12)

15. The tangent at a point P on a curve may be defined as the limiting position of a _____ PQ as Q approaches P along the curve.   (frame 13)

16. In polar coordinates the position of a point is determined by a _____ and a _____ .   (frame 15)

17. Write three other pairs of polar coordinates for the point $(2,30°)$.

_____

(frame 15)

18. Construct the table of values and plot the curve $r = 5 \cos 2\Theta$. (Use polar coordinate paper in Appendix.)   (frame 16)

19. Transform the equation $r = 5 - 8 \sin \Theta$ into rectangular coordinates.

_____

(frame 17)

20. Transform the equation $y^2 = 8x$ into polar coordinates.
(frame 17)

Answers to Self-Test

1. $x^2 + y^2 + 8x - 4y + 4 = 0$.

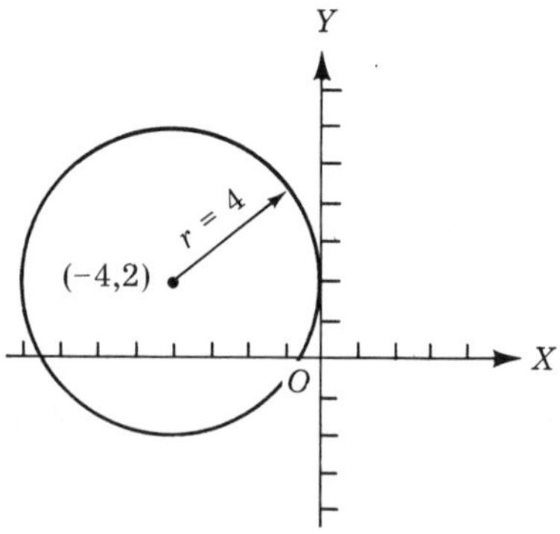

2. $(-\frac{4}{3}, -\frac{2}{3})$, $r = \frac{2}{3}\sqrt{5}$
3. $x^2 + y^2 - 6x - 6y + 9 = 0$, $x^2 + y^2 - 30x - 30y + 225 = 0$.
4. (a) focus;   (b) directrix;   (c) axis;,   (d) vertex

5. $(\frac{1}{2},0)$; $(\frac{1}{2},\pm1)$; $2x + 1 = 0$.

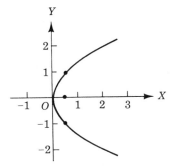

6. Standard equation of the parabola is $x^2 = 4py$. Since the parabola passes through the point $(500,100)$, we can substitute these values for x and y to obtain $(500)^2 = 4p(100)$, or $4p = \dfrac{250,000}{100} = 2,500$. There-fore our equation for the parabola is $x^2 = 2,500y$, or $y = .0004x^2$. Hence, when $x = 200$, the value of y becomes $y = .0004(200)^2 = 16$ feet as the height of the cable above the road bed at a distance of 300 feet from either end.

7. sum, constant

8. major, minor, eccentricity

9. $2\sqrt{2}$; 2; $(0,\pm2)$; $(0,\pm2\sqrt{2})$; $2\sqrt{2}$; $\frac{1}{2}\sqrt{2}$

10. difference, two fixed points

11. principal axis; transverse axis; conjugate axis.

12. 8; 8; $8\sqrt{2}$; $\sqrt{2}$; $(\pm8\sqrt{2},0)$; $(\pm8,0)$; $(\pm8\sqrt{2},\pm8)$; 16; $x \pm y = 0$.

13. True

14. length

15. secant

16. distance, direction

17. $(-2,210°)$, $(2,-330°)$, $(-2,-150°)$

18.

Θ	2Θ	$r = 5 \cos \Theta$
$0°$	$0°$	5
$15°$	$30°$	$\frac{5}{2}\sqrt{3}$
$22\frac{1}{2}°$	$45°$	$\frac{5}{2}\sqrt{2}$
$30°$	$60°$	$\frac{5}{2}$
$45°$	$90°$	0

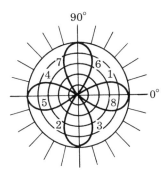

You can see that as Θ varies from $45°$ to $90°$, 2Θ will vary from $90°$ to $180°$, hence we will obtain the negative values of r in reverse order from the above table. Plotting the points we get half-loops 1 and 2. Note also that the curve is symmetric with respect to the polar axis. Continuing for all values of Θ up to $360°$ and using the symmetric property we obtain the entire curve shown above. Note the order of the half-loops.

19. $x^4 + y^4 + 39y^2 + 2x^2y^2 + 16x^2y + 16y^3 - 25x^2 = 0$.
20. $r = 8 \cot \Theta \csc \Theta$. Thus, $y^2 = 8x$ becomes $(r \sin \Theta)^2 = 8(r \cos \Theta)$ or

$r^2 \sin^2 \Theta = 8r \cos \Theta$, from which $r^2 = 8r \cdot \dfrac{\cos \Theta}{\sin \Theta} \cdot \dfrac{1}{\sin \Theta}$, and

$r = 8 \cot \Theta \csc \Theta$

CHAPTER NINE

Limits

Even though, in the last chapter, we touched on the definition of a tangent as the limiting position of a secant as it approaches a point along a curve, you may very well wonder what the subject of limits is doing in this book. It's a good question and one that deserves an answer.

As I'm sure you are aware, the major goal of this Self-Teaching Guide is to enable you to become generally familiar with the basic mathematical concepts you will need for the study of calculus. And calculus is, as we have mentioned before, very much concerned with the subject of limits. In fact calculus was invented by Leibnitz and Newton (separately, nor cooperatively) because of mathematicians' need to find some method of solving problems, including *calcula*tion, that simply couldn't be solved by any mathematics they knew. But, like many other useful inventions, the development of calculus methods had to await other mathematical developments. An important example of such a development is the analytic geometry of René Descartes we studied in the preceding two chapters. This was a necessary prelude to calculus. But just what were these puzzling problems that cried out for a method of solution?

In the half century B.C. the Greeks took an important step forward when they managed to separate mathematics from purely applied problems and began an abstract exploration of space based upon a study of points, lines, and figures such as triangles and circles. Interest in mathematics turned to logical reasoning rather than facts found in nature. It became a blend of mathematics and philosophy, since the Greeks were mainly interested in geometry as a means of advancing logical reasoning.

Even at this early date, however, these "philoso-maticians" ran into a number of puzzling problems. Some of these are embodied in the paradoxes of Zeno (495–435 B.C.). One of these involves a mythical race between Achilles and the tortoise. Even if the tortoise begins the race with a 100-yard start, if Achilles could run ten times as fast as the tortoise, it seemed perfectly apparent that he would overtake the tortoise. Not so, said Zeno.

The problem was to disprove Zeno's "proof" that the tortoise would always be ahead. He reasoned that while Achilles is covering the 100 yards that separates them at the start, the tortoise moves forward 10 yards. While Achilles dashes over this ten yards the tortoise plods on a yard and is still a yard ahead. When Achilles has covered *this* one yard, the tortoise is still one-tenth of a yard ahead. Thus, by dividing the distance run by Achilles into smaller and smaller amounts, Zeno argued that he would *never* pass the tortoise. The fact that an infinite set of distances could add up to a finite total distance was the unknown element that made Zeno's "proof" appear plausible. It was not until a better understanding of *limits* was developed that it became possible to demonstrate the fallacy in Zeno's logic, as we shall see later.

But there were other problems as well arising from this lack of an understanding of limits. Most of these involved calculating the measures of curved figures: the area of a circle or of the surface of a sphere, the volume of a sphere or of a cone, and similar problems. Problems of this kind were treated by what came to be known as the *Method of Exhaustions*, actually a method of limits wherein the circle was regarded as a limit of a series of inscribed polygons. This method enabled Archimedes (287–212 B.C.) to arrive at very close approximations of the correct values in many cases. A related method of limits, much more general in form, is one of the essential features of calculus today.

Not until the advent of calculus were these proximate methods replaced by a precise method. The problem of continuous motion was also the subject of much speculation. The Greeks made important conceptual contributions toward an understanding of motion (partly because of Zeno's paradoxes, no doubt). But not until the development of calculus was there available a workable, systematic method for describing in both qualitative and quantitative terms such things as velocity and acceleration, and for making analytical studies of various particular motions.

Perhpas you see, now, why you need to start doing some thinking about the subject of limits in preparation for your study of calculus. In this chapter, therefore, you will learn:

- some new things about limits;

- how our intuitive notions about limits can help us begin to understand and appreciate the mathematical concept of a limit;

- why, in a function, as one variable approaches zero, the other variable can approach some definite numerical value as a limit;

- how we find the limiting value of a function such as $f(x) = \dfrac{x^2 - 1}{x - 1}$ when x approaches 1 [i.e., $(x - 1)$ approaches zero] as a limit;

- the meaning of an expression such as $\lim\limits_{x \to a} f(x) = L$;

- about sequences, both finite and infinite, what arithmetic and geometric progressions are and how to solve problems involving progressions, and how to find the sum of an infinite progression;

- what we mean by the term *series* and why series are important in calculus;

- how we go about finding the instantaneous velocity of a free-falling body or the "instantaneous" slope of a curve at a particular point on a curve.

AN INTUITIVE APPROACH TO LIMITS

1. Let's begin with what you already know about limits. Did you ever feel you were reaching the "limit of your patience?" This thought is based on the notion (which we won't debate here) that each of us has only a fixed supply of patience and that circumstances can make a person feel he has just about used up his allotment. A mathematical way of saying this would be to say that one's reserve (remaining amount) of patience is approaching zero *as a limit*. And using standard mathematical symbols we could express this situation symbolically as:

    Patience ⟶ 0.

    (In case you have forgotten, the arrow means *approaches.*)

    Similarly, when we speak of reaching the "limit of our endurance" we really are referring to the fact that our supply of energy is fast approaching zero *as a limit*. Thus,

    Endurance ⟶ 0.

    Many of us have been faced with the dilemma of having the amount of gasoline remaining in our gas tank "approach zero" at an inopportune moment. We also know about military limits (being "off limits"), speed limits, the ground being the limit for a falling ball, and so on.

    All these examples have something in common. Can you tell what it is? Try putting it into words, then check your answer with the one given

    below. _____

    _____

    _ _ _ _ _ _ _ _ _ _ _ _ _ _ _ _

    The concept of a quantity or of the distance from some fixed position approaching zero as a limit.

2. Wasn't the distance between Achilles and the tortoise fast approaching zero as a limit? Examples of this kind are fine for developing an intuitive notion of limits. But in order to be able to *use* this concept to solve the kinds of problems that concerned Newton, Leibnitz, and other early mathematicians back to the days of the Greeks, we will need to examine it more closely.

If your normal weight is 150 pounds and you decide you are not going to let it exceed 160 pounds, then you have set a limit. When you weigh 151 pounds you will be nine pounds from your limit. When you get to 155 pounds you will be only five pounds from your limit. And as you reach 156, 157, and 158 pounds, the difference between what you weigh and the maximum weight *increase* you will accept is rapidly approaching zero. Thus, although your *actual weight* is approaching 160 pounds as a limit, the *weight increase you will accept* is approaching zero as a limit! There is a subtle difference between thinking in terms of *total* quantity versus thinking in terms of *difference* in quantity. We need to be aware of this in our discussion of limits. *In mathematics we are interested primarily in the difference between some quantity and the limit zero.*

Consider this idea with relation to a specific speed limit such as 35 mph. Usually we would say that as our car speed increases it approaches 35 mph as a limit. How could you express this situation in terms of the

difference between your speed and 35 mph? _____

_____

— — — — — — — — — — — — — — — —

As your speed increases, the difference between your speed and 35 mph approaches zero as a limit. (Do you see the difference?)

3. As we *commonly* use the word "limit" we are chiefly interested in the *magnitude* or size of a quantity as it gets nearer and nearer to some limit. In *mathematics* it is more efficient to discuss the *difference in* or *distance between* the quantity and its limit. Relating these two notions we can say that as a quantity approaches a limit, the *difference* between the quantity and its limit *approaches zero as a limit.*

In order to get used to seeing what this kind of relationship looks like in mathematical shorthand, let's try expressing it symbolically. We will suppose, for example, that you are filling your car's gas tank. As the Quantity of gasoline in the tank approaches 16 gallons (the tank's capacity), the Space remaining in the tank approaches zero. We can express this as follows:

as $Q_g \longrightarrow 16,\ S_r \longrightarrow 0,$

where Q_g represents the *Quantity* of *gas* (in gallons) and S_r represents the *Space remaining*. Obviously all we have done is to use the little arrows to mean "approaches" and invented a couple of letter symbols to represent the values involved. Not very technical and not very formal mathematics, certainly, but it says what we want it to say and that is the basic purpose of any mathematical symbol.

Now suppose you make up some symbols of your own and try representing the situation where your car speed is approaching the posted speed limit of 35 mph. When you have something that looks right to you, check it against the symbology shown below.

—————————————————————————

— — — — — — — — — — — — — —

as $S_c \longrightarrow 35$, $D_s \longrightarrow 0$.

Here S_c was used to represent the *speed* of the car as it approaches 35 mph, and D_s stood for the *difference* between the speed of the car and the speed *limit* of 35 mph. Whatever symbology you used is just as good as long as it represents the same values.

4. No doubt you could think of many other examples similar to those we have used here, but even these few are sufficiently typical to allow us to arrive at some kind of a general statement about such situations. We might say, for example, something like this: As the value of any quantity approaches some limit, the *difference* between the value and its limit approaches zero. Symbolically expressed our statement would look something like this:

    as $V_q \longrightarrow L$, $(V \sim L) \longrightarrow 0$.

The little sine-wave shaped symbol between V and L means "difference of." V_q stands for the value of the quantity (whatever its nature — gallons, miles per hour, inches, oranges, light years), L represents the limit which the value is approaching, and $V \sim L$ stands for the amount by which the value differs from its limit at any given moment.

Now suppose you were climbing a mountain and your objective was to reach a height (h) of 5,000 feet above sea level (L). How would you interpret, in words, the following symbolical representation — or mathematical model — of this situation?

   $h \longrightarrow L$, $(h \sim L) \longrightarrow 0$.

—————————————————————————

— — — — — — — — — — — — —

As your height above sea level approaches the goal (limit), the difference between your height above sea level and your goal (limit) approaches zero.

5. Your interpretation of the meaning of the symbols in the above exercise should have been generally similar to that shown in the answer. We could, of course, have used 5,000 (feet) in place of the symbol L (for *limit*, in this case) since we happen to know the numerical value of the limit in this instance. In this case we would write

$$h \longrightarrow 5{,}000, \ (h \sim 5{,}000) \longrightarrow 0,.$$

If you are saying to yourself that these examples are absurdly simple, you are quite right. However, you will perhaps remember this fact with gratitude when we get to some that are not quite so obvious. Let's consider, for example, the idea of a limit as applied to a function. A *function*, as you will recall from your study of algebra, is a set of ordered pairs such that no two ordered pairs have the same first element. A common notation for a function f is $f(x,y)$, where each ordered pair is of the form (x,y). Sometimes x, the first element, is called the *independent variable*, and the second element, y, the *dependent variable*. The element y also often is called the *value* of the function. Or the value may simply be represented by $f(x)$, which we read "f of x."

You have encountered a number of important kinds of functions thus far in your study of mathematics — polynomial, logarithmic, exponential, and trigonometric functions. However, we are concerned now only with relations between the elements x and y represented by linear or higher degree algebraic equations.

Consider, then, the function, $f(x) = \dfrac{x^2 - 1}{x - 1}$. Suppose we wish to evaluate this function for various values of x. If we let $x = 2$, we have no difficulty arriving at the value $f(x) = 3$. Nor do we have any difficulty if we substitute greater values of x than 2. But what happens when we substitute the value $x = 1$? _____

_____

- - - - - - - - - - - - - - - - -

$$f(x) = \frac{x^2 - 1}{x - 1} = \frac{1^2 - 1}{1 - 1} = \frac{0}{0} = ?$$

6. This meaningless result doesn't tell us much. And yet if we factor the numerator of our function and divide out the like binomial terms in the numerator and denominator, we get

$$f(x) = \frac{x^2 - 1}{x - 1} = \frac{(x - 1)(x + 1)}{(x - 1)} = x + 1.$$

This result would lead us to suspect that for $x = 1$ the value of the function *should* be $x + 1 = 2$. Why is it we can't get this result from our original equation?

Strictly speaking the function $\frac{x^2 - 1}{x - 1}$ has no definite value when $x = 1$, that is, it has no value that can be deduced from any of the principles of which we so far are aware. Obviously, however, we would like to see the rule maintained that a function has a definite value corresponding to every value of the dependent variable.

We are also guided by the principle of continuity that prompts us to seek a value of $\frac{x^2 - 1}{x - 1}$, when $x = 1$, that differs very slightly from the value of $\frac{x^2 - 1}{x - 1}$ when x differs only slightly from 1. With these thoughts in mind, let's prepare a table of values of the function as x varies from 2 towards 1.

Notice in the table at the right that as x approaches 1 *as a limit*, $f(x)$ appears to be approaching 2 *as a limit*. And as long as $x \neq 1$, no matter how little it differs from 1, we can perform the indicated division, and we have the identity

$$\frac{x^2 - 1}{x - 1} = x + 1.$$

x	$f(x)$
2	3
1.5	2.5
1.4	2.4
1.3	2.3
1.2	2.2
1.1	2.1
↓	↓
1	2

In the table above we let x vary from 2 towards 1. To substantiate our conclusions about what happens to the value of $f(x)$ as x approaches a value of 1, let's come toward 1 from the other side. That is, let's allow x to assume successive values between 0 and 1.

The table at the right shows the results of doing so. Again it is apparent that the closer x gets to a value of 1, the closer $f(x)$ approaches a value of 2.

We can see, therefore, that for values of x differing very little from 1, the value of $\frac{x^2 - 1}{x - 1}$ differs very little from 2. It is apparent, then, that by bringing x sufficiently near to 1, we can cause $\frac{x^2 - 1}{x - 1}$ to differ from 2 by as little as we please.

x	$f(x)$
0	1
0.5	1.5
0.6	1.6
0.7	1.7
0.8	1.8
0.9	1.9
↓	↓
1	2

The value of $\frac{x^2 - 1}{x - 1}$, as thus defined, is termed the *limiting value*, or the *limit* of $\frac{x^2 - 1}{x - 1}$ as x approaches 1 as a limit. Or (in light of our previous discussion) we could also say that the *difference* between the value of x and 1 approaches *zero* as a limit! We can write this, using symbols, as below, where "lim" stands for limit.

$$\lim_{x \to 1} \left(\frac{x^2 - 1}{x - 1} \right) = 2.$$

Generalizing a bit from the above example we can say that, when, by causing x to differ sufficiently little from a (i.e., the difference approaches zero as a limit), we can make the value of $f(x)$ approach as near as we please to L, then L is said to be the limiting value, or limit, of $f(x)$ when $x \to a$; and we write

$$\lim_{x \to a} f(x) = L.$$

Now we haven't really *proved* anything as a result of the foregoing exercise, but we have *intuited* several interesting concepts and defined some terms. Remember, we are not attempting to take a rigorous approach to the subject of limits. We mainly are trying to become acquainted with some of the basic concepts associated with limits and gain some "feel" for a few of the ways in which we can go about finding the limit of a function when our known methods of evaluation fail.

Before going on, we'd better make sure you remember what some of the basic terms and symbols mean.

(a)   A *function* is _____

_____ .

(b) $f(x)$ means _____ .

(c)   The abbreviation "lim" stands for _____ .

(d) $x \to a$ means _____ .

(e)   Put into words what $\lim\limits_{x \to a} f(x) = L$ means.

- - - - - - - - - - - - - - -

(a)   a set of ordered pairs such that no two ordered pairs have the same first element
(b)   a function of x; for example, in the equation $y = x^2$, y is a function of x, and we could substitute $f(x)$ for y
(c)   limit
(d) x approaches a as a limit
(e)   The limit of the function, f of x, approaches the value L as x approaches the value a as a limit.

SEQUENCES, PROGRESSIONS, AND SERIES

7. Having taken a first, intuitive look at the general concept of limits and considered a brief example of how to find the limiting value of a function, let's turn our attention now to another important application of limits — number sequences, progressions, and series.

Incidentally, even though we make no attempt in this chapter to supply any formal proofs, you must not get the idea that there is anything unworthy or improper about the intuitive approach. In daily life our thinking very often is intuitive. In fact, some of the terms — and even concepts — used in this book have been taken from everyday life, and both author and reader would be hard put to agree on a precise definition of many of them. Mathematicians have a high regard for the intuitive approach. The early development of calculus was, in fact, largely based on highly developed intuition. It required many more years of deep investigation and precise thinking about many of the things they had taken for granted before mathematicians were able to supply an analytic proof for some of the useful but fuzzy aspects of calculus.

Now, why are we going to discuss sequences and series? Because working with series involves finding the sum of sequences of numbers, and some of these series go out to infinity. Have you ever tried adding up something that extended to infinity? No? Then you may not yet appreciate the fact that this can be a very difficult task at times. And there are some series whose sum cannot be found except by the methods of calculus. We are not (rest assured) going to get into calculus in this chapter or in this book. But we are going to see if we can discover a little about how limits are involved in the study of sequences, progressions, and series.

If you have studied about sequences and progressions before, this will be a review for you. However, we may get into some aspects of the subject you didn't consider when you were introduced to it in intermediate or advanced algebra. On the other hand, if you haven't had occasion to learn anything about these topics before, this will be a small preparation for a deeper look into the subject of series when you begin the study of calculus. It will also, as we suggested earlier, help to expand your concept of limits. So let's proceed.

It often is desirable to order (arrange) a group of objects in such a way that there is a first object, a second object, and so on. When the property of *order* is imposed on the elements of a set (group) of objects, the result is a *sequence*. For example, if we arrange the positive integers in their natural order they form a sequence.

1, 2, 3, 4, . . .

Other examples are:

1^2, 2^2, 3^2, 4^2, . . .   the squares of the positive integers

1, 3, 5, 7, . . .   the odd positive integers

1, 2, 3, 5, 7, . . .   the prime integers

$\dfrac{1}{2}$, $\dfrac{2}{3}$, $\dfrac{3}{4}$, $\dfrac{4}{5}$, . . . , $\dfrac{n}{n+1}$, . . .

The terms of a sequence usually are separated by commas, as you see above.

See if you can write the first four terms of the set of the *even* positive

integers. _____

– – – – – – – – – – – – – –

2, 4, 6, 8, . . .

8. If our set of numbers is such that it goes on indefinitely, it gives rise to an *infinite sequence.* We can define this as follows:

> *A set of numbers arranged in order, so that there is a first number and every number is followed by another (its successor), is called an infinite sequence.*

The word *infinite* in this definition means that the sequence in question has no last term; the terms continue on and on in unbroken sequence. Take for example the fraction $\frac{5}{11}$. Dividing the numerator by the denominator to obtain the decimal fraction we get $\frac{5}{11}$ = 0.454545 . . . In this case the digits in the decimal constitute a sequence which, since the decimal is recurring (repeating), could carry on as far as we wish. Hence it has no last term. Therefore we would have to consider it an infinite sequence.

Similarly, the ratio π (from geometry and trigonometry) represents an infinite but non-repeating sequence since the digits continue indefinitely, each digit having a successor. Thus, π = 3.1415926 . . . The value of π has been computed (with the aid of an electronic computer) to more than 100,000 decimal places (that is, with more than 100,000 digits after the decimal point), and this could be continued indefinitely.

As you might suspect, a sequence that has a first and a last term and in which each term except the last has a successor is called a *finite sequence.* We are not going to be too concerned with finite sequences. When the word sequence alone is used it will refer to an infinite sequence.

Before we go on, let's see if you have caught the main concept of an infinite sequence. Complete the following.

An infinite sequence is _____

_____ .

— — — — — — — — — — — —

a set of numbers arranged in order, so that there is a first one and every number is followed by another, its successor.

9. A *progression* is a special kind of sequence. That is, it is a sequence of numbers formed according to some law. We are going to consider two types of progressions: arithmetic progressions and geometric progressions.

   An *arithmetic progression* is a sequence of numbers each of which differs from the one that precedes it by a constant amount, called the *common difference*.

   For example, if the first term is 2 and the common difference is 5, then the first eight terms of an arithmetic progression are 2, 7, 12, 17, 22, 27, 32, and 37. In this case since the common difference was positive the terms appear in *increasing* order. However, in an arithmetic progression such as 16, 14, 12, 10, . . . the terms appear in *decreasing* order, and it is apparent that the common difference is -2.

   State the common difference in the following progressions.

   (a)   1, 4, 7, 10, 13, . . .   _____

   (b)   27, 23, 19, 15, . . .   _____

   (c) $8\frac{1}{2}$, 10, $11\frac{1}{2}$, 13, . . .   _____

   — — — — — — — — — — — — — — —

   (a) 3;   (b) -4;   (c) $1\frac{1}{2}$

10. Most problems in arithmetic progressions deal with three or more of the following five quantities: the first term, the last term, the number of terms, the common difference, and the sum of all the terms. In order to derive formulas that will enable us to find any of these five quantities if we know the value of three of the others, we will let the following letters represent the five quantities.

    a = the first term of the progression
    l = the last term
    d = the common difference
    n = the number of terms
    s = the sum of all the terms

Using the above notation, the first four terms of an arithmetic progression are a, $a + d$, $a + 2d$, $a + 3d$. Notice that d appears with the implied coefficient 1 (one) in the second term and that this coefficient increases by 1 as we move from one term to the next. Therefore, the coefficient of d is one less than the number of that term in the progression. Thus, the sixth term is $a + 5d$, the ninth is $a + 8d$, and finally the last, or nth term, is $a + (n - 1)d$. So we now have as our formula for the last term

$$l = a + (n - 1)d \qquad (1)$$

Let's see how we can use this formula.

Example 1: If the first three terms of an arithmetic progression are 2, 6, and 10, find the eighth term.
Solution: Since the first and second terms, as well as the second and third, differ by 4, it is apparent that $d = 4$. Furthermore, $a = 2$ and $n = 8$. Therefore, if we substitute these values in (1) we get

$$\begin{aligned}
l &= 2 + (8 - 1)4 \\
&= 2 + 28 \\
&= 30
\end{aligned}$$

Example 2: If the first term of an arithmetic progression is -3 and the eighth term is 11, find d and write the eight terms of the progression.
Solution: In this problem, $a = -3$, $n = 8$, and $l = 11$. If these values are substituted in (1) we get

$$\begin{aligned}
11 &= -3 + (8 - 1)d \\
11 &= -3 + 7d \\
-7d &= -14 \\
d &= 2
\end{aligned}$$

Therefore, since $a = -3$, the first eight terms of the desired progression are $-3, -1, 1, 3, 5, 7, 9, 11$.

Now try these.

(a) If the first three terms of an arithmetic progression are 3, 8, and 13, find the sixth term.

(b) If the last term of an arithmetic progression is 16, the common difference is 2, and there are six terms in the progression, find the first term and write the six terms of the progression.

— — — — — — — — — — — —

(a) $l = 3 + (6 - 1)5$, or $l = 28$.
(b) $16 = a + (6 - 1)2$, or $a = 6$, hence the terms of the series are 6, 8, 10, 12, 14, 16.

11. Now suppose we wish to find a formula that will give us the sum, s, of the n terms of an arithmetic progression in which the first term is a and the common difference is d. As we found in frame 10, the terms in the progression are a, $a + d$, $a + 2d$, and so on until we reach the last term, which from (1) is $l = a + (n - 1)d$. Thus we can write

$$s = a + (a + d) + (a + 2d) + \ldots + [a + (n - 1)d] \qquad (2)$$

And since there are n terms (2) and each term contains a, we can rearrange the terms and write s as

$$s = na + [d + 2d + \ldots + (n - 1)d] \qquad (3)$$

Now, if we reverse the order of the terms in the progression by writing l as the first term, then the second term is $l - d$, the third $l - 2d$, and so on, to the nth term which, from (1), is $l + (n - 1)(-d)$. So we can write the sum as

$$s = l + (l - d) + (l - 2d) + \ldots + [l + (n - 1)(-d)]$$

Next, combining the l's and d's we get

$$s = nl - [d + 2d + \ldots + (n - 1)d] \qquad (4)$$

And finally, if we add the corresponding members of (3) and (4) and combine like terms we get

$$2s = na + nl$$
$$= n(a + l)$$

or, dividing both sides by 2,

$$s = \frac{n}{2}(a + l) \qquad (5)$$

Formulas (1) and (5) make it possible for us to find values for all five of the elements whenever any three of them are known.

Example: Find the sum of all the numbers between 1 and 100 that are divisible by 3.
Solution: These numbers form an arithmetic progression with the first term $a = 3$, $d = 3$, and $l = 99$. Using these values in (1) we get

$$99 = 3 + (n - 1)3, \text{ or}$$
$$= 3 + 3n - 3$$
$$= 3n$$
$$n = 33.$$

We can now obtain the sum from formula (5).

$$s = \frac{n}{2}(a + l)$$
$$= \frac{33}{2}(3 + 99)$$
$$= 33 \cdot 51$$
$$= 1{,}683.$$

Try this problem. If $a = 4$, $d = 5$, and $l = 49$, find n and s.

_ _ _ _ _ _ _ _ _ _ _ _ _

from (1): $49 = 4 + (n - 1)5$, or $n = 10$.

from (5): $s = \dfrac{10}{2}(4 + 49)$, or $s = 265$.

12. The terms between the first and last terms of an arithmetic progression
    are called *arithmetic means*. If the progression contains only three
    terms, the middle term is called *the arithmetic mean* of the first and
    last term. We can obtain the arithmetic means between two numbers
    by using (1) to find d, and the means can then be computed. If the
    progression consists of the three terms a, m, and l, then by formula (1)

   $$l = a + (3 - 1)d = a + 2d, \text{ hence}$$
   $$d = \frac{l - a}{2}, \text{ and}$$
   $$m = a + \frac{l - a}{2} = \frac{a + l}{2}.$$

   Therefore, the *arithmetic mean of two numbers is equal to one-half of
   their sum.*

   Example: Insert the five arithmetic means between 6 and -10.
   Solution: Since we want to find the five means between 6 and -10 we
   will have seven terms in all. Hence $n = 7$, $a = 6$, and $l = -10$. Therefore, from (1) we have

   $$-10 = 6 + (7 - 1)d, \text{ or}$$
   $$6d = -16, \text{ and}$$
   $$d = -\frac{16}{6} = -\frac{8}{3}.$$

   Thus, the progression is $6, \dfrac{10}{3}, \dfrac{2}{3}, -\dfrac{6}{3}, -\dfrac{14}{3}, -\dfrac{22}{3}, -\dfrac{30}{3}$.

   What are the five arithmetic means between 3 and 15?

   _ _ _ _ _ _ _ _ _ _ _ _ _

   $a = 3$, $l = 15$, $n = 7$ (the first and last terms plus the five means in
   between them). Therefore, from (1), $15 = 3 + (7 - 1)d$, or
   $6d = 12$, and $d = 2$. Hence the progression is 3, 5, 7, 9, 11, 13, 15 and
   5, 7, 9, 11, 13 are the five arithmetic means between 3 and 15.

13. The second type of progression we are going to consider is the geometric progression. A *geometric progression* is a sequence of numbers so related that each term after the first can be obtained from the preceding term by multiplying it by a fixed constant called the *common ratio*. A few such progressions are:

$-4, -2, -1, -\frac{1}{2}, -\frac{1}{4}, \ldots$ common ratio $\frac{1}{2}$

$3, -3, 3, -3, 3, \ldots$ common ratio -1

$2, 6, 18, 54, 162, \ldots$ common ratio 3

In order to obtain formulas for geometric progressions we again will use letter symbols as follows.

a = the first term
l = the last term
r = the common ratio
n = the number of terms
s = the sum of the terms

As you probably noticed, these are the same letters we used for the arithmetic progressions except for r, the common ratio, in place of d, the common difference.

Using the above notation, the first six terms of a geometric progression in which the first term is a and the common ratio is r are

$$a \quad ar \quad ar^2 \quad ar^3 \quad ar^4 \quad ar^5$$

Notice that the exponent of r in the second term is 1 and that this exponent increases by 1 as we proceed from each term to the next. Therefore, the exponent of r in any term is 1 less than the number of that term in the progression. Thus the nth term is ar^{n-1}. This gives us the formula

$$l = ar^{n-1} \tag{6}$$

Let's look at an example of how we can apply this formula.

Example: Find the seventh term of the geometric progression $36, -12, 4, \ldots$

Solution: The common ratio, r, is obtained by taking any two consecutive terms of the progression — for example, 36 and -12 — and dividing the second by the first. Thus, $-12 \div 36 = -\frac{1}{3}$. In this progression each term after the first is obtained by multiplying the preceding term by $-\frac{1}{3}$. We also know, from the information given, that $a = 36$, $n = 7$, and the seventh term is, of course, represented by the letter l. Substituting these values in formula (6) gives us

$$l = ar^{n-1} = 36(-\tfrac{1}{3})^{7-1}$$
$$= \frac{36}{(-3)^6}$$
$$= \frac{36}{729}$$
$$= \frac{4}{81}$$

Try one of these and see for yourself how the formula works. Find the fifth term of the geometric progression 4, 8, 16, . . .

— — — — — — — — — — — —

$l = ar^{n-1}$, and $r = 2$ (since $\tfrac{8}{4}$, for example, is 2), $a = 4$, and $n = 5$. Therefore, $l = 4(2)^{5-1} = 4(2)^4 = 4(16) = 64$, the fifth term of the geometric progression.

14. If we add the terms of a geometric progression, represented by a, ar, ar^2, . . . , ar^{n-2}, ar^{n-1}, we get

$$s = a + ar + ar^2 + \ldots + ar^{n-2} + ar^{n-1} \qquad (7)$$

However, by use of an algebraic device we can obtain a more compact formula for s. First, we multiply each member of (7) by r and get

$$rs = ar + ar^2 + ar^3 + \ldots + ar^{n-1} + ar^n \qquad (8)$$

Now notice that if we subtract the corresponding members of (7) and (8) and combine like terms we get

$$s - rs = a - ar^n, \text{ or } s(1 - r) = a - ar^n$$

Solving this equation for s we get

$$s = \frac{a - ar^n}{1 - r}, \qquad (9)$$

where $r \neq 1$. Now if we multiply each member of formula (6) by r we get $rl = ar^n$, and if we replace ar^n by rl in (9), we get

$$s = \frac{a - rl}{1 - r}, \qquad (10)$$

where $r \neq 1$.

Now let's find out how this formula for the sum of the terms of a geometric progression works.

Example 1: Find the sum of the first six terms of the progression 2, −6, 18, . . .

Solution: In this progression $a = 2$, $r = -3$, and $n = 6$. Therefore, if we substitute these values in (9) we get

$$s = \frac{2 - 2(-3)^6}{1 - (-3)}$$
$$= \frac{2 - 2(729)}{1 + 3}$$
$$= -364$$

Example 2: The first term of a geometric progression is 3; the fourth term is 24. Find the tenth term and the sum of the first 10 terms.
Solution: In order to find either the tenth term or the sum we must have the value of r. We can obtain this value by considering the progression as made up of the first four terms defined above. Then we have $a = 3$, $n = 4$, and $l = 24$. Substituting these values in (6) gives us $24 = 3r^{4-1}$, or $r^3 = 8$, and $r = 2$.

Now, by using (6) again with $a = 3$, $r = 2$, and $n = 10$, we get

$$l = 3(2^{10-1})$$
$$= 3(512)$$
$$= 1,536$$

Therefore, the tenth term is 1,536.

To obtain s we will use (9), with $a = 3$, $r = 2$, and $n = 10$. This gives us

$$s = \frac{3 - 3(2)^{10}}{1 - 2}$$
$$= \frac{3 - 3(1,024)}{-1}$$
$$= \frac{3 - 3,072}{-1}$$
$$= 3,069$$

Try this problem. Find the sum of the geometric progression in which $a = 32$, $r = \frac{1}{2}$, and $n = 6$.

- - - - - - - - - - - - - -

from (9), $s = \dfrac{a - ar^n}{1 - r}$, or $s = \dfrac{32 - 32(\frac{1}{2})^6}{1 - \frac{1}{2}} = \dfrac{32 - 32(\frac{1}{64})}{1 - \frac{1}{2}} = \dfrac{32 - \frac{1}{2}}{1 - \frac{1}{2}}$,

hence $s = 63$.

15. The terms between the first and last terms of a geometric progression are called *geometric means*. If the progression contains only three terms then the middle term is called *the geometric mean* of the other

two. In order to obtain the geometric means between a and l we use formula (6) to find the value of r, and the means can then be computed. If there are only three terms in the progression, use (6), which would become $l = ar^2$.

Solving $l = ar^2$ for r we get

$$r = \pm \sqrt{\frac{1}{a}}$$

Therefore the second, or the geometric mean between a and l, is (from frame 13) ar, or

$$a \pm \sqrt{\frac{l}{a}} = \pm \sqrt{\frac{a^2 l}{a}} = \pm \sqrt{al}$$

Putting this into words we would say that the *geometric mean between two quantities is either the positive or the negative square root of their product.*

Example: Find the five geometric means between 3 and 192.
Solution: A geometric progression starting with 3, ending with 192, and containing five intermediate terms has seven terms. Hence we know that $n = 7$, $a = 3$, and $l = 192$. And so, from (6)

$$192 = 3(r^{7-1})$$
$$r^6 = \frac{192}{3}$$
$$= 64, \text{ and}$$
$$r = \pm\sqrt[6]{64} = \pm 2$$

Therefore, the two sets of geometric means of five terms each between 3 and 192 are 6, 12, 24, 48, 96, and -6, 12, -24, 48, -96.

Now you try this one. Insert the two geometric means between 5 and 40.

$a = 5$, $l = 40$, and $n = 4$. Therefore, from (6), $40 = 5(r^{4-1})$, or $r^3 = 8$, and $r = 2$. Thus the second term, ar, would be $5 \cdot 2$, or 10, and the second term would be $ar^2 = 5 \cdot 2^2 = 20$. The four terms of the progression, then, are 5, 10, 20, 40.

16. Back at the beginning of this section (in frame 7, to be exact) we said that our main reason for studying a little about sequences, progressions, and series was to see how the idea of limits applied. From what we have

covered thus far about progressions you should be able to appreciate the significance of what we are going to discuss next, namely, infinite geometric progressions.

Our task will be to find the *limit* of the sum of a geometric progression where n (the number of terms) increases indefinitely and where the absolute value of r (the common ratio) is less than one. Using symbols we can express this value of r as $|r| < 1$. Or, putting the whole idea into symbols (using the symbol ∞, introduced in Chapter 6, frame 7, which means "infinity" or "without limit"),

$$\lim_{n \to \infty} s(n) = s$$

by which we mean that, by taking n sufficiently large (that is, using as many terms as we please), the value of $s(n)$ (the sum of n terms) will differ from s (the sum of an *infinite* number of terms) by an amount that is less than any positive number we wish to select in advance. Although this approach is new to you — and may not make much sense to you at this point — bear with us for a bit and the reason for it should begin to clarify itself in your mind.

To work out the formula we will need to find the sum of an infinite geometric progression. Let's go back for a moment to equation (9), which allows us to find the sum of a finite geometric progression. If we use $s(n)$ to represent the sum of the first n terms we can write, from (9)

$$s(n) = \frac{a - ar^n}{1 - r}$$

or, factoring the right hand side,

$$s(n) = \frac{a}{1 - r}(1 - r^n).$$

But since $|r| < 1$, then $|r| > |r^2| > \ldots > |r^n|$. That is, since the value of r is less than one, it will be greater than the value of any higher power of r, because as you raise a fraction to a higher power its value decreases. And it can be proved (though we're going to spare you the proof) that r^n can be made arbitrarily small by taking n sufficiently large. Therefore we can write

$$\lim_{n \longrightarrow \infty} s(n) = \frac{a}{1 - r}$$

since $r^n \to 0$ and thus $1 - r^n \to 1$, and we follow the usual procedure by writing

$$s = \frac{a}{1 - r} \text{ for } |r| < 1 \tag{11}$$

Let's look at an example to help clarify the concept.

Example 1: Find the sum of the geometric progression $1, \frac{1}{2}, \frac{1}{4}, \ldots$,
(the dots indicate that there is no end to the progression).
Solution: In this progression, $a = 1$ and $r = \frac{1}{2}$, hence from (11),

$$s = \frac{1}{1 - \frac{1}{2}} = \frac{1}{\frac{1}{2}} = 2.$$

This can also be seen geometrically from the figure below.

On a coordinate system we simply add the amounts of the terms. The
first term takes us from the origin to the point 1. The sum of the two
terms is $1\frac{1}{2}$; the sum of the first three is $1\frac{3}{4}$, and so on. It becomes
evident that the amount which is added each time is half the remaining
distance to 2. Hence by adding an infinite number of these we approach
two as a limit (sum).

Does this remind you of the race between Achilles and the tortoise?
Let's apply what we have learned to see if we can get a firm answer to
that paradox.

You remember that the tortoise had a head start of 100 yards and
that Achilles could run ten times as fast as the tortoise. Zeno's argument was that Achilles never would catch up with the tortoise because
no matter how much ground he covered in any one unit of time, the
tortoise would always be $\frac{1}{10}$ of that distance ahead of him. In other
words, Zeno was implying that an infinite set of distances could never
add up to a finite total distance. Let's see if that's true.

Using formula (11) we can see that $a = 100$ yards (their distance apart
initially) and $r = \frac{1}{10}$ (the ratio of their speeds), hence

$$s = \frac{100}{1 - \frac{1}{10}} = \frac{100}{\frac{9}{10}} = \frac{1000}{9} = 111.111 \ldots \text{ yards.}$$

Thus our formula tells us that the sum of the incremental distances
run (that is, the total distance run) before Achilles overtakes the tortoise is 111.111 . . . yards, that is, a definite limit. Looking at the
diagram above we can see that the amount added each time is $\frac{1}{10}$ the
distance between them previously. Thus, at point s_1 they were 100
yards apart, at s_2 ten yards apart, at s_3 one yard apart, at s_4 $\frac{1}{10}$ of a
yard apart, and at s_5 they were virtually together. If Achilles could run

10 yards per second (surely no trick for a god), then he would have overtaken the tortoise sometime between the eleventh and twelfth seconds of the race.

A nonterminating, repeating decimal fraction is an illustration of an infinite geometric progression with $-1 < r < 1$. For example,

$$.232323 \ldots = .23 + .0023 + .000023 + \ldots$$

The sequence of terms on the right is a geometric progression with $a = .23$ and $r = \frac{1}{100}$.

By the use of (11) we can express any repeating decimal fraction as a common fraction by the method illustrated in the next example.

Example 2: Show that $.333 \ldots = \frac{1}{3}$.
Solution: The decimal fraction $.333 \ldots$ can be expressed as the progression $.3 + .03 + .003 + \ldots$ in which $a = .3$ and $r = .1$. Hence, by (11), the sum s is

$$s = \frac{.3}{1 - .1} = \frac{.3}{.9} = \frac{3}{9} = \frac{1}{3}.$$

Find the sum of the infinite geometric progressions with elements listed in each of the following problems.

(a) $a = 3, r = \frac{1}{2}$

(b) $a = 4, r = \frac{1}{3}$

(c) $a = 4, r = \frac{1}{5}$

- - - - - - - - - - - - - -

(a) $s = \frac{3}{\frac{1}{2}} = 6$;   (b) 6;   (c) 5

Since the purpose of our investigation into the notion of numerical sequences and progressions is not intended to make you an expert on this subject but, rather, simply to furnish another illustration of the application of *limits* in mathematics, we will rest our efforts here. Thus, we will forego such matters as harmonic progressions, convergence or divergence of series, and so forth.

Incidentally, the terms "progression" and "series" basically are interchangeable. We have talked primarily about arithmetic and geometric *progressions*, but we could have used the word *series* just as properly. The only reason for bringing in the term "series" at all, really, is because it is used most commonly in calculus, where you will study different kinds of infinite series. Now, however, we are going to take our third (and last) journey of exploration into the land of limits.

THE PROBLEM OF TANGENTS

17. In Chapter 8, frame 13, we considered briefly the concept of the tangent to a curve at a point as the limiting position of a secant rotating about that point.

This is a *dynamic* concept because the secant constantly is changing direction as it approaches the tangent, with one of its end points moving along the curve. It therefore represents rather nicely a host of other dynamic situations which the older mathematics (i.e., before calculus) found it impossible to cope with in any precise way.

One of these other dynamic problems — and one with which Newton and his colleagues were very much concerned since it related to the phenomenon of gravity — was that of finding the instantaneous velocity of a free-falling object. In case this sounds like too simple a problem, bear in mind that an object such as a ball, thrown up into the air, is constantly changing its speed. Why? Because when it is moving upward the pull of gravity constantly is slowing it down (we say it is decelerating), and when it changes direction and starts to fall toward the earth, gravity is speeding it up. It's easy to find its *average* velocity over very short periods of time, but until the advent of calculus there was no known method of *calculat*ing the *instantaneous* velocity of the ball at any given moment in time.

In looking into how the concept of a limit helped solve many dynamic problems, we could work with the equation $h = 128t - 16t^2$, which represents the change of height with time of the ball thrown into the air (disregarding the resistance of the air). Here we would be concerned with the two variables h and t, where h is expressed as a function of t, the independent variable. But because this equation defines a parabola and the equation for the "unit" parabola is simpler and will produce the same result, we will use the simpler equation instead.

The equation for a parabola is, as you may recall, $x^2 = 4py$, where the curve is symmetrical about the Y-axis. And we will make it even more simple by letting $4p = 1$. Thus we get for our equation, $y = x^2$.

Using the values for x shown in the table at the right, find the corresponding values for y and plot the resulting curve on the coordinate system provided below, then check your results with those shown below.

y	x
	0
	1
	2
	3

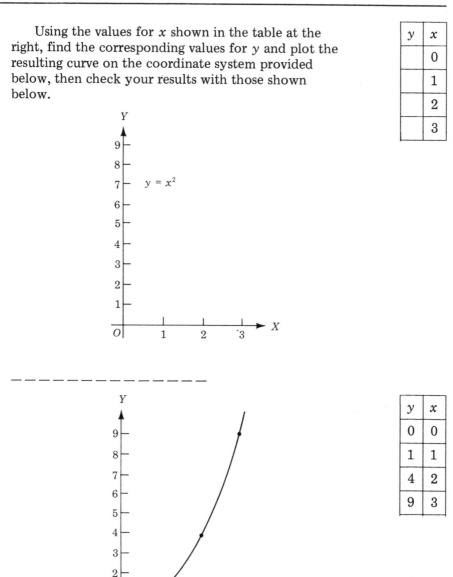

y	x
0	0
1	1
4	2
9	3

18. Note from your curve that, just as the *velocity* of the ball thrown into the air is constantly changing, the *direction* of the curve also is constantly changing, reflecting the rate at which one variable is changing with respect to the other.

So if we can find some way to determine the instantaneous rate of change of direction at any point on the curve, we should be able to use the same general approach to find the instantaneous rate of change in the velocity of the ball. Why? Because although the variables are different in each case and the physical situations they symbolize are different, mathematically the two equations involved are essentially the same! That is, they represent the same type of curve.

But how do we find the *direction* of the curve? The answer to this is that the direction of a curve at any point is, as you may recall, simply the *slope* of the curve at that point. Therefore, what we really are seeking is the slope, or angle of inclination, between the (positive direction of the) X-axis and a line tangent to the curve at the given point.

In the sketch below, identify the line T and give the ratio that represents the slope of T.

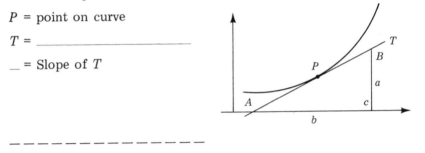

P = point on curve

T = _____

__ = Slope of T

T = tangent to the curve at point P; $\dfrac{a}{b}$ = slope of tangent line T

19. What we have found thus far is that it is convenient to represent the slope of the curve at the point P by means of a line, T, tangent to the curve at that point.

Now to help us in our analysis let's add another, random point on the curve at some indeterminate distance from P. This point we will designate as Q. Connecting this point to P by a straight line gives us the secant line, S. Notice this in the figure at the right.

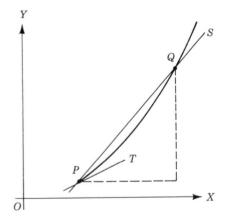

How do you think we should designate the coordinates of the point, P, bearing in mind that P is *any* point on the curve? _____

It probably would be best to designate the coordinates of P as (x,y) in order to illustrate the general nature of this point.

20. We also need to indicate the position of the point Q with relation to P. And since Q is a bit further from the X-axis and the Y-axis than P, we will designate the horizontal distance of Q from P as Δx (*delta x*, that is, a little bit of x), and the vertical distance as Δy (*delta y*, a little bit of y). With this information added, our graph now looks like this.

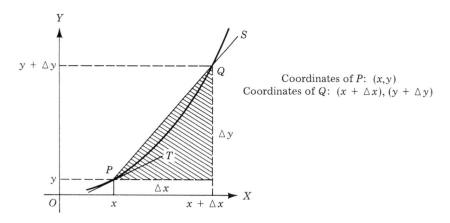

Coordinates of P: (x,y)
Coordinates of Q: $(x + \Delta x), (y + \Delta y)$

See if you can write the equation for the *slope* of the secant line S, keeping in mind that it simply will be the ratio of the vertical distance to the horizontal distance between the points P and Q.

Slope of S = _____

- - - - - - - - - - - - - - -

Slope of S = $\dfrac{\Delta y}{\Delta x}$

21.

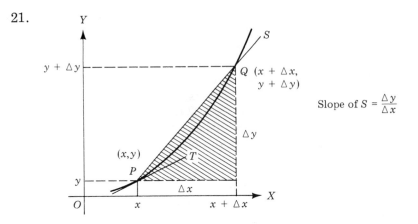

Slope of $S = \dfrac{\Delta y}{\Delta x}$

It may appear as though we had only succeeded in accumulating an odd assortment of letters. However, don't despair. They all are necessary and will be of great help shortly. You will notice also that we have shaded in the triangle of which the secant line S is the hypotenuse. This was done to help focus your attention on it.

Now, remembering that the equation for our curve is $y = x^2$, substitute the coordinates of the point Q for x and y in this equation and see what kind of an expression you get.

_____

_ _ _ _ _ _ _ _ _ _ _ _ _ _ _

You should get $(y + \Delta y) = (x + \Delta x)^2$. The equation of the curve is $y = x^2$, and the coordinates of the point are: x-coordinate $= x + \Delta x$, y-coordinate $= y + \Delta y$. Substituting these coordinates in the equation $y = x^2$ gives us $(y + \Delta y) = (x + \Delta x)^2$.

22. Now try the next step, namely, expanding the binomial $(x + \Delta x)^2$ and write your answer in the space.

$(x + \Delta x)(x + \Delta x) = $ _____

_ _ _ _ _ _ _ _ _ _ _ _ _ _ _

$(x + \Delta x)^2$ or $(x + \Delta x)(x + \Delta x) = x^2 + 2x \cdot \Delta x + \overline{\Delta x}^2$ (the little bar, or *vinculum*, over the Δx means that the exponent, 2, applies to the entire expression, not just to the x).

23. What we are seeking by this algebraic procedure is a relationship between x and y. Specifically, what we would like to find is the ratio of Δy to Δx (that is, the *slope* of the secant line S) based on what we know

about the equation for the curve. Once we find this, you will then learn how we can use it.

So, from the last two frames we now have this information:

$$y = x^2 \qquad (12)$$

and
$$y + \Delta y = x^2 + 2x \cdot \Delta x + \overline{\Delta x}^2. \qquad (13)$$

Since, from (12), we know the value of y in terms of x, we can substitute x^2 for y in the equation (13) and get: (you do it)

_____

— — — — — — — — — — — — —

$$x^2 + \Delta y = x^2 + 2x \cdot \Delta x + \overline{\Delta x}^2$$

24. Now look at the answer just above and notice that we can subtract x^2 from both members of the equation. Doing so gives us $\Delta y = 2x \cdot \Delta x + \overline{\Delta x}^2$, and dividing both sides by Δx gives us the new equation

$$\frac{\Delta y}{\Delta x} = 2x + \Delta x.$$

This looks a little simpler, doesn't it?

But what does it represent? See if you can complete the following sentence.

The quantity $2x + \Delta x$ represents _____

_____ .

— — — — — — — — — — — — —

the slope of the secant line S.

25. Let's state it again.

$$\frac{\Delta y}{\Delta x} = 2x + \Delta x = \text{slope of the secant line } S.$$

Look at it once more in the following figure. What the secant line S really represents, in effect, is the average slope of the tangent lines to the curve between the two points P and Q, much as the average velocity of the falling ball, taken between any two instants of time, would represent the *average velocity* of the ball. But just as we are seeking instantaneous velocity in the case of the falling ball, here we are seeking the exact slope of the curve at a specific point — not the *average* slope.

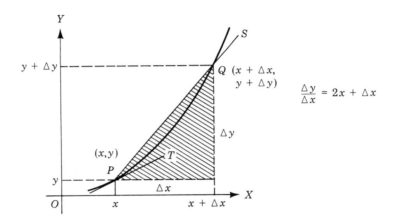

Very well then. Since what we *really* want is the slope of the curve $y = x^2$ at the precise point P, let's imagine the point Q to move slowly along the curve towards P. What we now get is a series of secants (shown as S_1, S_2, S_3, and S_4 in the figure below).

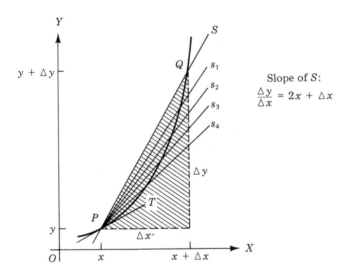

At the same time — since the secants are associated with (*define*, actually) the position of the point Q — the distances Δy and Δx grow shorter and shorter, and our shaded triangle diminishes in size.

Obviously Q is approaching a *limit* (sound familiar?), namely, the point P. What limit is the secant S approaching? _____

– – – – – – – – – – – – – – –

The tangent line T at point P.

26.   Here is our figure again.

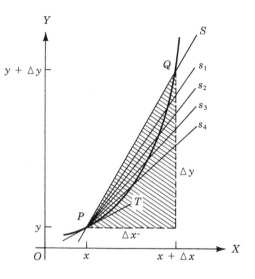

Do you see that the secant S is approaching the tangent line T as a limit?
By the time the point Q reaches point P, the secant S (one end of
which moves with Q) will coincide with the tangent T. Not only
coincide with it; it will *become* the tangent of the curve at the point P.
   It is important that you see these two things very clearly:

(1) Q is approaching the point P *as a limit!*
(2)   The horizontal distance, Δx, between Q and P, is approaching
      zero as a limit.

How do you think the expression for the slope of the secant, $\dfrac{\Delta y}{\Delta x} =$
$2x + \Delta x$, will change as Δx approaches zero as a limit?

_ _ _ _ _ _ _ _ _ _ _ _ _ _

As $\Delta x \rightarrow 0$, then $2x + \Delta x \rightarrow 2x$.

27.   The above answer is correct. But if Δx, approaching zero, comes so
infinitely small that it in effect drops out of the right-hand member of
the equation

$$\lim_{\Delta x \rightarrow 0} \frac{\Delta y}{\Delta x} = 2x + \Delta x,$$

leaving just $2x$, it seems reasonable to ask why it doesn't also drop out
of the expression $\dfrac{\Delta y}{\Delta x}$ on the left-hand side. The answer is: As Δx is

approaching zero as a limit, so is $\triangle y$. Hence (to oversimplify a matter that involves the theorems of infinitesimals), the *ratio*

$\dfrac{\triangle y}{\triangle x}$ remains intact.

Remember, $\dfrac{\triangle y}{\triangle x}$, interpreted graphically, is approaching the tangent to the curve at the point P. This is a *specific* number value! So while $\triangle x$ is approaching

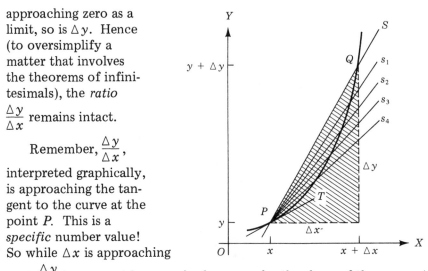

zero, $\dfrac{\triangle y}{\triangle x}$ is approaching a *real* value, namely, the slope of the curve at the point P. Hence the diminishing value of $\triangle x$ has a different effect on the two sides of the equation.

To summarize, then:

(1)   As Q approaches P as a limit, and
(2) $\triangle x$ approaches zero as a limit, then
(3)   The *secant* tends to become *tangent* to, and therefore its slope becomes the slope of, the curve at the point P.

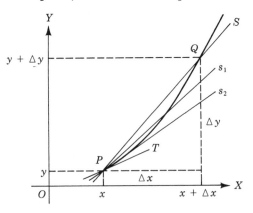

Using the arrow (symbol for "approaches") which we used earlier, and the abbreviation "lim" for *limit* (also familiar to you by now), we can express symbolically what is happening like this:

$$\lim_{\triangle x \to 0} \frac{\triangle y}{\triangle x} = 2x.$$

Try putting this symbolical expression into words just to make sure you understand its meaning. _____

_____

— — — — — — — — — — — —

The limit of $\frac{\Delta y}{\Delta x}$ as Δx approaches zero is equal to $2x$.

28. Let's repeat this entire limit result so we'll have it in front of us while we examine it a bit more.

$$\lim_{\Delta x \to 0} \frac{\Delta y}{\Delta x} = 2x$$

Put into words this result is saying: As Δx approaches zero as a limit, the limit of the *ratio* $\frac{\Delta y}{\Delta x}$ in the expression $\frac{\Delta y}{\Delta x} = 2x + \Delta x$ becomes $2x$.

Now, what useful information have we discovered that we didn't know before we started all this investigating? It is important that you be able to answer that question if you are going to get any value out of what we have covered. So see if you can select the *best* answer below.

___ (a) The secant becomes the tangent as Δx approaches zero as a limit.

___ (b) As the interval Δx of the independent variable approaches the limit zero, the ratio $\frac{\Delta y}{\Delta x}$ becomes the instantaneous rate of change of the function $y = x^2$ at the point P.

___ (c) In the expression $\frac{\Delta y}{\Delta x} = 2x + \Delta x$, the term Δx drops out as Δx approaches zero as a limit.

— — — — — — — — — — — —

Answers (a) and (c) both are correct statements, but neither is the *best* answer nor the most significant thing that occurs. Answer (b) actually is the best answer.

The really important piece of information is that we have found an expression for the *instantaneous rate of change* of a curve — that is, of the *function* which the curve represents — at a specific point, or instant!

To understand the real significance of this, realize that if $y = x^2$ happened to represent the relationship between the height and time increments of the ball thrown into the air, then $2x$ would represent the instantaneous velocity (time rate of change) of the ball at any given moment!

And this is exactly what we were trying to determine when we started out.

In other words, we essentially have done what we set out to do, namely, we have discovered a means of calculating instantaneous rate of change, or growth rate, of a function at a given instant.

CONCLUSION

You must not get the idea that you have studied calculus in this chapter on limits, although the so-called "delta approach" which we followed in this last section is one that frequently is used to introduce beginning calculus students to the concept of the *derivative*. All three of the approaches that we considered in this chapter in our exploration of the subject of limits have a bearing upon the methods of calculus.

But we leave you here, on the doorstep of calculus, since it is not in the province of this book to go further. You will find the study of calculus a fascinating and worthwhile one, which you certainly should pursue if you plan to continue your academic career in any aspect of science or engineering. It will give you a marvelously useful tool. And, as we stated at the outset, because the subject of growth rate, or instantaneous rate of change, appears in so many fields of endeavor today, the need for a knowledge of calculus is no longer confined (as it once was) to the fields of science and engineering. Marketing, statistics, survey techniques, psychology, economics, and many other vocational and professional fields use calculus as a tool.

You will find an excellent introduction to calculus in the Wiley Self-Teaching Guide *Quick Calculus*, by Daniel Kleppner and Norman Ramsey.

Now it is time for a final Self-Test to help you see if you have caught the highlights of this chapter.

SELF-TEST

1. You have been asked to fill a 100-gallon tank with water. Using symbols (letters, arrows, parentheses, etc.), express the fact that as the *quantity* of water in the tank (Q_w) approaches 100 gallons, the *difference* between that amount and the space remaining (S_r) approaches

    zero as a limit. _____

    (frames 3, 4)

2. A function is a set of _____ .

    (frame 5)

3. In the expression $y = x^2$, which is the independent variable? _____
(frame 5)

4. Put into words what $\lim\limits_{x \to a} f(x) = L$ means. _____
_____
(frame 6)

5. Write the first four terms of the set of alternate (that is, every other one) odd positive integers. _____
(frame 7)

6. A set of numbers that goes on indefinitely is called an _____ .
(frame 8)

7. See if you can give an example of an infinite sequence. _____
_____
(frame 8)

8. A progression is a special kind of _____ .
(frame 9)

9. Write an example of an arithmetic progression. _____
_____
(frame 9)

10. What is the common difference in this arithmetic progression? _____
(frame 9)

11. If the first three terms of an arithmetic progression are 2, 5, and 8, find the sixth term. _____   (frame 10)

12. Find the sum of the arithmetic progression where $a = 3$, $l = 15$, and $n = 7$. _____
(frame 11)

13. Insert the five arithmetic means between 29 and 5. _____

(frame 12)

14. Find the fourth term of the geometric progression $2, 5, \frac{25}{2}, \frac{125}{4}, \ldots$

_____

(frame 13)

15. Find the sum of the geometric progression in which $a = 2$, $r = -2$, and $l = 128$. _____

(frame 14)

16. Insert the three geometric means between 1 and 81. _____

(frame 15)

17. Find the sum of the infinite geometric progression having the elements $a = 5$, $r = -\frac{1}{4}$. _____

(frame 16)

18. See if you can put the following symbolical expression into words.

$$\lim_{\Delta x \to 0} \frac{\Delta y}{\Delta x} = 2x$$ _____

_____

(frame 27)

Answers to Self-Test

1. $Q_w \longrightarrow 100$, $S_r \longrightarrow 0$
2. ordered pairs such that no two ordered pairs have the same first element
3. x
4. The function $f(x)$ approaches L as a limiting value as x approaches a as a limit.
5. $1, 5, 9, 13$
6. infinite sequence
7. $\frac{1}{3}$; π; $\frac{5}{11}$; $\frac{1}{9}$; all the even numbers; all positive numbers; etc.
8. sequence
9. $2, 5, 8, 11, \ldots$ or any sequence of numbers in which each number differs from the one that precedes it by a constant amount (the common difference)

10. 3 (in the example above)
11. 17
12. 63
13. 25, 21, 17, 13, 9
14. $l = ar^{n-1} = 2(\frac{5}{2})^4 = 2(\frac{625}{16}) = \frac{625}{8}$
15. 86
16. 3, 9, 27; −3, 9, −27
17. 4
18. The limit of the ratio $\triangle y$ to $\triangle x$ approaches $2x$, as $\triangle x$ approaches zero as a limit.

Appendix

SYMBOLS AND ABBREVIATIONS

$\angle$	angle		$\doteq$	equals approximately
$\angle\!\!\!\angle$	angles		$\stackrel{\circ}{=}$	is measured by
$\frown$	arc		$\neq$	is not equal to
$\odot$	circle		$>$	is greater than
$\circledS$	circles		$\geq$	is greater than or equal to
$\cong$	is congruent to		$<$	is less than
$\circ$	degree		$\leq$	is less than or equal to
$\parallel$	is parallel to		$+$	plus
$\square\!\!\!\!\diagup$	parallelogram		$-$	minus
$\perp$	is perpendicular to		$\pm$	plus or minus
$\perp\!\!\!s$	perpendiculars		rt.	right
$\sim$	is similar to		st.	straight
$\square$	square		sin	sine
$\therefore$	therefore		cos	cosine
$\triangle$	triangle		tan	tangent
$\triangle\!\!\!\triangle$	triangles		csc	cosecant
$\mid\mid$	absolute value		sec	secant
$=$	is equal to		cot	cotangent

GREEK ALPHABET

Letters		Names		Letters		Names		Letters		Names
A	α	Alpha		I	ι	Iota		P	ρ	Rho
B	β	Beta		K	κ	Kappa		Σ	σ	Sigma
Γ	γ	Gamma		Λ	λ	Lambda		T	τ	Tau
Δ	δ	Delta		M	μ	Mu		Υ	υ	Upsilon
E	ϵ	Epsilon		N	ν	Nu		Φ	ϕ	Phi
Z	ζ	Zeta		Ξ	ξ	Xi		X	χ	Chi
H	η	Eta		O	o	Omicron		Ψ	ψ	Psi
Θ	θ	Theta		Π	π	Pi		Ω	ω	Omega

SOME IMPORTANT FORMULAS

Plane Geometry

Notation
a   side of a triangle
    leg of a right triangle
A   area
b   side of a triangle
    leg of a right triangle
    base
c   side of a triangle
    hypotenuse
C   circumference
d   diameter
    diagonal
h   altitude
l   length of an arc
n   number of arc degrees
p   perimeter
r   radius
    apothem of a polygon
s   side of a polygon
S   sum of angles

General Triangle
$$A = \tfrac{1}{2} bh$$

Right Triangle
$$A = \tfrac{1}{2} ab$$
$$c^2 = a^2 + b^2$$

Isosceles Right Triangle
$$c = a\sqrt{2}$$

Equilateral Triangle
$$h = \tfrac{1}{2} a\sqrt{3}$$

Parallelogram
$$A = bh$$

Square
$$A = s^2$$
$$d = s\sqrt{2}$$

Rhombus
$$a = \tfrac{1}{2} dd'$$

Trapezoid
$$A = \tfrac{1}{2} h(b + b')$$

Regular Polygon
$$p = ns$$
$$A = \tfrac{1}{2} pr$$

Polygon
$$S = (n - 2)180°$$

Circle
$$C = \pi d = 2\pi r$$
$$A = \tfrac{1}{2} Cr = \pi r^2 = \tfrac{1}{4}\pi d^2$$

Arc of Circle
$$l = \frac{n}{360}(2\pi r)$$

Sector of Circle
$$A = \tfrac{1}{2} rl = \frac{n}{360}(\pi r^2)$$

Trigonometry and Analytic Geometry

Notation
a   side of a triangle opposite $\angle A$
    leg of a right triangle
A   angle of a triangle
    acute angle of a right triangle
    area
b   side of a triangle opposite $\angle B$
    leg of a right triangle
B   angle of a triangle
c   side of a triangle opposite $\angle C$
    hypotenuse

C   angle of a triangle
    center of a circle
d   distance
(h,k) coordinates of center of a
    circle
m   slope of a line
x_1,y_1 coordinates of point P_1
x_2,y_2 coordinates of point P_2

Acute Angle in a Right Triangle

$$\sin A = \frac{a}{c} \qquad \csc A = \frac{c}{a}$$

$$\cos A = \frac{b}{c} \qquad \sec A = \frac{c}{b}$$

$$\tan A = \frac{a}{b} \qquad \cot A = \frac{b}{a}$$

Obtuse Angle
$$\sin x = \sin(180° - x)$$
$$\cos x = -\cos(180° - x)$$
$$\tan x = -\tan(180° - x)$$

Law of Sines
$$\frac{a}{\sin A} = \frac{b}{\sin B} = \frac{c}{\sin C}$$

Law of Cosines
$$a^2 = b^2 + c^2 - 2bc \cos A$$
$$b^2 = a^2 + c^2 - 2ac \cos B$$
$$c^2 = a^2 + b^2 - 2ab \cos C$$

Distance $P_1 P_2$
$$d = \sqrt{(x_2 - x_1)^2 + (y_2 - y_1)^2}$$

Midpoint of $P_1 P_2$
$$\frac{x_1 + x_2}{2}, \frac{y_1 + y_2}{2}$$

Parallel Lines
$$m_1 = m_2$$

Perpendicular Lines
$$m_1 = -\frac{1}{m_2}$$

Point-Slope Equation of a Line
$$y - y_1 = m(x - x_1)$$

Slope-Intercept Equation of a Line
$$y = mx + c$$

Two-Point Equation of a Line
$$\frac{y - y_1}{x - x_1} = \frac{y_1 - y_2}{x_1 - x_2}$$

Intercept Equation of a Line
$$\frac{x}{a} + \frac{y}{b} = 1$$

General Equation of a Straight Line
$$Ax + Bx + C = 0$$

Two Intersecting Lines
$$\tan \beta = \frac{m_2 - m_1}{1 + m_1 m_2}$$

General Equation of a Circle
$$(x - h)^2 + (y - k)^2 = r^2$$

Equation of Circle Whose Center is at Origin
$$x^2 + y^2 = r^2$$

Equation of a Parabola
$$y^2 = 4px$$

Equation of an Ellipse
$$\frac{x^2}{a^2} + \frac{y^2}{b^2} = 1$$

Equation of an Hyperbola
$$\frac{x^2}{a^2} - \frac{y^2}{b^2} = 1$$

TABLE OF TRIGONOMETRIC FUNCTIONS

0°

'	sin	tan	cot	cos	
0	00000	.00000	∞	1.0000	60
1	029	029	3437.7	000	59
2	058	058	1718.9	000	58
3	087	087	1145.9	000	57
4	116	116	859.44	000	56
5	.00145	.00145	687.55	1.0000	55
6	175	175	572.96	000	54
7	204	204	491.11	000	53
8	233	233	429.72	000	52
9	262	262	381.97	000	51
10	.00291	.00291	343.77	1.0000	50
11	320	320	312.52	.99999	49
12	349	349	286.48	999	48
13	378	378	264.44	999	47
14	407	407	245.55	999	46
15	.00436	.00436	229.18	.99999	45
16	465	465	214.86	999	44
17	495	495	202.22	999	43
18	524	524	190.98	999	42
19	553	553	180.93	998	41
20	.00582	.00582	171.89	.99998	40
21	611	611	163.70	998	39
22	640	640	156.26	998	38
23	669	669	149.47	998	37
24	698	698	143.24	998	36
25	.00727	.00727	137.51	.99997	35
26	756	756	132.22	997	34
27	785	785	127.32	997	33
28	814	815	122.77	997	32
29	844	844	118.54	996	31
30	.00873	.00873	114.59	.99996	30
31	902	902	110.89	996	29
32	931	931	107.43	996	28
33	960	960	104.17	995	27
34	.00989	.00989	101.11	995	26
35	.01018	.01018	98.218	.99995	25
36	047	047	95.489	995	24
37	076	076	92.908	994	23
38	105	105	90.463	994	22
39	134	135	88.144	994	21
40	.01164	.01164	85.940	.99993	20
41	193	193	83.844	993	19
42	222	222	81.847	993	18
43	251	251	79.943	992	17
44	280	280	78.126	992	16
45	.01309	.01309	76.390	.99991	15
46	338	338	74.729	991	14
47	367	367	73.139	991	13
48	396	396	71.615	990	12
49	425	425	70.153	990	11
50	.01454	.01455	68.750	.99989	10
51	483	484	67.402	989	9
52	513	513	66.105	989	8
53	542	542	64.858	988	7
54	571	571	63.657	988	6
55	.01600	.01600	62.499	.99987	5
56	629	629	61.383	987	4
57	658	658	60.306	986	3
58	687	687	59.266	986	2
59	716	716	58.261	985	1
60	.01745	.01746	57.290	.99985	0
	cos	cot	tan	sin	'

89°

1°

'	sin	tan	cot	cos	
0	.01745	.01746	57.290	.99985	60
1	774	775	56.351	984	59
2	803	804	55.442	984	58
3	832	833	54.561	983	57
4	862	862	53.709	983	56
5	.01891	.01891	52.882	.99982	55
6	920	920	52.081	982	54
7	949	949	51.303	981	53
8	.01978	.01978	50.549	980	52
9	.02007	.02007	49.816	980	51
10	.02036	.02036	49.104	.99979	50
11	065	066	48.412	979	49
12	094	095	47.740	978	48
13	123	124	47.085	977	47
14	152	153	46.449	977	46
15	.02181	.02182	45.829	.99976	45
16	211	211	45.226	976	44
17	240	240	44.639	975	43
18	269	269	44.066	974	42
19	298	298	43.508	974	41
20	.02327	.02328	42.964	.99973	40
21	356	357	42.433	972	39
22	385	386	41.916	972	38
23	414	415	41.411	971	37
24	443	444	40.917	970	36
25	.02472	.02473	40.436	.99969	35
26	501	502	39.965	969	34
27	530	531	39.506	968	33
28	560	560	39.057	967	32
29	589	589	38.618	966	31
30	.02618	.02619	38.188	.99966	30
31	647	648	37.769	965	29
32	676	677	37.358	964	28
33	705	706	36.956	963	27
34	734	735	36.563	963	26
35	.02763	.02764	36.178	.99962	25
36	792	793	35.801	961	24
37	821	822	35.431	960	23
38	850	851	35.070	959	22
39	879	881	34.715	959	21
40	.02908	.02910	34.368	.99958	20
41	938	939	34.027	957	19
42	967	968	33.694	956	18
43	.02996	.02997	33.366	955	17
44	.03025	.03026	33.045	954	16
45	.03054	.03055	32.730	.99953	15
46	083	084	32.421	952	14
47	112	114	32.118	952	13
48	141	143	31.821	951	12
49	170	172	31.528	950	11
50	.03199	.03201	31.242	.99949	10
51	228	230	30.960	948	9
52	257	259	30.683	947	8
53	286	288	30.412	946	7
54	316	317	30.145	945	6
55	.03345	.03346	29.882	.99944	5
56	374	376	29.624	943	4
57	403	405	29.371	942	3
58	432	434	29.122	941	2
59	461	463	28.877	940	1
60	.03490	.03492	28.636	.99939	0
	cos	cot	tan	sin	'

88°

2°

′	sin	tan	cot	cos	
0	.03490	.03492	28.636	.99939	60
1	519	521	.399	938	59
2	548	550	28.166	937	58
3	577	579	27.937	936	57
4	606	609	.712	935	56
5	.03635	.03638	27.490	.99934	55
6	664	667	.271	933	54
7	693	696	27.057	932	53
8	723	725	26.845	931	52
9	752	754	.637	930	51
10	.03781	.03783	26.432	.99929	50
11	810	812	.230	927	49
12	839	842	26.031	926	48
13	868	871	25.835	925	47
14	897	900	.642	924	46
15	.03926	.03929	25.452	.99923	45
16	955	958	.264	922	44
17	.03984	.03987	25.080	921	43
18	.04013	.04016	24.898	919	42
19	042	046	.719	918	41
20	.04071	.04075	24.542	.99917	40
21	100	104	.368	916	39
22	129	133	.196	915	38
23	159	162	24.026	913	37
24	188	191	23.859	912	36
25	.04217	04220	23.695	.99911	35
26	246	250	.532	910	34
27	275	279	.372	909	33
28	304	308	.214	907	32
29	333	337	23.058	906	31
30	.04362	.04366	22.904	.99905	30
31	391	395	.752	904	29
32	420	424	.602	902	28
33	449	454	.454	901	27
34	478	483	.308	900	26
35	.04507	.04512	22.164	.99898	25
36	536	541	22.022	897	24
37	565	570	21.881	896	23
38	594	599	.743	894	22
39	623	628	.606	893	21
40	.04653	.04658	21.470	.99892	20
41	682	687	.337	890	19
42	711	716	.205	889	18
43	740	745	21.075	888	17
44	769	774	20.946	886	16
45	.04798	.04803	20.819	.99885	15
46	827	833	.693	883	14
47	856	862	.569	882	13
48	885	891	.446	881	12
49	914	920	.325	879	11
50	.04943	.04949	20.206	.99878	10
51	.04972	.04978	20.087	876	9
52	.05001	.05007	19.970	875	8
53	030	037	.855	873	7
54	059	066	.740	872	6
55	.05088	.05095	19.627	.99870	5
56	117	124	.516	869	4
57	146	153	.405	867	3
58	175	182	.296	866	2
59	205	212	.188	864	1
60	.05234	.05241	19.081	.99863	0
	cos	cot	tan	sin	′

87°

3°

′	sin	tan	cot	cos	
0	.05234	.05241	19.081	.99863	60
1	263	270	18.976	861	59
2	292	299	.871	860	58
3	321	328	.768	858	57
4	350	357	.666	857	56
5	.05379	.05387	18.564	.99855	55
6	408	416	.464	854	54
7	437	445	.366	852	53
8	466	474	.268	851	52
9	495	503	.171	849	51
10	.05524	.05533	18.075	.99847	50
11	553	562	17.980	846	49
12	582	591	.886	844	48
13	611	620	.793	842	47
14	640	649	.702	841	46
15	.05669	.05678	17.611	.99839	45
16	698	708	.521	838	44
17	727	737	.431	836	43
18	755	766	.343	834	42
19	785	795	.256	833	41
20	.05814	.05824	17.169	.99831	40
21	844	854	17.084	829	39
22	873	883	16.999	827	38
23	902	912	.915	826	37
24	931	941	.832	824	36
25	.05960	.05970	16.750	.99822	35
26	.05989	.05999	.668	821	34
27	.06018	.06029	.587	819	33
28	047	058	.507	817	32
29	076	087	.428	815	31
30	.06105	.06116	16.350	.99813	30
31	134	145	.272	812	29
32	163	175	.195	810	28
33	192	204	.119	808	27
34	221	233	16.043	806	26
35	.06250	.06262	15.969	.99804	25
36	279	291	.895	803	24
37	308	321	.821	801	23
38	337	350	.748	799	22
39	366	379	.676	797	21
40	.06395	.06408	15.605	.99795	20
41	424	438	.534	793	19
42	453	467	.464	792	18
43	482	496	.394	790	17
44	511	525	.325	788	16
45	.06540	.06554	15.257	.99786	15
46	569	584	.189	784	14
47	598	613	.122	782	13
48	627	642	15.056	780	12
49	656	671	14.990	778	11
50	.06685	.06700	14.924	.99776	10
51	714	730	.860	774	9
52	743	759	.795	772	8
53	773	788	.732	770	7
54	802	817	.669	768	6
55	.06831	.06847	14.606	.99766	5
56	860	876	.544	764	4
57	889	905	.482	762	3
58	918	934	.421	760	2
59	947	963	.361	758	1
60	.06976	.06993	14.301	.99756	0
	cos	cot	tan	sin	′

86°

4°

′	sin	tan	cot	cos	
0	06976	.06993	14.301	.99756	60
1	.07005	.07022	.241	754	59
2	034	051	.182	752	58
3	063	080	.124	750	57
4	092	110	.065	748	56
5	.07121	.07139	14.008	.99746	55
6	150	168	13.951	744	54
7	179	197	.894	742	53
8	208	227	.838	740	52
9	237	256	.782	738	51
10	.07266	.07285	13.727	.99736	50
11	295	314	.672	734	49
12	324	344	.617	731	48
13	353	373	.563	729	47
14	382	402	.510	727	46
15	.07411	.07431	13.457	.99725	45
16	440	461	.404	723	44
17	469	490	.352	721	43
18	498	519	.300	719	42
19	527	548	.248	716	41
20	.07556	.07578	13.197	.99714	40
21	585	607	.146	712	39
22	614	636	.096	710	38
23	643	665	13.046	708	37
24	672	695	12.996	705	36
25	.07701	.07724	12.947	.99703	35
26	730	753	.898	701	34
27	759	782	.850	699	33
28	788	812	.801	696	32
29	817	841	.754	694	31
30	.07846	.07870	12.706	.99692	30
31	875	899	.659	689	29
32	904	929	.612	687	28
33	933	958	.566	685	27
34	962	.07987	.520	683	26
35	.07991	.08017	12.474	.99680	25
36	.08020	046	.429	678	24
37	049	075	.384	676	23
38	078	104	.339	673	22
39	107	134	.295	671	21
40	.08136	.08163	12.251	.99668	20
41	165	192	.207	666	19
42	194	221	.163	664	18
43	223	251	.120	661	17
44	252	280	.077	659	16
45	.08281	.08309	12.035	.99657	15
46	310	339	11.992	654	14
47	339	368	.950	652	13
48	368	397	.909	649	12
49	397	427	.867	647	11
50	.08426	.08456	11.826	.99644	10
51	455	485	.785	642	9
52	484	514	.745	639	8
53	513	544	.705	637	7
54	542	573	.664	635	6
55	.08571	.08602	11.625	.99632	5
56	600	632	.585	630	4
57	629	661	.546	627	3
58	658	690	.507	625	2
59	687	720	.468	622	1
60	08716	.08749	11.430	99619	0
	cos	cot	tan	sin	′

85°

5°

′	sin	tan	cot	cos	
0	.08716	.08749	11.430	.99619	60
1	745	778	.392	617	59
2	774	807	.354	614	58
3	803	837	.316	612	57
4	831	866	.279	609	56
5	.08860	.08895	11.242	.99607	55
6	889	925	.205	604	54
7	918	954	.168	602	53
8	947	.08983	.132	599	52
9	.08976	.09013	.095	596	51
10	.09005	.09042	11.059	.99594	50
11	034	071	11.024	591	49
12	063	101	10.988	588	48
13	092	130	.953	586	47
14	121	159	.918	583	46
15	.09150	.09189	10.883	.99580	45
16	179	218	.848	578	44
17	208	247	.814	575	43
18	237	277	.780	572	42
19	266	306	.746	570	41
20	.09295	.09335	10.712	.99567	40
21	324	365	.678	564	39
22	353	394	.645	562	38
23	382	423	.612	559	37
24	411	453	.579	556	36
25	.09440	.09482	10.546	.99553	35
26	469	511	.514	551	34
27	498	541	.481	548	33
28	527	570	.449	545	32
29	556	600	.417	542	31
30	.09585	.09629	10.385	.99540	30
31	614	658	.354	537	29
32	642	688	.322	534	28
33	671	717	.291	531	27
34	700	746	.260	528	26
35	.09729	.09776	10.229	.99526	25
36	758	805	.199	523	24
37	787	834	.168	520	23
38	816	864	.138	517	22
39	845	893	.108	514	21
40	.09874	.09923	10.078	.99511	20
41	903	952	.048	508	19
42	932	.09981	10.019	506	18
43	961	.10011	9.9893	503	17
44	.09990	040	.9601	500	16
45	.10019	.10069	9.9310	.99497	15
46	048	099	.9021	494	14
47	077	128	.8734	491	13
48	106	158	.8448	488	12
49	135	187	.8164	485	11
50	.10164	.10216	9.7882	.99482	10
51	192	246	.7601	479	9
52	221	275	.7322	476	8
53	250	305	.7044	473	7
54	279	334	.6768	470	6
55	.10308	.10363	9.6493	.99467	5
56	337	393	.6220	464	4
57	366	422	.5949	461	3
58	395	452	.5679	458	2
59	424	481	.5411	455	1
60	10453	.10510	9.5144	.99452	0
	cos	cot	tan	sin	′

84°

6°

'	sin	tan	cot	cos	
0	.10453	.10510	9.5144	.99452	60
1	482	540	.4878	449	59
2	511	569	.4614	446	58
3	540	599	.4352	443	57
4	569	628	.4090	440	56
5	.10597	.10657	9.3831	.99437	55
6	626	687	.3572	434	54
7	655	716	.3315	431	53
8	684	746	.3060	428	52
9	713	775	.2806	424	51
10	.10742	.10805	9.2553	.99421	50
11	771	834	.2302	418	49
12	800	863	.2052	415	48
13	829	893	.1803	412	47
14	858	922	.1555	409	46
15	.10887	.10952	9.1309	.99406	45
16	916	.10981	.1065	402	44
17	945	.11011	.0821	399	43
18	.10973	040	.0579	396	42
19	.11002	070	.0338	393	41
20	.11031	.11099	9.0098	.99390	40
21	060	128	8.9860	386	39
22	089	158	.9623	383	38
23	118	187	.9387	380	37
24	147	217	.9152	377	36
25	.11176	.11246	8.8919	.99374	35
26	205	276	.8686	370	34
27	234	305	.8455	367	33
28	263	335	.8225	364	32
29	291	364	.7996	360	31
30	.11320	.11394	8.7769	.99357	30
31	349	423	.7542	354	29
32	378	452	.7317	351	28
33	407	482	.7093	347	27
34	436	511	.6870	344	26
35	.11465	.11541	8.6648	.99341	25
36	494	570	.6427	337	24
37	523	600	.6208	334	23
38	552	629	.5989	331	22
39	580	659	.5772	327	21
40	.11609	.11688	8.5555	.99324	20
41	638	718	.5340	320	19
42	667	747	.5126	317	18
43	696	777	.4913	314	17
44	725	806	.4701	310	16
45	.11754	.11836	8.4490	.99307	15
46	783	865	.4280	303	14
47	812	895	.4071	300	13
48	840	924	.3863	297	12
49	869	954	.3656	293	11
50	.11898	.11983	8.3450	.99290	10
51	927	.12013	.3245	286	9
52	956	042	.3041	283	8
53	.11985	072	.2838	279	7
54	.12014	101	.2636	276	6
55	.12043	.12131	8.2434	.99272	5
56	071	160	.2234	269	4
57	100	190	.2035	265	3
58	129	219	.1837	262	2
59	158	249	.1640	258	1
60	.12187	.12278	8.1443	.99255	0
	cos	cot	tan	sin	'

83°

7°

'	sin	tan	cot	cos	
0	.12187	.12278	8.1443	.99255	60
1	216	308	.1248	251	59
2	245	338	.1054	248	58
3	274	367	.0860	244	57
4	302	397	.0667	240	56
5	.12331	.12426	8.0476	.99237	55
6	360	456	.0285	233	54
7	389	485	8.0095	230	53
8	418	515	7.9906	226	52
9	447	544	.9718	222	51
10	.12476	.12574	7.9530	.99219	50
11	504	603	.9344	215	49
12	533	633	.9158	211	48
13	562	662	.8973	208	47
14	591	692	.8789	204	46
15	.12620	.12722	7.8606	.99200	45
16	649	751	8424	197	44
17	678	781	.8243	193	43
18	706	810	.8062	189	42
19	735	840	.7882	186	41
20	.12764	.12869	7.7704	.99182	40
21	793	899	.7525	178	39
22	822	929	.7348	175	38
23	851	958	.7171	171	37
24	880	.12988	.6996	167	36
25	.12908	.13017	7.6821	.99163	35
26	937	047	.6647	160	34
27	966	076	.6473	156	33
28	.12995	106	.6301	152	32
29	.13024	136	.6129	148	31
30	.13053	.13165	7.5958	.99144	30
31	081	195	.5787	141	29
32	110	224	.5618	137	28
33	139	254	.5449	133	27
34	168	284	.5281	129	26
35	.13197	.13313	7.5113	.99125	25
36	226	343	.4947	122	24
37	254	372	.4781	118	23
38	283	402	.4615	114	22
39	312	432	.4451	110	21
40	.13341	.13461	7.4287	.99106	20
41	370	491	.4124	102	19
42	399	521	.3962	098	18
43	427	550	.3800	094	17
44	456	580	.3639	091	16
45	.13485	.13609	7.3479	.99087	15
46	514	639	.3319	083	14
47	543	669	.3160	079	13
48	572	698	.3002	075	12
49	600	728	.2844	071	11
50	.13629	.13758	7.2687	.99067	10
51	658	787	.2531	063	9
52	687	817	.2375	059	8
53	716	846	.2220	055	7
54	744	876	.2066	051	6
55	.13773	.13906	7.1912	.99047	5
56	802	935	.1759	043	4
57	831	965	.1607	039	3
58	860	.13995	.1455	035	2
59	889	.14024	.1304	031	1
60	.13917	.14054	7.1154	.99027	0
	cos	cot	tan	sin	'

82°

8°

′	sin	tan	cot	cos	
0	.13917	.14054	7.1154	.99027	60
1	946	084	.1004	023	59
2	.13975	113	.0855	019	58
3	.14004	143	.0706	015	57
4	033	173	.0558	011	56
5	.14061	.14202	7.0410	.99006	55
6	090	232	.0264	.99002	54
7	119	262	7.0117	.98998	53
8	148	291	6.9972	994	52
9	177	321	.9827	990	51
10	.14205	.14351	6.9682	.98986	50
11	234	381	.9538	982	49
12	263	410	.9395	978	48
13	292	440	.9252	973	47
14	320	470	.9110	969	46
15	.14349	.14499	6.8969	.98965	45
16	378	529	.8828	961	44
17	407	559	.8687	957	43
18	436	588	.8548	953	42
19	464	618	.8408	948	41
20	.14493	.14648	6.8269	.98944	40
21	522	678	.8131	940	39
22	551	707	.7994	936	38
23	580	737	.7856	931	37
24	608	767	.7720	927	36
25	.14637	.14796	6.7584	.98923	35
26	666	826	.7448	919	34
27	695	856	.7313	914	33
28	723	886	.7179	910	32
29	752	915	.7045	906	31
30	.14781	.14945	6.6912	.98902	30
31	810	.14975	.6779	897	29
32	838	.15005	.6646	893	28
33	867	034	.6514	889	27
34	896	064	.6383	884	26
35	.14925	.15094	6.6252	.98880	25
36	954	124	.6122	876	24
37	.14982	153	.5992	871	23
38	.15011	183	.5863	867	22
39	040	213	.5734	863	21
40	.15069	.15243	6.5606	.98858	20
41	097	272	.5478	854	19
42	126	302	.5350	849	18
43	155	332	.5223	845	17
44	184	362	.5097	841	16
45	.15212	.15391	6.4971	.98836	15
46	241	421	.4846	832	14
47	270	451	.4721	827	13
48	299	481	.4596	823	12
49	327	511	.4472	818	11
50	.15356	.15540	6.4348	.98814	10
51	385	570	.4225	809	9
52	414	600	.4103	805	8
53	442	630	.3980	800	7
54	471	660	.3859	796	6
55	.15500	.15689	6.3737	.98791	5
56	529	719	.3617	787	4
57	557	749	.3496	782	3
58	586	779	.3376	778	2
59	615	809	.3257	773	1
60	.15643	.15838	6.3138	.98769	0
	cos	cot	tan	sin	′

81°

9°

′	sin	tan	cot	cos	
0	.15643	.15838	6.3138	.98769	60
1	672	868	.3019	764	59
2	701	898	.2901	760	58
3	730	928	.2783	755	57
4	758	958	.2666	751	56
5	.15787	.15988	6.2549	.98746	55
6	816	.16017	.2432	741	54
7	845	047	.2316	737	53
8	873	077	.2200	732	52
9	902	107	.2085	728	51
10	.15931	.16137	6.1970	.98723	50
11	959	167	.1856	718	49
12	.15988	196	.1742	714	48
13	.16017	226	.1628	709	47
14	046	256	.1515	704	46
15	.16074	.16286	6.1402	.98700	45
16	103	316	.1290	695	44
17	132	346	.1178	690	43
18	160	376	.1066	686	42
19	189	405	.0955	681	41
20	.16218	.16435	6.0844	.98676	40
21	246	465	.0734	671	39
22	275	495	.0624	667	38
23	304	525	.0514	662	37
24	333	555	.0405	657	36
25	.16361	.16585	6.0296	.98652	35
26	390	615	.0188	648	34
27	419	645	6.0080	643	33
28	447	674	5.9972	638	32
29	476	704	.9865	633	31
30	.16505	.16734	5.9758	.98629	30
31	533	764	.9651	624	29
32	562	794	.9545	619	28
33	591	824	.9439	614	27
34	620	854	.9333	609	26
35	.16648	.16884	5.9228	.98604	25
36	677	914	.9124	600	24
37	706	944	.9019	595	23
38	734	.16974	.8915	590	22
39	763	.17004	.8811	585	21
40	.16792	.17033	5.8708	.98580	20
41	820	063	.8605	575	19
42	849	093	.8502	570	18
43	878	123	.8400	565	17
44	906	153	.8298	561	16
45	.16935	.17183	5.8197	.98556	15
46	964	213	.8095	551	14
47	16992	243	.7994	546	13
48	.17021	273	.7894	541	12
49	050	303	.7794	536	11
50	.17078	.17333	5.7694	.98531	10
51	107	363	.7594	526	9
52	136	393	.7495	521	8
53	164	423	.7396	516	7
54	193	453	.7297	511	6
55	.17222	.17483	5.7199	.98506	5
56	250	513	.7101	501	4
57	279	543	.7004	496	3
58	308	573	.6906	491	2
59	336	603	.6809	486	1
60	.17365	.17633	5.6713	.98481	0
	cos	cot	tan	sin	′

80°

10°

′	sin	tan	cot	cos	
0	.17365	.17633	5.6713	.98481	60
1	393	663	.6617	476	59
2	422	693	.6521	471	58
3	451	723	.6425	466	57
4	479	753	.6329	461	56
5	.17508	.17783	5.6234	.98455	55
6	537	813	.6140	450	54
7	565	843	.6045	445	53
8	594	873	.5951	440	52
9	623	903	.5857	435	51
10	.17651	.17933	5.5764	.98430	50
11	680	963	.5671	425	49
12	708	.17993	.5578	420	48
13	737	.18023	.5485	414	47
14	766	053	.5393	409	46
15	.17794	.18083	5.5301	.98404	45
16	823	113	.5209	399	44
17	852	143	.5118	394	43
18	880	173	.5026	389	42
19	909	203	.4936	383	41
20	.17937	.18233	5.4845	.98378	40
21	966	263	.4755	373	39
22	.17995	293	.4665	368	38
23	.18023	323	.4575	362	37
24	052	353	.4486	357	36
25	.18081	.18384	5.4397	.98352	35
26	109	414	.4308	347	34
27	138	444	.4219	341	33
28	166	474	.4131	336	32
29	195	504	.4043	331	31
30	.18224	.18534	5.3955	.98325	30
31	252	564	.3868	320	29
32	281	594	.3781	315	28
33	309	624	.3694	310	27
34	338	654	.3607	304	26
35	.18367	.18684	5.3521	.98299	25
36	395	714	.3435	294	24
37	424	745	.3349	288	23
38	452	775	.3263	283	22
39	481	805	.3178	277	21
40	.18509	.18835	5.3093	.98272	20
41	538	865	.3008	267	19
42	567	895	.2924	261	18
43	595	925	.2839	256	17
44	624	955	.2755	250	16
45	.18652	.18986	5.2672	.98245	15
46	681	.19016	.2588	240	14
47	710	046	.2505	234	13
48	738	076	.2422	229	12
49	767	106	.2339	223	11
50	.18795	.19136	5.2257	.98218	10
51	824	166	.2174	212	9
52	852	197	.2092	207	8
53	881	227	.2011	201	7
54	910	257	.1929	196	6
55	.18938	.19287	5.1848	.98190	5
56	967	317	.1767	185	4
57	.18995	347	.1686	179	3
58	.19024	378	.1606	174	2
59	052	408	.1526	168	1
60	.19081	.19438	5.1446	.98163	0
	cos	cot	tan	sin	′

79°

11°

′	sin	tan	cot	cos	
0	.19081	.19438	5.1446	.98163	60
1	109	468	.1366	157	59
2	138	498	.1286	152	58
3	167	529	.1207	146	57
4	195	559	.1128	140	56
5	.19224	.19589	5.1049	.98135	55
6	252	619	.0970	129	54
7	281	649	.0892	124	53
8	309	680	.0814	118	52
9	338	710	.0736	112	51
10	.19366	.19740	5.0658	.98107	50
11	395	770	.0581	101	49
12	423	801	.0504	096	48
13	452	831	.0427	090	47
14	481	861	.0350	084	46
15	.19509	.19891	5.0273	.98079	45
16	538	921	.0197	073	44
17	566	952	.0121	067	43
18	595	.19982	5.0045	061	42
19	623	.20012	4.9969	056	41
20	.19652	.20042	4.9894	.98050	40
21	680	073	.9819	044	39
22	709	103	.9744	039	38
23	737	133	.9669	033	37
24	766	164	.9594	027	36
25	.19794	.20194	4.9520	.98021	35
26	823	224	.9446	016	34
27	851	254	.9372	010	33
28	880	285	.9298	.98004	32
29	908	315	.9225	.97998	31
30	.19937	.20345	4.9152	.97992	30
31	965	376	.9078	987	29
32	.19994	406	.9006	981	28
33	.20022	436	.8933	975	27
34	051	466	.8860	969	26
35	.20079	.20497	4.8788	.97963	25
36	108	527	.8716	958	24
37	136	557	.8644	952	23
38	165	588	.8573	946	22
39	193	618	.8501	940	21
40	.20222	.20648	4.8430	.97934	20
41	250	679	.8359	928	19
42	279	709	.8288	922	18
43	307	739	.8218	916	17
44	336	770	.8147	910	16
45	.20364	.20800	4.8077	.97905	15
46	393	830	.8007	899	14
47	421	861	.7937	893	13
48	450	891	.7867	887	12
49	478	921	.7798	881	11
50	.20507	.20952	4.7729	.97875	10
51	535	.20982	.7659	869	9
52	563	.21013	.7591	863	8
53	592	043	.7522	857	7
54	620	073	.7453	851	6
55	.20649	.21104	4.7385	.97845	5
56	677	134	.7317	839	4
57	706	164	.7249	833	3
58	734	195	.7181	827	2
59	763	225	.7114	821	1
60	.20791	.21256	4.7046	.97815	0
	cos	cot	tan	sin	′

78°

12°

'	sin	tan	cot	cos	
0	.20791	.21256	4.7046	.97815	60
1	820	286	.6979	809	59
2	848	316	.6912	803	58
3	877	347	.6845	797	57
4	905	377	.6779	791	56
5	.20933	.21408	4.6712	.97784	55
6	962	438	.6646	778	54
7	.20990	469	.6580	772	53
8	.21019	499	.6514	766	52
9	047	529	.6448	760	51
10	.21076	.21560	4.6382	.97754	50
11	104	590	.6317	748	49
12	132	621	.6252	742	48
13	161	651	.6187	735	47
14	189	682	.6122	729	46
15	.21218	.21712	4.6057	.97723	45
16	246	743	.5993	717	44
17	275	773	.5928	711	43
18	303	804	.5864	705	42
19	331	834	.5800	698	41
20	.21360	.21864	4.5736	.97692	40
21	388	895	.5673	686	39
22	417	925	.5609	680	38
23	445	956	.5546	673	37
24	474	.21986	.5483	667	36
25	.21502	.22017	4.5420	.97661	35
26	530	047	.5357	655	34
27	559	078	.5294	648	33
28	587	108	.5232	642	32
29	616	139	.5169	636	31
30	.21644	.22169	4.5107	.97630	30
31	672	200	.5045	623	29
32	701	231	.4983	617	28
33	729	261	.4922	611	27
34	758	292	.4860	604	26
35	.21786	.22322	4.4799	.97598	25
36	814	353	.4737	592	24
37	843	383	.4676	585	23
38	871	414	.4615	579	22
39	899	444	.4555	573	21
40	.21928	.22475	4.4494	.97566	20
41	956	505	.4434	560	19
42	.21985	536	.4373	553	18
43	.22013	567	.4313	547	17
44	041	597	.4253	541	16
45	.22070	.22628	4.4194	.97534	15
46	098	658	.4134	528	14
47	126	689	.4075	521	13
48	155	719	.4015	515	12
49	183	750	.3956	508	11
50	.22212	.22781	4.3897	.97502	10
51	240	811	.3838	496	9
52	268	842	.3779	489	8
53	297	872	.3721	483	7
54	325	903	.3662	476	6
55	.22353	.22934	4.3604	.97470	5
56	382	964	.3546	463	4
57	410	.22995	.3488	457	3
58	438	.23026	.3430	450	2
59	467	056	.3372	444	1
60	.22495	.23087	4.3315	.97437	0
	cos	cot	tan	sin	'

77°

13°

'	sin	tan	cot	cos	
0	.22495	.23087	4.3315	.97437	60
1	523	117	.3257	430	59
2	552	148	.3200	424	58
3	580	179	.3143	417	57
4	608	209	.3086	411	56
5	.22637	.23240	4.3029	.97404	55
6	665	271	.2972	398	54
7	693	301	.2916	391	53
8	722	332	.2859	384	52
9	750	363	.2803	378	51
10	.22778	.23393	4.2747	.97371	50
11	807	424	.2691	365	49
12	835	455	.2635	358	48
13	863	485	.2580	351	47
14	892	516	.2524	345	46
15	.22920	.23547	4.2468	.97338	45
16	948	578	.2413	331	44
17	.22977	608	.2358	325	43
18	.23005	639	.2303	318	42
19	033	670	.2248	311	41
20	.23062	.23700	4.2193	.97304	40
21	090	731	.2139	298	39
22	118	762	.2084	291	38
23	146	793	.2030	284	37
24	175	823	.1976	278	36
25	.23203	.23854	4.1922	.97271	35
26	231	885	.1868	264	34
27	260	916	.1814	257	33
28	288	946	.1760	251	32
29	316	.23977	.1706	244	31
30	.23345	.24008	4.1653	.97237	30
31	373	039	.1600	230	29
32	401	069	.1547	223	28
33	429	100	.1493	217	27
34	458	131	.1441	210	26
35	.23486	.24162	4.1388	.97203	25
36	514	193	.1335	196	24
37	542	223	.1282	189	23
38	571	254	.1230	182	22
39	599	285	.1178	176	21
40	.23627	.24316	4.1126	.97169	20
41	656	347	.1074	162	19
42	684	377	.1022	155	18
43	712	408	.0970	148	17
44	740	439	.0918	141	16
45	.23769	.24470	4.0867	.97134	15
46	797	501	.0815	127	14
47	825	532	.0764	120	13
48	853	562	.0713	113	12
49	882	593	.0662	106	11
50	.23910	.24624	4.0611	.97100	10
51	938	655	.0560	093	9
52	966	686	.0509	086	8
53	.23995	717	.0459	079	7
54	.24023	747	.0408	072	6
55	.24051	.24778	4.0358	.97063	5
56	079	809	.0308	058	4
57	108	840	.0257	051	3
58	136	871	.0207	044	2
59	164	902	.0158	037	1
60	.24192	.24933	4.0108	.97030	0
	cos	cot	tan	sin	'

76°

14°

′	sin	tan	cot	cos	
0	.24192	.24933	4.0108	.97030	60
1	220	964	.0058	023	59
2	249	.24995	4.0009	015	58
3	277	.25026	3.9959	008	57
4	305	056	.9910	.97001	56
5	.24333	.25087	3.9861	.96994	55
6	362	118	.9812	987	54
7	390	149	.9763	980	53
8	418	180	.9714	973	52
9	446	211	.9665	966	51
10	.24474	.25242	3.9617	.96959	50
11	503	273	.9568	952	49
12	531	304	.9520	945	48
13	559	335	.9471	937	47
14	587	366	.9423	930	46
15	.24615	.25397	3.9375	.96923	45
16	644	428	.9327	916	44
17	672	459	.9279	909	43
18	700	490	.9232	902	42
19	728	521	.9184	894	41
20	.24756	.25552	3.9136	.96887	40
21	784	583	.9089	880	39
22	813	614	.9042	873	38
23	841	645	.8995	866	37
24	869	676	.8947	858	36
25	.24897	.25707	3.8900	.96851	35
26	925	738	.8854	844	34
27	954	769	.8807	837	33
28	.24982	800	.8760	829	32
29	.25010	831	.8714	822	31
30	.25038	.25862	3.8667	.96815	30
31	066	893	.8621	807	29
32	094	924	.8575	800	28
33	122	955	.8528	793	27
34	151	.25986	.8482	786	26
35	.25179	.26017	3.8436	.96778	25
36	207	048	.8391	771	24
37	235	079	.8345	764	23
38	263	110	.8299	756	22
39	291	141	.8254	749	21
40	.25320	.26172	3.8208	.96742	20
41	348	203	.8163	734	19
42	376	235	.8118	727	18
43	404	266	.8073	719	17
44	432	297	.8028	712	16
45	.25460	.26328	3.7983	.96705	15
46	488	359	.7938	697	14
47	516	390	.7893	690	13
48	545	421	.7848	682	12
49	573	452	.7804	675	11
50	.25601	.26483	3.7760	.96667	10
51	629	515	.7715	660	9
52	657	546	.7671	653	8
53	685	577	.7627	645	7
54	713	608	.7583	638	6
55	.25741	.26639	3.7539	.96630	5
56	769	670	.7495	623	4
57	798	701	.7451	615	3
58	826	733	.7408	608	2
59	854	764	.7364	600	1
60	25882	26795	3.7321	.96593	0
	cos	cot	tan	sin	′

75°

15°

′	sin	tan	cot	cos	
0	.25882	.26795	3.7321	.96593	60
1	910	826	.7277	585	59
2	938	857	.7234	578	58
3	966	888	.7191	570	57
4	.25994	920	.7148	562	56
5	.26022	.26951	3.7105	.96555	55
6	050	.26982	.7062	547	54
7	079	.27013	.7019	540	53
8	107	044	.6976	532	52
9	135	076	.6933	524	51
10	.26163	.27107	3.6891	.96517	50
11	191	138	.6848	509	49
12	219	169	.6806	502	48
13	247	201	.6764	494	47
14	275	232	.6722	486	46
15	.26303	.27263	3.6680	.96479	45
16	331	294	.6638	471	44
17	359	326	.6596	463	43
18	387	357	.6554	456	42
19	415	388	.6512	448	41
20	.26443	.27419	3.6470	.96440	40
21	471	451	.6429	433	39
22	500	482	.6387	425	38
23	528	513	.6346	417	37
24	556	545	.6305	410	36
25	.26584	.27576	3.6264	.96402	35
26	612	607	.6222	394	34
27	640	638	.6181	386	33
28	668	670	.6140	379	32
29	696	701	.6100	371	31
30	.26724	.27732	3.6059	.96363	30
31	752	764	.6018	355	29
32	780	795	.5978	347	28
33	808	826	.5937	340	27
34	836	858	.5897	332	26
35	.26864	.27889	3.5856	.96324	25
36	892	921	.5816	316	24
37	920	952	.5776	308	23
38	948	.27983	.5736	301	22
39	.26976	.28015	.5696	293	21
40	.27004	.28046	3.5656	.96285	20
41	032	077	.5616	277	19
42	060	109	.5576	269	18
43	088	140	.5536	261	17
44	116	172	.5497	253	16
45	.27144	.28203	3.5457	.96246	15
46	172	234	.5418	238	14
47	200	266	.5379	230	13
48	228	297	.5339	222	12
49	256	329	.5300	214	11
50	.27284	.28360	3.5261	.96206	10
51	312	391	.5222	198	9
52	340	423	.5183	190	8
53	368	454	.5144	182	7
54	396	486	.5105	174	6
55	.27424	.28517	3.5067	.96166	5
56	452	549	.5028	158	4
57	480	580	.4989	150	3
58	508	612	.4951	142	2
59	536	643	.4912	134	1
60	.27564	.28675	3.4874	.96126	0
	cos	cot	tan	sin	′

74°

16°

′	sin	tan	cot	cos	
0	.27564	.28675̄	3.4874	.96126	60
1	592	706	.4836	118	59
2	620	738	.4798	110	58
3	648	769	.4760	102	57
4	676	801	.4722	094	56
5	.27704	.28832	3.4684	.96086	55
6	731	864	.4646	078	54
7	759	895	.4608	070	53
8	787	927	.4570	062	52
9	815	958	.4533	054	51
10	.27843	.28990	3.4495	.96046	50
11	871	.29021	.4458	037	49
12	899	053	.4420	029	48
13	927	084	.4383	021	47
14	955̄	116	.4346	013	46
15	.27983	.29147	3.4308	.96005̄	45
16	.28011	179	.4271	.95997	44
17	039	210	.4234	989	43
18	067	242	.4197	981	42
19	095̄	274	.4160	972	41
20	.28123	.29305	3.4124	.95964	40
21	150	337	.4087	956	39
22	178	368	.4050	948	38
23	206	400	.4014	940	37
24	234	432	.3977	931	36
25	.28262	.29463	3.3941	.95923	35
26	290	495̄	.3904	915	34
27	318	526	.3868	907	33
28	346	558	.3832	898	32
29	374	590	.3796	890	31
30	.28402	.29621	3.3759	.95882	30
31	429	653	.3723	874	29
32	457	685̄	.3687	865	28
33	485	716	.3652	857	27
34	513	748	.3616	849	26
35	.28541	.29780	3.3580	.95841	25
36	569	811	.3544	832	24
37	597	843	.3509	824	23
38	625̄	875̄	.3473	816	22
39	652	906	.3438	807	21
40	.28680	.29938	3.3402	.95799	20
41	708	.29970	.3367	791	19
42	736	.30001	.3332	782	18
43	764	033	.3297	774	17
44	792	065̄	.3261	766	16
45	.28820	.30097	3.3226	.95757	15
46	847	128	.3191	749	14
47	875	160	.3156	740	13
48	903	192	.3122	732	12
49	931	224	.3087	724	11
50	.28959	.30255	3.3052	.95715	10
51	.28987	287	.3017	707	9
52	.29015̄	319	.2983	698	8
53	042	351	.2948	690	7
54	070	382	.2914	681	6
55	.29098	.30414	3.2879	.95673	5
56	126	446	.2845	664	4
57	154	478	.2811	656	3
58	182	509	.2777	647	2
59	209	541	.2743	639	1
60	.29237	.30573	3.2709	.95630	0
	cos	cot	tan	sin	′

73°

17°

′	sin	tan	cot	cos	
0	.29237	.30573	3.2709	.95630	60
1	265̄	605̄	.2675	622	59
2	293	637	.2641	613	58
3	321	669	.2607	605̄	57
4	348	700	.2573	596	56
5	.29376	.30732	3.2539	.95588	55
6	404	764	.2506	579	54
7	432	796	.2472	571	53
8	460	828	.2438	562	52
9	487	860	.2405̄	554	51
10	.29515	.30891	3.2371	.95545	50
11	543	923	.2338	536	49
12	571	955	.2305̄	528	48
13	599	.30987	.2272	519	47
14	626	.31019	.2238	511	46
15	.29654	.31051	3.2205	.95502	45
16	682	083	.2172	493	44
17	710	115̄	.2139	485̄	43
18	737	147	.2106	476	42
19	765	178	.2073	467	41
20	.29793	.31210	3.2041	.95459	40
21	821	242	.2008	450	39
22	849	274	.1975	441	38
23	876	306	.1943	433	37
24	904	338	.1910	424̄	36
25	.29932	.31370	3.1878	.95415	35
26	960	402	.1845	407	34
27	.29987	434	.1813	398	33
28	.30015	466	.1780	389	32
29	043	498	.1748	380	31
30	.30071	.31530	3.1716	.95372	30
31	098	562	.1684	363	29
32	126	594	.1652	354	28
33	154	626	.1620	345	27
34	182	658	.1588	337	26
35	.30209	.31690	3.1556	.95328	25
36	237	722	.1524	319	24
37	265̄	754	.1492	310	23
38	292	786	.1460	301	22
39	320	818	.1429	293	21
40	.30348	.31850̄	3.1397	.95284	20
41	376	882	.1366	275̄	19
42	403	914	.1334	260	18
43	431	946	.1303	257	17
44	459	.31978	.1271	248	16
45	.30486	.32010	3.1240	.95240	15
46	514	042	.1209	231	14
47	542	074	.1178	222	13
48	570	106	.1146	213	12
49	597	139	.1115	204	11
50	.30625̄	.32171	3.1084	.95195	10
51	653	203	.1053	186	9
52	680	235̄	.1022	177	8
53	708	267	.0991	168	7
54	736	299	.0961	159	6
55	.30763	.32331	3.0930	.95150	5
56	791	363	.0899	142	4
57	819	396	.0868	133	3
58	846	428	.0838	124̄	2
59	874	460	.0807	115̄	1
60	.30902	.32492	3.0777	.95106	0
	cos	cot	tan	sin	′

72°

18°

	sin	tan	cot	cos	
0	.30902	.32492	3.0777	.95106	60
1	929	524	.0746	097	59
2	957	556	.0716	088	58
3	.30985	588	.0686	079	57
4	.31012	621	.0655	070	56
5	.31040	.32653	3.0625	.95061	55
6	068	685	.0595	052	54
7	095	717	.0565	043	53
8	123	749	.0535	033	52
9	151	782	.0505	024	51
10	.31178	.32814	3.0475	.95015	50
11	206	846	.0445	.95006	49
12	233	878	.0415	.94997	48
13	261	911	.0385	988	47
14	289	943	.0356	979	46
15	.31316	.32975	3.0326	.94970	45
16	344	.33007	.0296	961	44
17	372	040	.0267	952	43
18	399	072	.0237	943	42
19	427	104	.0208	933	41
20	.31454	.33136	3.0178	.94924	40
21	482	169	.0149	915	39
22	510	201	.0120	906	38
23	537	233	.0090	897	37
24	565	266	.0061	888	36
25	.31593	.33298	3.0032	.94878	35
26	620	330	3.0003	869	34
27	648	363	2.9974	860	33
28	675	395	.9945	851	32
29	703	427	.9916	842	31
30	.31730	.33460	2.9887	.94832	30
31	758	492	.9858	823	29
32	786	524	.9829	814	28
33	813	557	.9800	805	27
34	841	589	.9772	795	26
35	.31868	.33621	2.9743	.94786	25
36	896	654	.9714	777	24
37	923	686	.9686	768	23
38	951	718	.9657	758	22
39	.31979	751	.9629	749	21
40	.32006	33783	2.9600	.94740	20
41	034	816	.9572	730	19
42	061	848	.9544	721	18
43	089	881	.9515	712	17
44	116	913	.9487	702	16
45	.32144	.33945	2.9459	.94693	15
46	171	.33978	.9431	684	14
47	199	.34010	.9403	674	13
48	227	043	.9375	665	12
49	254	075	.9347	656	11
50	.32282	.34108	2.9319	.94646	10
51	309	140	.9291	637	9
52	337	173	.9263	627	8
53	364	205	.9235	618	7
54	392	238	.9208	609	6
55	.32419	.34270	2.9180	.94599	5
56	447	303	.9152	590	4
57	474	335	.9125	580	3
58	502	368	.9097	571	2
59	529	400	.9070	561	1
60	.32557	.34433	2.9042	.94552	0
	cos	cot	tan	sin	′

71°

19°

	sin	tan	cot	cos	
0	.32557	.34433	2.9042	.94552	60
1	584	465	.9015	542	59
2	612	498	.8987	533	58
3	639	530	.8960	523	57
4	667	563	.8933	514	56
5	.32694	.34596	2.8905	.94504	55
6	722	628	.8878	495	54
7	749	661	.8851	485	53
8	777	693	.8824	476	52
9	804	726	.8797	466	51
10	.32832	.34758	2.8770	.94457	50
11	859	791	.8743	447	49
12	887	824	.8716	438	48
13	914	856	.8689	428	47
14	942	889	.8662	418	46
15	.32969	.34922	2.8636	.94409	45
16	.32997	954	.8609	399	44
17	.33024	.34987	.8582	390	43
18	051	.35020	.8556	380	42
19	079	052	.8529	370	41
20	.33106	.35085	2.8502	.94361	40
21	134	118	.8476	351	39
22	161	150	.8449	342	38
23	189	183	.8423	332	37
24	216	216	.8397	322	36
25	.33244	.35248	2.8370	.94313	35
26	271	281	.8344	303	34
27	298	314	.8318	293	33
28	326	346	.8291	284	32
29	353	379	.8265	274	31
30	.33381	.35412	2.8239	.94264	30
31	408	445	.8213	254	29
32	436	477	.8187	245	28
33	463	510	.8161	235	27
34	490	543	.8135	225	26
35	.33518	.35576	2.8109	.94215	25
36	545	608	.8083	206	24
37	573	641	.8057	196	23
38	600	674	.8032	186	22
39	627	707	.8006	176	21
40	.33655	.35740	2.7980	.94167	20
41	682	772	.7955	157	19
42	710	805	.7929	147	18
43	737	838	.7903	137	17
44	764	871	.7878	127	16
45	.33792	.35904	2.7852	.94118	15
46	819	937	.7827	108	14
47	846	.35969	.7801	098	13
48	874	.36002	.7776	088	12
49	901	035	.7751	078	11
50	.33929	.36068	2.7725	.94068	10
51	956	101	.7700	058	9
52	.33983	134	.7675	049	8
53	.34011	167	.7650	039	7
54	038	199	.7625	029	6
55	.34065	.36232	2.7600	.94019	5
56	093	265	.7575	.94009	4
57	120	298	.7550	.93999	3
58	147	331	.7525	989	2
59	175	364	.7500	979	1
60	.34202	.36397	2.7475	.93969	0
	cos	cot	tan	sin	′

70°

20°

'	sin	tan	cot	cos	
0	.34202	.36397	2.7475	.93969	60
1	229	430	.7450	959	59
2	257	463	.7425	949	58
3	284	496	.7400	939	57
4	311	529	.7376	929	56
5	.34339	.36562	2.7351	.93919	55
6	366	595	.7326	909	54
7	393	628	.7302	899	53
8	421	661	.7277	889	52
9	448	694	.7253	879	51
10	.34475	.36727	2.7228	.93869	50
11	503	760	.7204	859	49
12	530	793	.7179	849	48
13	557	826	.7155	839	47
14	584	859	.7130	829	46
15	.34612	.36892	2.7106	.93819	45
16	639	925	.7082	809	44
17	666	958	.7058	799	43
18	694	.36991	.7034	789	42
19	721	.37024	.7009	779	41
20	.34748	.37057	2.6985	.93769	40
21	775	090	.6961	759	39
22	803	123	.6937	748	38
23	830	157	.6913	738	37
24	857	190	.6889	728	36
25	.34884	.37223	2.6865	.93718	35
26	912	256	.6841	708	34
27	939	289	.6818	698	33
28	966	322	.6794	688	32
29	.34993	355	.6770	677	31
30	.35021	.37388	2.6746	.93667	30
31	048	422	.6723	657	29
32	075	455	.6699	647	28
33	102	488	.6675	637	27
34	130	521	.6652	626	26
35	.35157	.37554	2.6628	.93616	25
36	184	588	.6605	606	24
37	211	621	.6581	596	23
38	239	654	.6558	585	22
39	266	687	.6534	575	21
40	.35293	.37720	2.6511	.93565	20
41	320	754	.6488	555	19
42	347	787	.6464	544	18
43	375	820	.6441	534	17
44	402	853	.6418	524	16
45	.35429	.37887	2.6395	.93514	15
46	456	920	.6371	503	14
47	484	953	.6348	493	13
48	511	.37986	.6325	483	12
49	538	.38020	.6302	472	11
50	.35565	.38053	2.6279	.93462	10
51	592	086	.6256	452	9
52	619	120	.6233	441	8
53	647	153	.6210	431	7
54	674	186	.6187	420	6
55	.35701	.38220	2.6165	.93410	5
56	728	253	.6142	400	4
57	755	286	.6119	389	3
58	782	320	.6096	379	2
59	810	353	.6074	368	1
60	.35837	.38386	2.6051	.93358	0
	cos	cot	tan	sin	'

69°

21°

'	sin	tan	cot	cos	
0	.35837	.38386	2.6051	.93358	60
1	864	420	.6028	348	59
2	891	453	.6006	337	58
3	918	487	.5983	327	57
4	945	520	.5961	316	56
5	.35973	.38553	2.5938	.93306	55
6	.36000	587	.5916	295	54
7	027	620	.5893	285	53
8	054	654	.5871	274	52
9	081	687	.5848	264	51
10	.36108	.38721	2.5826	.93253	50
11	135	754	.5804	243	49
12	162	787	.5782	232	48
13	190	821	.5759	222	47
14	217	854	.5737	211	46
15	.36244	.38888	2.5715	.93201	45
16	271	921	.5693	190	44
17	298	955	.5671	180	43
18	325	.38988	.5649	169	42
19	352	.39022	.5627	159	41
20	.36379	.39055	2.5605	.93148	40
21	406	089	.5583	137	39
22	434	122	.5561	127	38
23	461	156	.5539	116	37
24	488	190	.5517	106	36
25	.36515	.39223	2.5495	.93095	35
26	542	257	.5473	084	34
27	569	290	.5452	074	33
28	596	324	.5430	063	32
29	623	357	.5408	052	31
30	.36650	.39391	2.5386	.93042	30
31	677	425	.5365	031	29
32	704	458	.5343	020	28
33	731	492	.5322	.93010	27
34	758	526	.5300	.92999	26
35	.36785	.39559	2.5279	.92988	25
36	812	593	.5257	978	24
37	839	626	.5236	967	23
38	867	660	.5214	956	22
39	894	694	.5193	945	21
40	.36921	.39727	2.5172	.92935	20
41	948	761	.5150	924	19
42	.36975	795	.5129	913	18
43	.37002	829	.5108	902	17
44	029	862	.5086	892	16
45	.37056	.39896	2.5065	.92881	15
46	083	930	.5044	870	14
47	110	963	.5023	859	13
48	137	.39997	.5002	849	12
49	164	.40031	.4981	838	11
50	.37191	.40065	2.4960	.92827	10
51	218	098	.4939	816	9
52	245	132	.4918	805	8
53	272	166	.4897	794	7
54	299	200	.4876	784	6
55	.37326	.40234	2.4855	.92773	5
56	353	267	.4834	762	4
57	380	301	.4813	751	3
58	407	335	.4792	740	2
59	434	369	.4772	729	1
60	.37461	.40403	2.4751	.92718	0
	cos	cot	tan	sin	'

68°

22°

′	sin	tan	cot	cos	
0	.37461	.40403	2.4751	.92718	60
1	488	436	.4730	707	59
2	515̄	470	.4709	697	58
3	542	504	.4689	686	57
4	569	538	.4668	675̄	56
5	.37595	.40572	2.4648	.92664	55
6	622	606	.4627	653	54
7	649	640	.4606	642	53
8	676	674	.4586	631	52
9	703	707	.4566	620	51
10	.37730	.40741	2.4545	.92609	50
11	·757	775	.4525̄	598	49
12	784	809	.4504	587	48
13	811	843	.4484	576	47
14	838	877	.4464	565	46
15	.37865̄	.40911	2.4443	.92554	45
16	892	945	.4423	543	44
17	919	.40979	.4403	532	43
18	946	.41013	.4383	521	42
19	973	047	.4362	510	41
20	.37999	.41081	2.4342	.92499	40
21	.38026	115	.4322	488	39
22	053	149	.4302	477	38
23	080	183	.4282	466	37
24	107	217	.4262	455̄	36
25	.38134	.41251	2.4242	.92444	35
26	161	285	.4222	432	34
27	188	319	.4202	421	33
28	215	353	.4182	410	32
29	241	387	.4162	399	31
30	.38268	.41421	2.4142	.92388	30
31	295	455	.4122	377	29
32	322	490	.4102	366	28
33	349	524	.4083	355̄	27
34	376	558	.4063	343	26
35	.38403	.41592	2.4043	.92332	25
36	430	626	.4023	321	24
37	456	660	.4004	310	23
38	483	694	.3984	299	22
39	510	728	.3964	287	21
40	.38537	.41763	2.3945̄	.92276	20
41	564	797	.3925	265̄	19
42	591	831	.3906	254	18
43	617	865	.3886	243	17
44	644	899	.3867	231	16
45	.38671	.41933	2.3847	.92220	15
46	698	.41968	.3828	209	14
47	725̄	.42002	.3808	198	13
48	752	036	.3789	186	12
49	778	070	.3770	175̄	11
50	.38805	.42105̄	2.3750	.92164	10
51	832	139	.3731	152	9
52	859	173	.3712	141	8
53	886	207	.3693	130	7
54	912	242	.3673	119	6
55	.38939	.42276	2.3654	.92107	5
56	966	310	.3635̄	096	4
57	.38993	345̄	.3616	085̄	3
58	.39020	379	.3597	073	2
59	046	413	.3578	062	1
60	.39073	.42447	2.3559	.92050	0
	cos	cot	tan	sin	′

67°

23°

′	sin	tan	cot	cos	
0	.39073	.42447	2.3559	.92050	60
1	100	482	.3539	039	59
2	127	516	.3520	028	58
3	153	551	.3501	016	57
4	180	585̄	.3483	.92005̄	56
5	.39207	.42619	2.3464	.91994	55
6	234	654	.3445̄	982	54
7	260	688	.3426	971	53
8	287	722	.3407	959	52
9	314	757	.3388	948	51
10	.39341	.42791	2.3369	.91936	50
11	367	826	.3351	925̄	49
12	394	860	.3332	914	48
13	421	894	.3313	902	47
14	448	929	.3294	891	46
15	.39474	.42963	2.3276	.91879	45
16	501	.42998	.3257	868	44
17	528	.43032	.3238	856	43
18	555̄	067	.3220	845̄	42
19	581	101	.3201	833	41
20	.39608	.43136	2.3183	.91822	40
21	635̄	170	.3164	810	39
22	661	205̄	.3146	799	38
23	688	239	.3127	787	37
24	715̄	274	.3109	775̄	36
25	.39741	.43308	2.3090	.91764	35
26	768	343	.3072	752	34
27	795̄	378	.3053	741	33
28	822	412	.3035̄	729	32
29	848	447	.3017	718	31
30	.39875̄	.43481	2.2998	.91706	30
31	902	516	.2980	694	29
32	928	550	.2962	683	28
33	955̄	585̄	.2944	671	27
34	.39982	620	.2925	660	26
35	.40008	.43654	2.2907	.91648	25
36	035̄	689	.2889	636	24
37	062	724	.2871	625̄	23
38	088	758	.2853	613	22
39	115̄	793	.2835̄	601	21
40	.40141	.43828	2.2817	.91590	20
41	168	862	.2799	578	19
42	195̄	897	.2781	566	18
43	221	932	.2763	555̄	17
44	248	.43966	.2745̄	543	16
45	.40275̄	.44001	2.2727	.91531	15
46	301	036	.2709	519	14
47	328	071	.2691	508	13
48	355̄	105̄	.2673	496	12
49	381	140	.2655̄	484	11
50	.40408	.44175̄	2.2637	.91472	10
51	434	210	.2620	461	9
52	461	244	.2602	449	8
53	488	279	.2584	437	7
54	514	314	.2566	425̄	6
55	.40541	.44349	2.2549	.91414	5
56	567	384	.2531	402	4
57	594	418	.2513	390	3
58	621	453	.2496	378	2
59	647	488	.2478	366	1
60	.40674	.44523	2.2460	.91355̄	0
	cos	cot	tan	sin	′

66°

24°

′	sin	tan	cot	cos	
0	.40674	.44523	2.2460	.91355	60
1	700	558	.2443	343	59
2	727	593	.2425	331	58
3	753	627	.2408	319	57
4	780	662	.2390	307	56
5	.40806	.44697	2.2373	.91295	55
6	833	732	.2355	283	54
7	860	767	.2338	272	53
8	886	802	.2320	260	52
9	913	837	.2303	248	51
10	.40939	.44872	2.2286	.91236	50
11	966	907	.2268	224	49
12	.40992	942	.2251	212	48
13	.41019	.44977	.2234	200	47
14	045	.45012	.2216	188	46
15	.41072	.45047	2.2199	.91176	45
16	098	082	.2182	164	44
17	125	117	.2165	152	43
18	151	152	.2148	140	42
19	178	187	.2130	128	41
20	.41204	.45222	2.2113	.91116	40
21	231	257	.2096	104	39
22	257	292	.2079	092	38
23	284	327	.2062	080	37
24	310	362	.2045	068	36
25	.41337	.45397	2.2028	.91056	35
26	363	432	.2011	044	34
27	390	467	.1994	032	33
28	416	502	.1977	020	32
29	443	538	.1960	.91008	31
30	.41469	.45573	2.1943	.90996	30
31	496	608	.1926	984	29
32	522	643	.1909	972	28
33	549	678	.1892	960	27
34	575	713	.1876	948	26
35	.41602	.45748	2.1859	.90936	25
36	628	784	.1842	924	24
37	655	819	.1825	911	23
38	681	854	.1808	899	22
39	707	889	.1792	887	21
40	.41734	.45924	2.1775	.90875	20
41	760	960	.1758	863	19
42	787	.45995	.1742	851	18
43	813	.46030	.1725	839	17
44	840	065	.1708	826	16
45	.41866	.46101	2.1692	.90814	15
46	892	136	.1675	802	14
47	919	171	.1659	790	13
48	945	206	.1642	778	12
49	972	242	.1625	766	11
50	.41998	.46277	2.1609	.90753	10
51	.42024	312	.1592	741	9
52	051	348	.1576	729	8
53	077	383	.1560	717	7
54	104	418	.1543	704	6
55	.42130	.46454	2.1527	.90692	5
56	156	489	.1510	680	4
57	183	525	.1494	668	3
58	209	560	.1478	655	2
59	235	595	.1461	643	1
60	.42262	.46631	2.1445	.90631	0
	cos	cot	tan	sin	′

65°

25°

′	sin	tan	cot	cos	
0	.42262	.46631	2.1445	.90631	60
1	288	666	.1429	618	59
2	315	702	.1413	606	58
3	341	737	.1396	594	57
4	367	772	.1380	582	56
5	.42394	.46808	2.1364	.90569	55
6	420	843	.1348	557	54
7	446	879	.1332	545	53
8	473	914	.1315	532	52
9	499	950	.1299	520	51
10	42525	.46985	2.1283	.90507	50
11	552	.47021	.1267	495	49
12	578	056	.1251	483	48
13	604	092	.1235	470	47
14	631	128	.1219	458	46
15	42657	.47163	2.1203	.90446	45
16	683	199	.1187	433	44
17	709	234	.1171	421	43
18	736	270	.1155	408	42
19	762	305	.1139	396	41
20	42788	.47341	2.1123	.90383	40
21	815	377	.1107	371	39
22	841	412	.1092	358	38
23	867	448	.1076	346	37
24	894	483	.1060	334	36
25	42920	.47519	2.1044	.90321	35
26	946	555	.1028	309	34
27	972	590	.1013	296	33
28	42999	626	0997	284	32
29	43025	662	.0981	271	31
30	43051	.47698	2.0965	.90259	30
31	077	733	.0950	246	29
32	104	769	.0934	233	28
33	130	805	.0918	221	27
34	156	840	.0903	208	26
35	43182	.47876	2.0887	.90196	25
36	209	912	.0872	183	24
37	235	948	.0856	171	23
38	261	.47984	.0840	158	22
39	287	.48019	.0825	146	21
40	43313	.48055	2.0809	.90133	20
41	340	091	.0794	120	19
42	366	127	.0778	108	18
43	392	163	.0763	095	17
44	418	198	.0748	082	16
45	43443	.48234	2.0732	.90070	15
46	471	270	.0717	057	14
47	497	306	.0701	045	13
48	523	342	.0686	032	12
49	549	378	.0671	019	11
50	43575	.48414	2.0655	.90007	10
51	602	450	.0640	.89994	9
52	628	486	.0625	981	8
53	654	521	.0609	968	7
54	680	557	.0594	956	6
55	43706	.48593	2.0579	.89943	5
56	733	629	.0564	930	4
57	759	665	.0549	918	3
58	785	701	.0533	905	2
59	811	737	.0518	892	1
60	43837	.48773	2.0503	.89879	0
	cos	cot	tan	sin	′

64°

26°

′	sin	tan	cot	cos	
0	.43837	.48773	2.0503	.89879	60
1	863	809	.0488	867	59
2	889	845	.0473	854	58
3	916	881	.0458	841	57
4	942	917	.0443	828	56
5	.43968	.48953	2.0428	.89816	55
6	.43994	.48989	.0413	803	54
7	.44020	.49026	.0398	790	53
8	046	062	.0383	777	52
9	072	098	.0368	764	51
10	.44098	.49134	2.0353	.89752	50
11	124	170	.0338	739	49
12	151	206	.0323	726	48
13	177	242	.0308	713	47
14	203	278	.0293	700	46
15	.44229	.49315	2.0278	.89687	45
16	255	351	.0263	674	44
17	281	387	.0248	662	43
18	307	423	.0233	649	42
19	333	459	.0219	636	41
20	.44359	.49495	2.0204	.89623	40
21	385	532	.0189	610	39
22	411	568	.0174	597	38
23	437	604	.0160	584	37
24	464	640	.0145	571	36
25	.44490	.49677	2.0130	.89558	35
26	516	713	.0115	545	34
27	542	749	.0101	532	33
28	568	786	.0086	519	32
29	594	822	.0072	506	31
30	.44620	.49858	2.0057	.89493	30
31	646	894	.0042	480	29
32	672	931	.0028	467	28
33	698	.49967	2.0013	454	27
34	724	.50004	1.9999	441	26
35	.44750	.50040	1.9984	.89428	25
36	776	076	.9970	415	24
37	802	113	.9955	402	23
38	828	149	.9941	389	22
39	854	185	.9926	376	21
40	.44880	.50222	1.9912	.89363	20
41	906	258	.9897	350	19
42	932	295	.9883	337	18
43	958	331	.9868	324	17
44	.44984	368	.9854	311	16
45	.45010	.50404	1.9840	.89298	15
46	036	441	.9825	285	14
47	062	477	.9811	272	13
48	088	514	.9797	259	12
49	114	550	.9782	245	11
50	.45140	.50587	1.9768	.89232	10
51	166	623	.9754	219	9
52	192	660	.9740	206	8
53	218	696	9725	193	7
54	243	733	.9711	180	6
55	.45269	.50769	1.9697	.89167	5
56	295	806	.9683	153	4
57	321	843	.9669	140	3
58	347	879	.9654	127	2
59	373	916	.9640	114	1
60	.45399	.50953	1.9626	.89101	0
	cos	cot	tan	sin	′

63°

27°

′	sin	tan	cot	cos	
0	.45399	.50953	1.9626	.89101	60
1	425	.50989	.9612	087	59
2	451	.51026	.9598	074	58
3	477	063	.9584	061	57
4	503	.099	.9570	048	56
5	.45529	.51136	1.9556	.89035	55
6	554	173	.9542	021	54
7	580	209	.9528	.89008	53
8	606	246	.9514	.88995	52
9	632	283	.9500	981	51
10	.45658	.51319	1.9486	.88968	50
11	684	356	.9472	955	49
12	710	393	.9458	942	48
13	736	430	.9444	928	47
14	762	467	.9430	915	46
15	.45787	.51503	1.9416	.88902	45
16	813	540	.9402	888	44
17	839	577	.9388	875	43
18	865	614	.9375	862	42
19	891	651	.9361	848	41
20	.45917	.51688	1.9347	.88835	40
21	942	724	.9333	822	39
22	968	761	.9319	808	38
23	.45994	798	.9306	795	37
24	.46020	835	.9292	782	36
25	.46046	.51872	1.9278	.88768	35
26	072	909	.9265	755	34
27	097	946	.9251	741	33
28	123	.51983	.9237	728	32
29	149	.52020	.9223	715	31
30	.46175	.52057	1.9210	.88701	30
31	201	094	.9196	688	29
32	226	131	.9183	674	28
33	252	168	.9169	661	27
34	278	205	.9155	647	26
35	.46304	.52242	1.9142	.88634	25
36	330	279	.9128	620	24
37	355	316	.9115	607	23
38	381	353	.9101	593	22
39	407	390	.9088	580	21
40	.46433	.52427	1.9074	.88566	20
41	458	464	.9061	553	19
42	484	501	9047	539	18
43	510	538	9034	526	17
44	536	575	.9020	512	16
45	.46561	.52613	1.9007	.88499	15
46	587	650	.8993	485	14
47	613	687	.8980	472	13
48	639	724	.8967	458	12
49	664	761	.8953	445	11
50	.46690	.52798	1.8940	.88431	10
51	716	836	.8927	417	9
52	742	873	.8913	404	8
53	767	910	.8900	390	7
54	793	947	.8887	377	6
55	.46819	.52985	1.8873	.88363	5
56	844	.53022	.8860	349	4
57	870	059	.8847	336	3
58	896	096	.8834	322	2
59	921	134	.8820	308	1
60	.46947	.53171	1.8807	.88295	0
	cos	cot	tan	sin	′

62°

28°

′	sin	tan	cot	cos	
0	.46947	.53171	1.8807	.88295	60
1	973	208	.8794	281	59
2	.46999	246	.8781	267	58
3	.47024	283	.8768	254	57
4	050	320	.8755	240	56
5	.47076	.53358	1.8741	.88226	55
6	101	395	.8728	213	54
7	127	432	.8715	199	53
8	153	470	.8702	185	52
9	178	507	.8689	172	51
10	.47204	.53545	1.8676	.88158	50
11	229	582	.8663	144	49
12	255	620	.8650	130	48
13	281	657	.8637	117	47
14	306	694	.8624	103	46
15	.47332	.53732	1.8611	.88089	45
16	358	769	.8598	075	44
17	383	807	.8585	062	43
18	409	844	.8572	048	42
19	434	882	.8559	034	41
20	.47460	.53920	1.8546	.88020	40
21	486	957	.8533	.88006	39
22	511	.53995	.8520	.87993	38
23	537	.54032	.8507	979	37
24	562	070	.8495	965	36
25	.47588	.54107	1.8482	.87951	35
26	614	145	.8469	937	34
27	639	183	.8456	923	33
28	665	220	.8443	909	32
29	690	258	.8430	896	31
30	.47716	.54296	1.8418	.87882	30
31	741	333	.8405	868	29
32	767	371	.8392	854	28
33	793	409	.8379	840	27
34	818	446	.8367	826	26
35	.47844	.54484	1.8354	.87812	25
36	869	522	.8341	798	24
37	895	560	.8329	784	23
38	920	597	.8316	770	22
39	946	635	.8303	756	21
40	.47971	.54673	1.8291	.87743	20
41	.47997	711	.8278	729	19
42	.48022	748	.8265	715	18
43	048	786	.8253	701	17
44	073	824	.8240	687	16
45	.48099	.54862	1.8228	.87673	15
46	124	900	.8215	659	14
47	150	938	.8202	645	13
48	175	.54975	.8190	631	12
49	201	.55013	.8177	617	11
50	.48226	.55051	1.8165	.87603	10
51	252	089	.8152	589	9
52	277	127	.8140	575	8
53	303	165	.8127	561	7
54	328	203	.8115	546	6
55	.48354	.55241	1.8103	.87532	5
56	379	279	.8090	518	4
57	405	317	.8078	504	3
58	430	355	.8065	490	2
59	456	393	.8053	476	1
60	.48481	.55431	1.8040	.87462	0
	cos	cot	tan	sin	′

61°

29°

′	sin	tan	cot	cos	
0	.48481	.55431	1.8040	.87462	60
1	506	469	.8028	448	59
2	532	507	.8016	434	58
3	557	545	.8003	420	57
4	583	583	.7991	406	56
5	.48608	.55621	1.7979	.87391	55
6	634	659	.7966	377	54
7	659	697	.7954	363	53
8	684	736	.7942	349	52
9	710	774	.7930	335	51
10	.48735	.55812	1.7917	.87321	50
11	761	850	.7905	306	49
12	786	888	.7893	292	48
13	811	926	.7881	278	47
14	837	.55964	.7868	264	46
15	.48862	.56003	1.7856	.87250	45
16	888	041	.7844	235	44
17	913	079	.7832	221	43
18	938	117	.7820	207	42
19	964	156	.7808	193	41
20	.48989	.56194	1.7796	.87178	40
21	.49014	232	.7783	164	39
22	040	270	.7771	150	38
23	065	309	.7759	136	37
24	090	347	.7747	121	36
25	.49116	.56385	1.7735	.87107	35
26	141	424	.7723	093	34
27	166	462	.7711	079	33
28	192	501	.7699	064	32
29	217	539	.7687	050	31
30	.49242	.56577	1.7675	.87036	30
31	268	616	.7663	021	29
32	293	654	.7651	.87007	28
33	318	693	.7639	.86993	27
34	344	731	7627	978	26
35	.49369	.56769	1.7615	.86964	25
36	394	808	.7603	949	24
37	419	846	.7591	935	23
38	445	885	.7579	921	22
39	470	923	.7567	906	21
40	.49495	.56962	1.7556	.86892	20
41	521	.57000	.7544	878	19
42	546	039	.7532	863	18
43	571	078	.7520	849	17
44	596	116	.7508	834	16
45	.49622	.57155	1.7496	.86820	15
46	647	193	.7485	805	14
47	672	232	.7473	791	13
48	697	271	.7461	777	12
49	723	309	.7449	762	11
50	.49748	.57348	1.7437	.86748	10
51	773	386	.7426	733	9
52	798	425	.7414	719	8
53	824	464	.7402	704	7
54	849	503	.7391	690	6
55	.49874	.57541	1.7379	.86675	5
56	899	580	.7367	661	4
57	924	619	.7355	646	3
58	950	657	.7344	632	2
59	.49975	696	.7332	617	1
60	.50000	.57735	1.7321	.86603	0
	cos	cot	tan	sin	′

60°

30°

′	sin	tan	cot	cos	
0	.50000	.57735	1.7321	.86603	60
1	025	774	.7309	588	59
2	050	813	.7297	573	58
3	076	851	.7286	559	57
4	101	890	.7274	544	56
5	.50126	.57929	1.7262	.86530	55
6	151	.57968	.7251	515	54
7	176	.58007	.7239	501	53
8	201	046	.7228	486	52
9	227	085	.7216	471	51
10	.50252	.58124	1.7205	.86457	50
11	277	162	.7193	442	49
12	302	201	.7182	427	48
13	327	240	.7170	413	47
14	352	279	.7159	398	46
15	.50377	.58318	1.7147	.86384	45
16	403	357	.7136	369	44
17	428	396	.7124	354	43
18	453	435	.7113	340	42
19	478	474	.7102	325	41
20	.50503	.58513	1.7090	.86310	40
21	528	552	.7079	295	39
22	553	591	.7067	281	38
23	578	631	.7056	266	37
24	603	670	.7045	251	36
25	.50628	.58709	1.7033	.86237	35
26	654	748	.7022	222	34
27	679	787	.7011	207	33
28	704	826	.6999	192	32
29	729	865	.6988	178	31
30	.50754	.58905	1.6977	.86163	30
31	779	944	.6965	148	29
32	804	.58983	.6954	133	28
33	829	.59022	.6943	119	27
34	854	061	.6932	104	26
35	.50879	.59101	1.6920	.86089	25
36	904	140	.6909	074	24
37	929	179	.6898	059	23
38	954	218	.6887	045	22
39	.50979	258	.6875	030	21
40	.51004	.59297	1.6864	.86015	20
41	029	336	.6853	.86000	19
42	054	376	.6842	.85985	18
43	079	415	.6831	970	17
44	104	454	.6820	956	16
45	.51129	.59494	1.6808	.85941	15
46	154	533	.6797	926	14
47	179	573	.6786	911	13
48	204	612	.6775	896	12
49	229	651	.6764	881	11
50	.51254	.59691	1.6753	.85866	10
51	279	730	.6742	851	9
52	304	770	.6731	836	8
53	329	809	.6720	821	7
54	354	849	.6709	806	6
55	.51379	.59888	1.6698	.85792	5
56	404	928	.6687	777	4
57	429	.59967	.6676	762	3
58	454	.60007	.6665	747	2
59	479	046	.6654	732	1
60	.51504	.60086	1.6643	.85717	0
′	cos	cot	tan	sin	

59°

31°

′	sin	tan	cot	cos	′
0	.51504	.60086	1.6643	.85717	60
1	529	126	.6632	702	59
2	554	165	.6621	687	58
3	579	205	.6610	672	57
4	604	245	.6599	657	56
5	.51628	.60284	1.6588	.85642	55
6	653	324	.6577	627	54
7	678	364	.6566	612	53
8	703	403	.6555	597	52
9	728	443	.6545	582	51
10	51753	.60483	1.6534	.85567	50
11	778	522	.6523	551	49
12	803	562	.6512	536	48
13	828	602	.6501	521	47
14	852	642	.6490	506	46
15	51877	.60681	1.6479	.85491	45
16	902	721	.6469	476	44
17	927	761	.6458	461	43
18	952	801	.6447	446	42
19	.51977	841	.6436	431	41
20	52002	.60881	1.6426	.85416	40
21	026	921	.6415	401	39
22	051	.60960	.6404	385	38
23	076	.61000	.6393	370	37
24	101	040	.6383	355	36
25	52126	.61080	1.6372	.85340	35
26	151	120	.6361	325	34
27	175	160	.6351	310	33
28	200	200	.6340	294	32
29	225	240	.6329	279	31
30	52250	.61280	1.6319	.85264	30
31	275	320	.6308	249	29
32	299	360	.6297	234	28
33	324	400	.6287	218	27
34	349	440	.6276	203	26
35	52374	.61480	1.6265	.85188	25
36	399	520	.6255	173	24
37	423	561	.6244	157	23
38	448	601	.6234	142	22
39	473	641	.6223	127	21
40	.52498	.61681	1.6212	.85112	20
41	522	721	.6202	096	19
42	547	761	.6191	081	18
43	572	801	.6181	066	17
44	597	842	.6170	051	16
45	.52621	.61882	1.6160	.85035	15
46	646	922	.6149	020	14
47	671	.61962	.6139	.85005	13
48	696	.62003	.6128	.84989	12
49	720	043	.6118	974	11
50	.52745	.62083	1.6107	.84959	10
51	770	124	.6097	943	9
52	794	164	.6087	928	8
53	819	204	.6076	913	7
54	844	245	.6066	897	6
55	.52869	.62285	1.6055	.84882	5
56	893	325	.6045	866	4
57	918	366	.6034	851	3
58	943	406	.6024	836	2
59	967	446	.6014	820	1
60	.52992	.62487	1.6003	.84805	0
′	cos	cot	tan	sin	

58°

32°

′	sin	tan	cot	cos	
0	.52992	.62487	1.6003	.84805	60
1	.53017	527	.5993	789	59
2	041	568	.5983	774	58
3	066	608	.5972	759	57
4	091	649	.5962	743	56
5	.53115	.62689	1.5952	.84728	55
6	140	730	.5941	712	54
7	164	770	.5931	697	53
8	189	811	.5921	681	52
9	214	852	.5911	666	51
10	.53238	.62892	1.5900	.84650	50
11	263	933	.5890	635	49
12	288	.62973	.5880	619	48
13	312	.63014	.5869	604	47
14	337	055	.5859	588	46
15	.53361	.63095	1.5849	.84573	45
16	386	136	.5839	557	44
17	411	177	.5829	542	43
18	435	217	.5818	526	42
19	460	258	.5808	511	41
20	.53484	.63299	1.5798	.84495	40
21	509	340	.5788	480	39
22	534	380	.5778	464	38
23	558	421	.5768	448	37
24	583	462	.5757	433	36
25	.53607	.63503	1.5747	.84417	35
26	632	544	.5737	402	34
27	656	584	.5727	386	33
28	681	625	.5717	370	32
29	705	666	.5707	355	31
30	.53730	.63707	1.5697	.84339	30
31	754	748	.5687	324	29
32	779	789	.5677	308	28
33	804	830	.5667	292	27
34	828	871	.5657	277	26
35	.53853	.63912	1.5647	.84261	25
36	877	953	.5637	245	24
37	902	.63994	.5627	230	23
38	926	.64035	.5617	214	22
39	951	076	.5607	198	21
40	.53975	.64117	1.5597	.84182	20
41	.54000	158	.5587	167	19
42	024	199	.5577	151	18
43	049	240	.5567	135	17
44	073	281	.5557	120	16
45	.54097	.64322	1.5547	.84104	15
46	122	363	.5537	088	14
47	146	404	.5527	072	13
48	171	446	.5517	057	12
49	195	487	.5507	041	11
50	.54220	.64528	1.5497	.84025	10
51	244	569	.5487	.84009	9
52	269	610	.5477	.83994	8
53	293	652	.5468	978	7
54	317	693	.5458	962	6
55	.54342	.64734	1.5448	.83946	5
56	366	775	.5438	930	4
57	391	817	.5428	915	3
58	415	858	.5418	899	2
59	440	899	.5408	883	1
60	.54464	.64941	1.5399	.83867	0
	cos	cot	tan	sin	′

57°

33°

′	sin	tan	cot	cos	
0	.54464	.64941	1.5399	.83867	60
1	488	.64982	.5389	851	59
2	513	.65024	.5379	835	58
3	537	065	.5369	819	57
4	561	106	.5359	804	56
5	.54586	.65148	1.5350	.83788	55
6	610	189	.5340	772	54
7	635	231	.5330	756	53
8	659	272	.5320	740	52
9	683	314	.5311	724	51
10	.54708	.65355	1.5301	.83708	50
11	732	397	.5291	692	49
12	756	438	.5282	676	48
13	781	480	.5272	660	47
14	805	521	.5262	645	46
15	.54829	.65563	1.5253	.83629	45
16	854	604	.5243	613	44
17	878	646	.5233	597	43
18	902	688	.5224	581	42
19	927	729	.5214	565	41
20	.54951	.65771	1.5204	.83549	40
21	975	813	.5195	533	39
22	.54999	854	.5185	517	38
23	.55024	896	.5175	501	37
24	048	938	.5166	485	36
25	.55072	.65980	1.5156	.83469	35
26	097	.66021	.5147	453	34
27	121	063	.5137	437	33
28	145	105	.5127	421	32
29	169	147	.5118	405	31
30	.55194	.66189	1.5108	.83389	30
31	218	230	.5099	373	29
32	242	272	.5089	356	28
33	266	314	.5080	340	27
34	291	356	.5070	324	26
35	.55315	.66398	1.5061	.83308	25
36	339	440	.5051	292	24
37	363	482	.5042	276	23
38	388	524	.5032	260	22
39	412	566	.5023	244	21
40	.55436	.66608	1.5013	.83228	20
41	460	650	.5004	212	19
42	484	692	.4994	195	18
43	509	734	.4985	179	17
44	533	776	.4975	163	16
45	.55557	.66818	1.4966	.83147	15
46	581	860	.4957	131	14
47	605	902	.4947	115	13
48	630	944	.4938	098	12
49	654	.66986	.4928	082	11
50	.55678	.67028	1.4919	.83066	10
51	702	071	.4910	050	9
52	726	113	.4900	034	8
53	750	155	.4891	017	7
54	775	197	.4882	.83001	6
55	.55799	.67239	1.4872	.82985	5
56	823	282	.4863	969	4
57	847	324	.4854	953	3
58	871	366	.4844	936	2
59	895	409	.4835	920	1
60	.55919	.67451	1.4826	.82904	0
	cos	cot	tan	sin	′

56°

34°

′	sin	tan	cot	cos	
0	.55919	.67451	1.4826	.82904	60
1	943	493	.4816	887	59
2	968	536	.4807	871	58
3	.55992	578	.4798	855̄	57
4	.56016	620	.4788	839	56
5	.56040	.67663	1.4779	.82822	55
6	064	705	.4770	806	54
7	088	748	.4761	790	53
8	112	790	.4751	773	52
9	136	832	.4742	757	51
10	.56160	.67875	1.4733	.82741	50
11	184	917	.4724	724	49
12	208	.67960	.4715	708	48
13	232	.68002	.4705	692	47
14	256	045	.4696	675	46
15	.56280	.68088	1.4687	.82659	45
16	305̄	130	.4678	643	44
17	329	173	.4669	626	43
18	353	215	.4659	610	42
19	377	258	.4650	593	41
20	.56401	.68301	1.4641	.82577	40
21	425̄	343	.4632	561	39
22	449	386	.4623	544	38
23	473	429	.4614	528	37
24	497	471	.4605̄	511	36
25	.56521	.68514	1.4596	.82495̄	35
26	545̄	557	.4586	478	34
27	569	600	.4577	462	33
28	593	642	.4568	446	32
29	617	685	.4559	429	31
30	.56641	.68728	1.4550	.82413	30
31	665̄	771	.4541	396	29
32	689	814	.4532	380	28
33	713	857	.4523	363	27
34	736	900	.4514	347	26
35	.56760	.68942	1.4505̄	.82330	25
36	784	.68985	.4496	314	24
37	808	.69028	.4487	297	23
38	832	071	.4478	281	22
39	856	114	.4469	264	21
40	.56880	.69157	1.4460	.82248	20
41	904	200	.4451	231	19
42	928	243	.4442	214	18
43	952	286	.4433	198	17
44	.56976	329	.4424	181	16
45	.57000	.69372	1.4415̄	.82165̄	15
46	024	416	.4406	148	14
47	047	459	.4397	132	13
48	071	502	.4388	115̄	12
49	095	545̄	.4379	098	11
50	.57119	.69588	1.4370	.82082	10
51	143	631	.4361	065	9
52	167	675̄	.4352	048	8
53	191	718	.4344	032	7
54	215̄	761	.4335̄	.82015̄	6
55	.57238	.69804	1.4326	.81999	5
56	262	847	.4317	982	4
57	286	891	.4308	965	3
58	310	934	.4299	949	2
59	334	.69977	.4290	932	1
60	.57358	.70021	1.4281	.81915	0
	cos	cot	tan	sin	′

55°

35°

′	sin	tan	cot	cos	
0	.57358	.70021	1.4281	.81915	60
1	381	064	.4273	899	59
2	405̄	107	.4264	882	58
3	429	151	.4255̄	865̄	57
4	453	194	.4246	848	56
5	.57477	.70238	1.4237	.81832	55
6	501	281	.4229	815̄	54
7	524	325̄	.4220	798	53
8	548	368	.4211	782	52
9	572	412	.4202	765̄	51
10	.57596	.70455̄	1.4193	.81748	50
11	619	499	.4185̄	731	49
12	643	542	.4176	714	48
13	667	586	.4167	698	47
14	691	629	.4158	681	46
15	.57715̄	.70673	1.4150̄	.81664	45
16	738	717	.4141	647	44
17	762	760	.4132	631	43
18	786	804	.4124	614	42
19	810	848	.4115̄	597	41
20	.57833	.70891	1.4106	.81580	40
21	857	935̄	.4097	563	39
22	881	.70979	.4089	546	38
23	904	.71023	.4080	530	37
24	928	066	.4071	513	36
25	.57952	.71110	1.4063	.81496	35
26	976	154	.4054	479	34
27	.57999	198	.4045̄	462	33
28	.58023	242	.4037	445̄	32
29	047	285̄	.4028	428	31
30	.58070	.71329	1.4019	.81412	30
31	094	373	.4011	395̄	29
32	118	417	.4002	378	28
33	141	461	.3994	361	27
34	165̄	505̄	.3985̄	344	26
35	.58189	.71549	1.3976	.81327	25
36	212	593	.3968	310	24
37	236	637	.3959	293	23
38	260	681	.3951	276	22
39	283	725̄	.3942	259	21
40	.58307	.71769	1.3934	.81242	20
41	330	813	.3925̄	225̄	19
42	354	857	.3916	208	18
43	378	901	.3908	191	17
44	401	946	.3899	174	16
45	.58425̄	.71990	1.3891	.81157	15
46	449	.72034	.3882	140	14
47	472	078	.3874	123	13
48	496	122	.3865̄	106	12
49	519	167	.3857	089	11
50	.58543	.72211	1.3848	.81072	10
51	567	255̄	.3840	055̄	9
52	590	299	.3831	038	8
53	614	344	.3823	021	7
54	637	388	.3814	.81004	6
55	.58661	.72432	1.3806	80987	5
56	684	477	.3798	970	4
57	708	521	.3789	953	3
58	731	565̄	.3781	936	2
59	755̄	610	.3772	919	1
60	.58779	.72654	1.3764	.80902	0
	cos	cot	tan	sin	′

54°

36°

'	sin	tan	cot	cos	
0	.58779	.72654	1.3764	.80902	60
1	802	699	.3755	885	59
2	826	743	.3747	867	58
3	849	788	.3739	850	57
4	873	832	.3730	833	56
5	.58896	.72877	1.3722	.80816	55
6	920	921	.3713	799	54
7	943	.72966	.3705	782	53
8	967	.73010	.3697	765	52
9	.58990	055	.3688	748	51
10	.59014	.73100	1.3680	.80730	50
11	037	144	.3672	713	49
12	061	189	.3663	696	48
13	084	234	.3655	679	47
14	108	278	.3647	662	46
15	.59131	.73323	1.3638	.80644	45
16	154	368	.3630	627	44
17	178	413	.3622	610	43
18	201	457	.3613	593	42
19	225	502	.3605	576	41
20	.59248	.73547	1.3597	.80558	40
21	272	592	.3588	541	39
22	295	637	.3580	524	38
23	318	681	.3572	507	37
24	342	726	.3564	489	36
25	.59365	.73771	1.3555	.80472	35
26	389	816	.3547	455	34
27	412	861	.3539	438	33
28	436	906	.3531	420	32
29	459	951	.3522	403	31
30	.59482	.73996	1.3514	.80386	30
31	506	74041	.3506	368	29
32	529	086	.3498	351	28
33	552	131	.3490	334	27
34	576	176	.3481	316	26
35	.59599	.74221	1.3473	.80299	25
36	622	267	.3465	282	24
37	646	312	.3457	264	23
38	669	357	.3449	247	22
39	693	402	.3440	230	21
40	.59716	.74447	1.3432	.80212	20
41	739	492	.3424	195	19
42	763	538	.3416	178	18
43	786	583	.3408	160	17
44	809	628	.3400	143	16
45	.59832	.74674	1.3392	.80125	15
46	856	719	.3384	108	14
47	879	764	.3375	091	13
48	902	810	.3367	073	12
49	926	855	.3359	056	11
50	59949	.74900	1.3351	.80038	10
51	972	946	.3343	021	9
52	.59995	.74991	.3335	.80003	8
53	.60019	.75037	.3327	.79986	7
54	042	082	.3319	968	6
55	.60065	.75128	1.3311	.79951	5
56	089	173	.3303	934	4
57	112	219	.3295	916	3
58	135	264	.3287	899	2
59	158	310	.3278	881	1
60	.60182	.75355	1.3270	.79864	0
	cos	cot	tan	sin	'

53°

37°

'	sin	tan	cot	cos	
0	.60182	.75355	1.3270	.79864	60
1	205	401	.3262	846	59
2	228	447	.3254	829	58
3	251	492	.3246	811	57
4	274	538	.3238	793	56
5	.60298	.75584	1.3230	.79776	55
6	321	629	.3222	758	54
7	344	675	.3214	741	53
8	367	721	.3206	723	52
9	390	767	.3198	706	51
10	.60414	.75812	1.3190	.79688	50
11	437	858	.3182	671	49
12	460	904	.3175	653	48
13	483	950	.3167	635	47
14	506	.75996	.3159	618	46
15	.60529	.76042	1.3151	.79600	45
16	553	088	.3143	583	44
17	576	134	.3135	565	43
18	599	180	.3127	547	42
19	622	226	.3119	530	41
20	.60645	.76272	1.3111	.79512	40
21	668	318	.3103	494	39
22	691	364	.3095	477	38
23	714	410	.3087	459	37
24	738	456	.3079	441	36
25	.60761	.76502	1.3072	.79424	35
26	784	548	.3064	406	34
27	807	594	.3056	388	33
28	830	640	.3048	371	32
29	853	686	.3040	353	31
30	.60876	.76733	1.3032	.79335	30
31	899	779	.3024	318	29
32	922	825	.3017	300	28
33	945	871	.3009	282	27
34	968	918	.3001	264	26
35	.60991	.76964	1.2993	.79247	25
36	.61015	.77010	.2985	229	24
37	038	057	.2977	211	23
38	061	103	.2970	193	22
39	084	149	.2962	176	21
40	.61107	.77196	1.2954	.79158	20
41	130	242	.2946	140	19
42	153	289	.2938	122	18
43	176	335	.2931	105	17
44	199	382	.2923	087	16
45	.61222	.77428	1.2915	.79069	15
46	245	475	.2907	051	14
47	268	521	.2900	033	13
48	291	568	.2892	.79016	12
49	314	615	.2884	.78998	11
50	.61337	.77661	1.2876	.78980	10
51	360	708	.2869	962	9
52	383	754	.2861	944	8
53	406	801	.2853	926	7
54	429	848	.2846	908	6
55	.61451	.77895	1.2838	.78891	5
56	474	941	.2830	873	4
57	497	.77988	.2822	855	3
58	520	.78035	.2815	837	2
59	543	082	.2807	819	1
60	.61566	.78129	1.2799	.78801	0
	cos	cot	tan	sin	'

52°

38°

′	sin	tan	cot	cos	
0	.61566	.78129	1.2799	.78801	60
1	589	175	.2792	783	59
2	612	222	.2784	765	58
3	635	269	.2776	747	57
4	658	316	.2769	729	56
5	.61681	.78363	1.2761	.78711	55
6	704	410	.2753	694	54
7	726	457	.2746	676	53
8	749	504	.2738	658	52
9	772	551	.2731	640	51
10	.61795	.78598	1.2723	.78622	50
11	818	645	.2715	604	49
12	841	692	.2708	586	48
13	864	739	.2700	568	47
14	887	786	.2693	550	46
15	.61909	.78834	1.2685	.78532	45
16	932	881	.2677	514	44
17	955	928	.2670	496	43
18	.61978	.78975	.2662	478	42
19	.62001	.79022	.2655	460	41
20	.62024	.79070	1.2647	.78442	40
21	046	117	.2640	424	39
22	069	164	.2632	405	38
23	092	212	.2624	387	37
24	115	259	.2617	369	36
25	.62138	.79306	1.2609	.78351	35
26	160	354	.2602	333	34
27	183	401	.2594	315	33
28	206	449	.2587	297	32
29	229	496	.2579	279	31
30	.62251	.79544	1.2572	.78261	30
31	274	591	.2564	243	29
32	297	639	.2557	225	28
33	320	686	.2549	206	27
34	342	734	.2542	188	26
35°	.62365	.79781	1.2534	.78170	25
36	388	829	.2527	152	24
37	411	877	.2519	134	23
38	433	924	.2512	116	22
39	456	.79972	.2504	098	21
40	.62479	.80020	1.2497	.78079	20
41	502	067	.2489	061	19
42	524	115	.2482	043	18
43	547	163	.2475	025	17
44	570	211	.2467	.78007	16
45	.62592	.80258	1.2460	.77988	15
46	615	306	.2452	970	14
47	638	354	.2445	952	13
48	660	402	.2437	934	12
49	683	450	.2430	916	11
50	.62706	.80498	1.2423	.77897	10
51	728	546	.2415	879	9
52	751	594	.2408	861	8
53	774	642	.2401	843	7
54	796	690	.2393	824	6
55	.62819	.80738	1.2386	.77806	5
56	842	786	.2378	788	4
57	864	834	.2371	769	3
58	887	882	.2364	751	2
59	909	930	.2356	733	1
60	.62932	.80978	1.2349	.77715	0
	cos	cot	tan	sin	′

51°

39°

′	sin	tan	cot	cos	
0	.62932	.80978	1.2349	.77715	60
1	955	.81027	.2342	696	59
2	.62977	075	.2334	678	58
3	.63000	123	.2327	660	57
4	022	171	.2320	641	56
5	.63045	.81220	1.2312	.77623	55
6	068	268	.2305	605	54
7	090	316	.2298	586	53
8	113	364	.2290	568	52
9	135	413	.2283	550	51
10	.63158	.81461	1.2276	.77531	50
11	180	510	.2268	513	49
12	203	558	.2261	494	48
13	225	606	.2254	476	47
14	248	655	.2247	458	46
15	.63271	.81703	1.2239	.77439	45
16	293	752	.2232	421	44
17	316	800	.2225	402	43
18	338	849	.2218	384	42
19	361	898	.2210	366	41
20	.63383	.81946	1.2203	.77347	40
21	406	.81995	.2196	329	39
22	428	.82044	.2189	310	38
23	451	092	.2181	292	37
24	473	141	.2174	273	36
25	.63496	.82190	1.2167	.77255	35
26	518	238	.2160	236	34
27	540	287	.2153	218	33
28	563	336	.2145	199	32
29	585	385	.2138	181	31
30	.63608	.82434	1.2131	.77162	30
31	630	483	.2124	144	29
32	653	531	.2117	125	28
33	675	580	.2109	107	27
34	698	629	.2102	088	26
35	.63720	.82678	1.2095	.77070	25
36	742	727	.2088	051	24
37	765	776	.2081	033	23
38	787	825	.2074	.77014	22
39	810	874	.2066	.76996	21
40	.63832	.82923	1.2059	.76977	20
41	854	.82972	.2052	959	19
42	877	.83022	.2045	940	18
43	899	071	.2038	921	17
44	922	120	.2031	903	16
45	.63944	.83169	1.2024	.76884	15
46	966	218	.2017	866	14
47	.63989	268	.2009	847	13
48	.64011	317	.2002	828	12
49	033	366	.1995	810	11
50	.64056	.83415	1.1988	.76791	10
51	078	465	.1981	772	9
52	100	514	.1974	754	8
53	123	564	.1967	735	7
54	145	613	.1960	717	6
55	.64167	.83662	1.1953	.76698	5
56	190	712	.1946	679	4
57	212	761	.1939	661	3
58	234	811	.1932	642	2
59	256	860	.1925	623	1
60	.64279	.83910	1.1918	.76604	0
	cos	cot	tan	sin	′

50°

40°

′	sin	tan	cot	cos	
0	.64279	.83910	1.1918	.76604	60
1	301	.83960	.1910	586	59
2	323	.84009	.1903	567	58
3	346	059	.1896	548	57
4	368	108	.1889	530	56
5	.64390	.84158	1.1882	.76511	55
6	412	208	.1875	492	54
7	435	258	.1868	473	53
8	457	307	.1861	455	52
9	479	357	.1854	436	51
10	.64501	.84407	1.1847	.76417	50
11	524	457	.1840	398	49
12	546	507	.1833	380	48
13	568	556	.1826	361	47
14	590	606	.1819	342	46
15	.64612	.84656	1.1812	.76323	45
16	635	706	.1806	304	44
17	657	756	.1799	286	43
18	679	806	.1792	267	42
19	701	856	.1785	248	41
20	.64723	.84906	1.1778	.76229	40
21	746	.84956	.1771	210	39
22	768	.85006	.1764	192	38
23	790	057	.1757	173	37
24	812	107	.1750	154	36
25	.64834	.85157	1.1743	.76135	35
26	856	207	.1736	116	34
27	878	257	.1729	097	33
28	901	308	.1722	078	32
29	923	358	.1715	059	31
30	.64945	.85408	1.1708	.76041	30
31	967	458	.1702	022	29
32	.64989	509	.1695	.76003	28
33	.65011	559	.1688	.75984	27
34	033	609	.1681	965	26
35	.65055	.85660	1.1674	.75946	25
36	077	710	.1667	927	24
37	100	761	.1660	908	23
38	122	811	.1653	889	22
39	144	862	.1647	870	21
40	.65166	.85912	1.1640	.75851	20
41	188	.85963	.1633	832	19
42	210	.86014	.1626	813	18
43	232	064	.1619	794	17
44	254	115	.1612	775	16
45	.65276	.86166	1.1606	.75756	15
46	298	216	.1599	738	14
47	320	267	.1592	719	13
48	342	318	.1585	700	12
49	364	368	.1578	680	11
50	.65386	.86419	1.1571	.75661	10
51	408	470	.1565	642	9
52	430	521	.1558	623	8
53	452	572	.1551	604	7
54	474	623	.1544	585	6
55	.65496	.86674	1.1538	.75566	5
56	518	725	.1531	547	4
57	540	776	.1524	528	3
58	562	827	.1517	509	2
59	584	878	.1510	490	1
60	.65606	86929	1.1504	.75471	0
	cos	cot	tan	sin	′

49°

41°

′	sin	tan	cot	cos	
0	.65606	.86929	1.1504	.75471	60
1	628	.86980	.1497	452	59
2	650	.87031	.1490	433	58
3	672	082	.1483	414	57
4	694	133	.1477	395	56
5	.65716	.87184	1.1470	.75375	55
6	738	236	.1463	356	54
7	759	287	.1456	337	53
8	781	338	.1450	318	52
9	803	389	.1443	299	51
10	.65825	.87441	1.1436	.75280	50
11	847	492	.1430	261	49
12	869	543	.1423	241	48
13	891	595	.1416	222	47
14	913	646	.1410	203	46
15	.65935	.87698	1.1403	.75184	45
16	956	749	.1396	165	44
17	978	801	.1389	146	43
18	.66000	852	.1383	126	42
19	022	904	.1376	107	41
20	.66044	.87955	1.1369	.75088	40
21	066	.88007	.1363	069	39
22	088	059	.1356	050	38
23	109	110	.1349	030	37
24	131	162	.1343	.75011	36
25	.66153	.88214	1.1336	.74992	35
26	175	265	.1329	973	34
27	197	317	.1323	953	33
28	218	369	.1316	934	32
29	240	421	.1310	915	31
30	.66262	.88473	1.1303	.74896	30
31	284	524	.1296	876	29
32	306	576	.1290	857	28
33	327	628	.1283	838	27
34	349	680	.1276	818	26
35	.66371	.88732	1.1270	.74799	25
36	393	784	.1263	780	24
37	414	836	.1257	760	23
38	436	888	.1250	741	22
39	458	940	.1243	722	21
40	.66480	.88992	1.1237	.74703	20
41	501	.89045	.1230	683	19
42	523	097	.1224	664	18
43	545	149	.1217	644	17
44	566	201	.1211	625	16
45	.66588	.89253	1.1204	.74606	15
46	610	306	.1197	586	14
47	632	358	.1191	567	13
48	653	410	.1184	548	12
49	675	463	.1178	528	11
50	.66697	.89515	1.1171	.74509	10
51	718	567	.1165	489	9
52	740	620	.1158	470	8
53	762	672	.1152	451	7
54	783	725	.1145	431	6
55	.66805	.89777	1.1139	.74412	5
56	827	830	.1132	392	4
57	848	883	.1126	373	3
58	870	935	.1119	353	2
59	891	.89988	.1113	334	1
60	.66913	.90040	1.1106	.74314	0
	cos	cot	tan	sin	′

48°

42°

′	sin	tan	cot	cos	
0	.66913	.90040	1.1106	.74314	60
1	935	093	.1100	295	59
2	956	146	.1093	276	58
3	978	199	.1087	256	57
4	.66999	251	.1080	237	56
5	.67021	.90304	1.1074	.74217	55
6	043	357	.1067	198	54
7	064	410	.1061	178	53
8	086	463	.1054	159	52
9	107	516	.1048	139	51
10	.67129	.90569	1.1041	.74120	50
11	151	621	.1035	100	49
12	172	674	.1028	080	48
13	194	727	.1022	061	47
14	215	781	.1016	041	46
15	.67237	.90834	1.1009	.74022	45
16	258	887	.1003	.74002	44
17	280	940	.0996	.73983	43
18	301	.90993	.0990	963	42
19	323	.91046	.0983	944	41
20	.67344	.91099	1.0977	.73924	40
21	366	153	.0971	904	39
22	387	206	.0964	885	38
23	409	259	.0958	865	37
24	430	313	.0951	846	36
25	.67452	.91366	1.0945	.73826	35
26	473	419	.0939	806	34
27	495	473	.0932	787	33
28	516	526	.0926	767	32
29	538	580	.0919	747	31
30	.67559	.91633	1.0913	.73728	30
31	580	687	.0907	708	29
32	602	740	.0900	688	28
33	623	794	.0894	669	27
34	645	847	.0888	649	26
35	.67666	.91901	1.0881	.73629	25
36	688	.91955	.0875	610	24
37	709	.92008	.0869	590	23
38	730	062	.0862	570	22
39	752	116	.0856	551	21
40	.67773	.92170	1.0850	.73531	20
41	795	224	.0843	511	19
42	816	277	.0837	491	18
43	837	331	.0831	472	17
44	859	385	.0824	452	16
45	.67880	.92439	1.0818	.73432	15
46	901	493	.0812	413	14
47	923	547	.0805	393	13
48	944	601	.0799	373	12
49	965	655	.0793	353	11
50	.67987	.92709	1.0786	.73333	10
51	.68008	763	.0780	314	9
52	029	817	.0774	294	8
53	051	872	.0768	274	7
54	072	926	.0761	254	6
55	.68093	.92980	1.0755	.73234	5
56	115	.93034	.0749	215	4
57	136	088	.0742	195	3
58	157	143	.0736	175	2
59	179	197	.0730	155	1
60	.68200	.93252	1.0724	.73135	0
	cos	cot	tan	sin	′

47°

43°

′	sin	tan	cot	cos	
0	.68200	.93252	1.0724	.73135	60
1	221	306	.0717	116	59
2	242	360	.0711	096	58
3	264	415	.0705	076	57
4	285	469	.0699	056	56
5	.68306	.93524	1.0692	.73036	55
6	327	578	.0686	.73016	54
7	349	633	.0680	.72996	53
8	370	688	.0674	976	52
9	391	742	.0668	957	51
10	.68412	.93797	1.0661	.72937	50
11	434	852	.0655	917	49
12	455	906	.0649	897	48
13	476	.93961	.0643	877	47
14	497	.94016	.0637	857	46
15	.68518	.94071	1.0630	.72837	45
16	539	125	.0624	817	44
17	561	180	.0618	797	43
18	582	235	.0612	777	42
19	603	290	.0606	757	41
20	.68624	.94345	1.0599	.72737	40
21	645	400	.0593	717	39
22	666	455	.0587	697	38
23	688	510	.0581	677	37
24	709	565	.0575	657	36
25	.68730	.94620	1.0569	.72637	35
26	751	676	.0562	617	34
27	772	731	.0556	597	33
28	793	786	.0550	577	32
29	814	841	.0544	557	31
30	.68835	.94896	1.0538	.72537	30
31	857	.94952	.0532	517	29
32	878	.95007	.0526	497	28
33	899	062	.0519	477	27
34	920	118	.0513	457	26
35	.68941	.95173	1.0507	.72437	25
36	962	229	.0501	417	24
37	.68983	284	.0495	397	23
38	.69004	340	.0489	377	22
39	025	395	.0483	357	21
40	.69046	.95451	1.0477	.72337	20
41	067	506	.0470	317	19
42	088	562	.0464	297	18
43	109	618	.0458	277	17
44	130	673	.0452	257	16
45	.69151	.95729	1.0446	.72236	15
46	172	785	.0440	216	14
47	193	841	.0434	196	13
48	214	897	.0428	176	12
49	235	.95952	.0422	156	11
50	.69256	.96008	1.0416	.72136	10
51	277	064	.0410	116	9
52	298	120	.0404	095	8
53	319	176	.0398	075	7
54	340	232	.0392	055	6
55	.69361	.96288	1.0385	.72035	5
56	382	344	.0379	.72015	4
57	403	400	.0373	.71995	3
58	424	457	.0367	974	2
59	445	513	.0361	954	1
60	.69466	.96569	1.0355	.71934	0
	cos	cot	tan	sin	′

46°

44°

′	sin	tan	cot	cos	
0	.69466	.96569	1.0355	.71934	60
1	487	625	.0349	914	59
2	508	681	.0343	894	58
3	529	738	.0337	873	57
4	549	794	.0331	853	56
5	.69570	.96850	1.0325	.71833	55
6	591	907	.0319	813	54
7	612	.96963	.0313	792	53
8	633	.97020	.0307	772	52
9	654	076	.0301	752	51
10	.69675	.97133	1.0295	.71732	50
11	696	189	.0289	711	49
12	717	246	.0283	691	48
13	737	302	.0277	671	47
14	758	359	.0271	650	46
15	.69779	.97416	1.0265	.71630	45
16	800	472	.0259	610	44
17	821	529	.0253	590	43
18	842	586	.0247	569	42
19	862	643	.0241	549	41
20	.69883	.97700	1.0235	.71529	40
21	904	756	.0230	508	39
22	925	813	.0224	488	38
23	946	870	.0218	468	37
24	966	927	.0212	447	36
25	.69987	.97984	1.0206	.71427	35
26	.70008	.98041	.0200	407	34
27	029	098	.0194	386	33
28	049	155	.0188	366	32
29	070	213	.0182	345	31
30	.70091	.98270	1.0176	.71325	30
31	112	327	.0170	305	29
32	132	384	.0164	284	28
33	153	441	.0158	264	27
34	174	499	.0152	243	26
35	.70195	.98556	1.0147	.71223	25
36	215	613	.0141	203	24
37	236	671	.0135	182	23
38	257	728	.0129	162	22
39	277	786	.0123	141	21
40	.70298	.98843	1.0117	.71121	20
41	319	901	.0111	100	19
42	339	.98958	.0105	080	18
43	360	.99016	.0099	059	17
44	381	073	.0094	039	16
45	.70401	.99131	1.0088	.71019	15
46	422	189	.0082	.70998	14
47	443	247	.0076	978	13
48	463	304	.0070	957	12
49	484	362	.0064	937	11
50	.70505	.99420	1.0058	.70916	10
51	525	478	.0052	896	9
52	546	536	.0047	875	8
53	567	594	.0041	855	7
54	587	652	.0035	834	6
55	.70608	.99710	1.0029	.70813	5
56	628	768	.0023	793	4
57	649	826	.0017	772	3
58	670	884	.0012	752	2
59	690	.99942	.0006	731	1
60	.70711	1.0000	1.0000	.70711	0
	cos	cot	tan	sin	

45°

POLAR COORDINATE PAPER

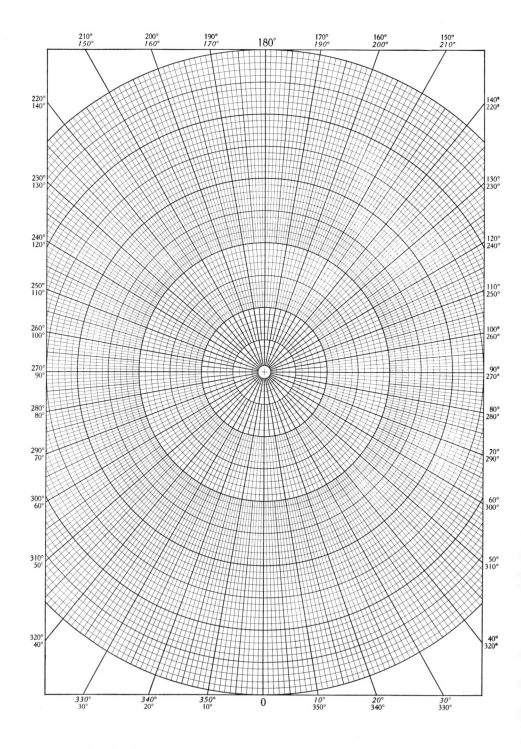

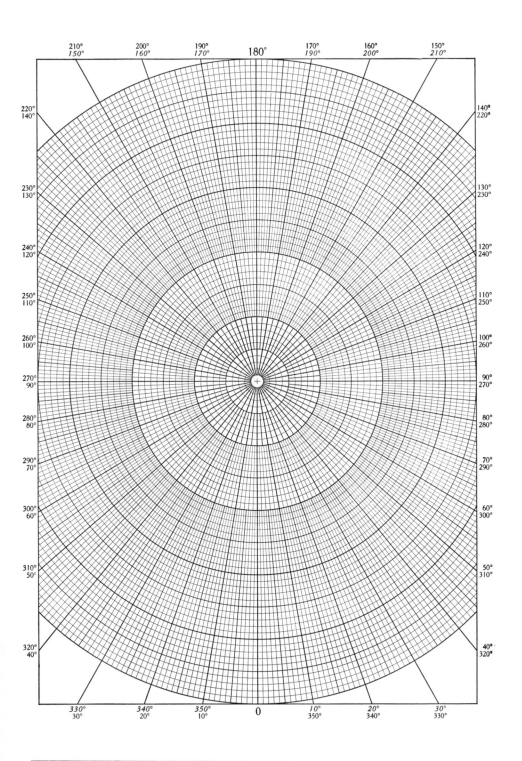

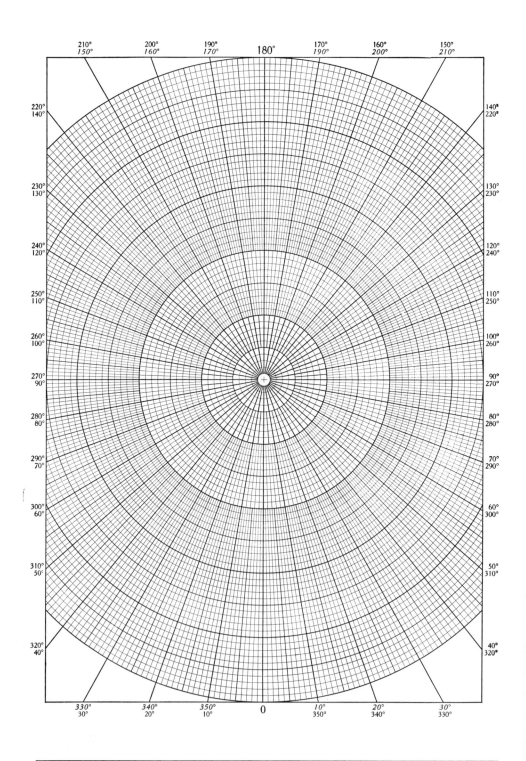

Index